Root Cause Failure Analysis:
A Guide to Improve Plant Reliability

Root Cause Failure Analysis:
A Guide to Improve Plant Reliability

Dr. Trinath Sahoo

WILEY

This edition first published 2021

Registered Office
John Wiley & Sons, Inc., 111 River Street, Hoboken, NJ 07030, USA

Editorial Office
111 River Street, Hoboken, NJ 07030, USA

For details of our global editorial offices, customer services, and more information about Wiley products visit us at www.wiley.com.

Wiley also publishes its books in a variety of electronic formats and by print-on-demand. Some content that appears in standard print versions of this book may not be available in other formats.

For general information on our other products and services or for technical support, please contact our Customer Care Department within the United States at (800) 762-2974, outside the United States at (317) 572-3993 or fax (317) 572-4002.
Wiley also publishes its books in a variety of electronic formats. Some content that appears in print may not be available in electronic formats. For more information about Wiley products, visit our web site at www.wiley.com.

Library of Congress Cataloging-in-Publication Data

Names: Sahoo, Trinath, author.
Title: Root cause failure analysis : a guide to improve plant reliability / Trinath Sahoo.
Description: Hoboken, New Jersey : Wiley, 2021. | Includes bibliographical references and index.
Identifiers: LCCN 2020053092 (print) | LCCN 2020053093 (ebook) | ISBN 9781119615545 (hardback) | ISBN 9781119615590 (adobe pdf) | ISBN 9781119615613 (epub)
Subjects: LCSH: Root cause analysis. | Piping. | Industrial equipment.
Classification: LCC TA169.55.R66 S25 2021 (print) | LCC TA169.55.R66 (ebook) | DDC 658.2–dc23
LC record available at https://lccn.loc.gov/2020053092
LC ebook record available at https://lccn.loc.gov/2020053093

Cover Design: Wiley
Cover Images: © ch123/Shutterstock, Yakov Oskanov/Shutterstock

Set in 9.5/12.5pt STIXTwoText by SPi Global, Pondicherry, India

MIX
Paper from responsible sources
FSC
www.fsc.org
FSC® C013604

Contents

Preface

Process industries are home to a huge number of machines, piping, structures, most of them critical to the industry's mission. Failure of these items can cause loss of life, unscheduled shutdowns, increased maintenance and repair costs, and damaging litigation disputes. Experience shows that all too often, process machinery problems are never defined sufficiently; they are merely "solved" to "get back on stream." Production pressures often override the need to analyze a situation thoroughly, and the problem and its underlying cause come back and haunt us later. Equipment downtime and component failure risk can be reduced only if potential problems are anticipated and avoided. To prevent future recurrence of the problem, it is essential to carry out an investigation aimed at detecting the root cause of failure.

The ability to identify this weakest link and propose remedial measures is the key for a successful failure analysis investigation. This requires a multidisciplinary approach, which forms the basis of this book. The results of the investigation can also be used as the basis for insurance claims, for marketing purposes, and to develop new materials or improve the properties of existing ones.

The objective of this book is to help anyone involved with machinery reliability, be it in the design of new plants or the maintenance and operation of existing ones, to understand why the process machine fails, so some preventive measures can be taken to avoid another failure of the same kind.

An important feature of this book is that it not only demonstrates the methodology for conducting a successful failure analysis investigation, but also provides the necessary background.

The book is divided in two parts:

1) The first part discusses the benefit of failure analysis, including some definitions and examples. Here, we examine the failure analysis procedure, including some approaches suitable for different types of problems. We also look at how plant-wide failure prevention efforts should be conducted, including a discussion about the importance of the role of the top management in the prevention of failure.
2) In the second part, different types of failure mechanisms that affect process equipment are discussed with several examples of bearings, seals, and other components' failures.

Because it is simply impossible to deal with every conceivable type of failure, this book is structured to teach failure identification and analysis methods that can be applied to virtually all problem situations that might arise.

Trinath Sahoo

About the Author

Trinath Sahoo, Ph.D., is the chief general manager at M/S Indian Oil Corporation Ltd. Dr. Sahoo has 30 years of experience in various fields such as engineering design, project management, asset management, maintenance management, lubrication, and reliability. He has published many papers in journals like *Hydrocarbon Processing, Chemical Engineering, Chemical Engineering Progress,* and *World Pumps*. Some of his articles were adjudged best articles and published as the cover page story in the magazines. He has also spoken in many international conferences. He was the convener for reliability enhancement projects for different refinery and petrochemical sites of M/S Indian Oil Corporation Ltd. Dr. Sahoo is the author of bestselling book *Process Plants: Shutdown and Turnaround Management*. He holds a Ph.D. degree from Indian Institute of Technology (ISM), Dhanbad, Jharkhand, India.

Acknowledgment

First and foremost, I would like to thank God, the Almighty, for His showers of blessings throughout to complete the book successfully. In the process of putting this book together, I realized how true this gift of writing is for me. You have given me the power to believe in my passion and pursue my dreams. I could never have done this without the faith I have in you, the Almighty.

I have to thank my parents for their love and support throughout my life. Thank you both for giving me strength to reach for the stars and chase my dreams.

For my wife Chinoo, all the good that comes from this book I look forward to sharing with you! Thanks for not just believing, but knowing that I could do this! I Love You Always and Forever!

To my children Sonu and Soha: You may outgrow my lap, but you will never outgrow my heart. Your growth provides a constant source of joy and pride to me and helped me to complete the book.

Without the experiences and support from my peers and team at Indian Oil, this book would not exist. You have given me the opportunity to lead a great group of individuals.

"Thanks to everyone on my publishing team."

Only those who dare to fail greatly can ever achieve greatly.

Robert F. Kennedy.

1

FAILURE: How to Understand It, Learn from It and Recover from It

Failure and fault are virtually inseparable in households, organizations, and cultures. But the wisdom of learning from failure is much more than from success. Many a time we discover what works well, by finding out what will not work; and "probably he who have never made a mistake never made a discovery."

Thomas Edison's associate, Walter S. Mallory, while discussing inventions, once said to him, "Isn't it a shame that with the tremendous amount of work you have done you haven't been able to get any results?" Edison replied, with a smile, "Results! Why, my dear, I have gotten a lot of results! I know several thousand things that won't work."

People see success as positive and failure as negative phenomena. Edison's quote emphasizes that failure isn't a bad thing. You can learn and evolve from your past mistakes. But in organizations executives believe that failure is bad. These widely held beliefs are misguided. Understanding of failure's causes and contexts will help to avoid the blame game and create an atmosphere of learning in the organization. Failure may sometimes considered bad, sometimes inevitable, and sometimes even good in organizations. In most companies, the system and procedures required to effectively detect and analyze failures are in short supply. Even the context-specific learning strategies are not appreciated many times. In many organizations, managers often want to learn from failures to improve future performance. In the process, they and their teams used to devote many hours in after-action reviews, post-mortems, etc. But time after time these painstaking efforts led to no real change. The reason: being, managers think about failure in a wrong way.

To be able to learn from our failures, we need to develop a methodology to decode the "teachable moments" hidden within them. We need to find out what exactly those lessons are and how they can improve our chances of future success.

Failure Type

Although an infinite number of things can go wrong in machinery, systems, and process, mistakes fall into three broad categories: preventable failure, failure in complex system, and intelligent failure.

Root Cause Failure Analysis: A Guide to Improve Plant Reliability, First Edition. Trinath Sahoo.

Preventable Failures

Most failures in this category are considered as "bad." These could have been foreseen but weren't. This is the worst kind of failure, and it usually occurs because an employee didn't follow best practices, didn't have the right talent, or didn't pay attention to detail. They usually deviate from specification in the closely defined processes or deviate from routine operations and maintenance practices. But in such cases, the causes can be readily identified and solutions can be developed.

If you've experienced a preventable failure, it's time to more deeply analyze the effort's weaknesses and stick to what works in future. Employees can follow those new processes learned from past mistakes consistently, with proper training and support.

Human error used to be an area that was associated with high-risk industries like aviation, rail, petrochemical and the nuclear industry. The high consequences of failure in these industries meant that there was a real obligation on companies to try to reduce the likelihood of all failure causes. Human error is also a high-priority, preventable issue.

Unavoidable Failures in Complex Systems

In complex organizations such as aircraft carriers, nuclear power plants, and petrochemical plants, system failure is a perpetual risk. A large number of failures are due to the inherent uncertainty of working of such systems.

The lesson from this type of failure is to create systems to try to spot small failures resulting from complex factors, and take corrective action before it snowballs and destroys the whole system. These type of failure may not be considered bad but reviewed how complex systems work. Most accidents in these systems result from a series of small failures that went unnoticed and unfortunately lined up in just the wrong way.

The complex systems are heavily and successfully defended against failure by construction of multiple layers of defense against failure. These defenses include obvious technical components (e.g. backup systems, "safety" features of equipment) and human components (e.g. training, knowledge) but also a variety of organizational, institutional, and regulatory defenses (e.g. policies and procedures, certification, work rules, team training). The effect of these measures is to provide a series of shields that normally divert operations away from accidents.

Intelligent Failures

Intelligent failures occur when answers are not known in advance because this exact situation hasn't been encountered before and experimentation is necessary in these cases. For example testing a prototype, designing a new type of machinery or operating a machine in different operating condition. In these settings, "trial and error" is the common term used for the kind of experimentation needed. These type of failures can be considered "good," because they provide valuable insight and new knowledge that can help an organization to learn from past mistakes for its future growth. The lesson here is clear: If something works, do more of it. If it doesn't, go back to the drawing board

Building a Learning Culture

Leaders can create and reinforce a culture that makes people feel comfortable for surfacing and learning from failures to avoid blame game. When things go wrong, they should insist to find out what happened – rather than "who did it." This requires consistently reporting failures, small, and large; systematically analyzing them; and proactively taking steps to avoid reoccurrence.

Most organizations engage in all three kinds of work discussed above – routine, complex, and intelligent. Leaders must ensure that the right approach to learning from failure is applied in each of them. All organizations learn from failure through following essential activities: detection, analysis, learning, and sharing.

Detecting Failure

Spotting big, painful, expensive failures are easy. But failure that are hidden are hidden as long as it's unlikely to cause immediate or obvious harm. The goal should be to surface it early, before it can create disaster when accompanied by other lapses in the system. High-reliability-organization (HRO) helps prevent catastrophic failures in complex systems like nuclear power plants, aircraft through early detection.

In a big petrochemical plant, the top management is religiously interested to tracks each plant for anything even slightly out of the ordinary, immediately investigates whatever turns up, and informs all its other plants of any anomalies. But many a time, these methods are not widely employed because senior executives – remain reluctant to convey bad news to bosses and colleagues.

Analyzing Failure

Most people avoid analyzing the failure altogether because many a time it is emotionally unpleasant and can chip away at our self-esteem. Another reason is that analyzing organizational failures requires inquiry and openness, patience, and a tolerance for causal ambiguity. Hence, managers should be rewarded for thoughtful reflection. That is why the right culture can percolate in the organization.

Once a failure has been detected, it's essential to find out the root causes not just relying on the obvious and superficial reasons. This requires the discipline to use sophisticated analysis to ensure that the right lessons are learned and the right remedies are employed. Engineers need to see that their organizations don't just move on after a failure but stop to dig in and discover the wisdom contained in it.

A team of leading physicists, engineers, aviation experts, naval leaders, and even astronauts devoted months to an analysis of the Columbia disaster. They conclusively established not only the first-order cause – a piece of foam had hit the shuttle's leading edge during launch – but also second-order causes: A rigid hierarchy and schedule-obsessed culture at NASA made it especially difficult for engineers to speak up about anything but the most rock-solid concerns.

Motivating people to go beyond first-order reasons (procedures weren't followed) to understanding the second- and third-order reasons can be a major challenge. One way to do this is to use interdisciplinary teams with diverse skills and perspectives. Complex failures in particular are the result of multiple events that occurred in different departments or disciplines or at different levels of the organization. Understanding what happened and how to prevent it from happening again requires detailed, team-based discussion, and analysis.

Here are some common root causes and their corresponding corrective actions:

- Design deficiency caused failure → Revisit in-service loads and environmental effects, modify design appropriately.
- Manufacturing defect caused failure → Revisit manufacturing processes (e.g. casting, forging, machining, heat treat, coating, assembly) to ensure design requirements are met.
- Material defect caused failure → Implement raw material quality control plan.
- Misuse or abuse caused failure → Educate user in proper installation, use, care, and maintenance.
- Useful life exceeded → Educate user in proper overhaul/replacement intervals.
- There are various methods that failure analysts use – for example, Ishikawa "fishbone" diagrams, failure modes and effects analysis (FMEA), or fault tree analysis (FTA). Methods vary in approach, but all seek to determine the root cause of failure by looking at the characteristics and clues left behind.

Once the root cause of the failure has been determined, it is possible to develop a corrective action plan to prevent recurrence of the same failure mode. Understanding what caused one failure may allow us to improve upon our design process, manufacturing processes, material properties, or actual service conditions. This valuable insight may allow us to foresee and avoid potential problems before they occur in the future.

Share the Lessons

Failure is less painful when you extract the maximum value from it. If you learn from each mistake, large and small, share those lessons, and periodically check that these processes are helping your organization move more efficiently in the right direction, your return on failure will skyrocket. While it's useful to reflect on individual failures, the real payoff comes when you spread the lessons across the organization. As one executive commented, "You need to build a review cycle where this is fed into a broader conversation." When the information, ideas, and opportunities for improvement gained from an failure incident are passed on to another, their benefits are magnified. The information on root cause failure analysis should be made available to others in the organization so that they can learn too.

Benefits of Failure Analysis

The best way to get risk-averse managers and employees to learn to accept higher risks and their associated failures are to educate them on the many positive aspects and benefits of failure. Some of those many benefits include:

- **Failure tells you what to stop doing** – Obviously, failure reveals what doesn't work, so you can avoid using similar unmodified approaches in the future. And over time, by continually eliminating failure factors, you obviously increase the probability of future success.

- **Failure is the best teacher** – Failure is only valuable if you use it to identify what worked and what didn't work and to use that information to minimize future failures. In the corporate and engineering worlds, learning from failure starts with failure analysis. This is a process that helps you identify specifically what failed and then to understand the "root causes" of that failure (i.e. critical failure factors). But since failure and success factors are often closely related, the identification of the failure factors will likely aid you in identifying the critical success factors that cause an approach to succeed. The famous auto innovator Henry Ford revealed his understanding of learning from failure in this quote: "The only real mistake is the one from which we learn nothing."
- **A failure factor in one area may apply to another area** – Failure analysis tells you what failed and why. But the best corporations develop processes that "spread the word" and warn others in your organization about what clearly doesn't work so that others don't need to learn the hard way. On the positive side, lessons learned from both successes and failures in one discipline may be able to be applied to another discipline or functional area.
- **Experience builds your capability to handle future major failures** – When a major failure does occur, your "rusty" employees and your out of date processes simply won't be able to handle it. Both the military and healthcare managers have proven that the more often you train for and work through actual major failures, the better prepared you will be when an unplanned failure occurs in the future.

Conclusion

Many companies and organizations have been on the reliability journey for a number of years. There are many elements of a solid reliability program – establishing a reliability-centered culture, tracking key metrics, bad actor elimination programs and establishing equipment reliability plans – to name a few. But, one key element to a solid reliability program, and one that is very important to improving unit reliability metrics, is root cause failure analysis (RCFA). One of the interesting benefits of organizations that have fully embraced the RCFA work process across the entire organization is that over time the RCFA methodology starts to impact how people approach everyday problems – it becomes how they think about even the smallest failure, problems, or defects. Now the organization starts to evolve into a culture that does not accept failure and provides a mindset to help eliminate failures across the organization.

2

What Is Root Cause Analysis

It is not uncommon to see industries caught in the vicious cycle of failure, repair, blame, failure, repair, blame, etc. When there is premature failure of equipment, people involved often asked the question, whose fault it is. Many a time you will get the answer "it is other guy's fault."

If one were to ask a operator why the equipment fail, the immediate answer will be it was the fault of maintenance mechanic who had not fixed it properly. In the same line, a maintenance mechanic likely answer to that question would be "operator error." At times, there is some validity to both these answers, but the honest and complete answer is much more complex. This chapter briefly introduces the concepts of failure analysis, root cause analysis, and the role of failure analysis as a general engineering tool for enhancing failure prevention.

Failure analysis is a process that is performed in order to determine the causes that may have attributed to the loss of functionality. These defects may come from a deficient design, poor material, mistakes in manufacturing or wrong operation and maintenance. Many a time there is no single cause and no single train of events that lead to a failure. Rather, there are factors that combine at a particular time to allow a failure to occur. Failure analysis involves a logical sequence of steps that lead the investigator through identifying the root causes of faults or problems.

Look at any well-studied major disaster and ask if there was only one cause. Was there only one cause for the TITANIC? Three Mile Island? The Exxon Valdez mess? Bhopal? Chernobyl? It would be nice if there were only one cause per failure, because correcting the problem would then be easy. However, in reality, there are multiple causes to every equipment failure. Let us take the case of TITANIC failure.

The Causes of TITANIC disaster

The TITANIC passengers included some of the wealthiest and most prestigious people at that time. Captain Edward John Smith, one of the most experienced shipmasters on the Atlantic, was navigating the TITANIC. On the night of 14 April, although the wireless operators had received several ice warnings from others ships in the area, the TITANIC continued to rush through the darkness at nearly full steam. Suddenly, the captain spotted a massive iceberg less than a quarter of a mile off the bow of the ship. Immediately, the engines were thrown into reverse and the rudder turned hard left. Because of the tremendous mass of the ship, slowing and turning took an incredible distance, more than that available. Without

Root Cause Failure Analysis: A Guide to Improve Plant Reliability, First Edition. Trinath Sahoo.

enough distance to alter her course, the TITANIC sideswiped the iceberg, damaging nearly 300 feet of the right side of the hull above and below the waterline.

The two official investigations back in 1912 started with a conclusion – the TITANIC hit an iceberg and sank. They made somewhat of an attempt to answer why that happened without attaching too much blame. The result was not so much as getting to the root cause but found out the immediate cause.

Richard Corfield writes in a Physics World retrospective on the disaster that caused 1514 deaths on 14–15 April 1912. He described it was an event cascade followed by a perfect storm of circumstances conspired the TITANIC to fail. The iceberg that the TITANIC struck on its way from Southampton to New York is No. 1 on a top-9 list of circumstances. Here are eight other suggested circumstances from Richard Corfield's article and other sources:

Climate caused more icebergs: Weather conditions in the North Atlantic were particularly conducive for corralling icebergs at the intersection of the Labrador Current and the Gulf Stream, due to warmer-than-usual waters in the Gulf Stream. As a result, there were icebergs and sea ice concentrated in the very position where the collision happened

The iron rivets were too weak: Metallurgists Tim Foecke and Jennifer Hooper McCarty looked into the materials used for the building of the TITANIC at its Belfast shipyard and found that the steel plates toward the bow and the stern were held together with low-grade iron rivets. Those rivets may have been used because higher-grade rivets were in short supply, or because the better rivets couldn't be inserted in those areas using the shipyard's crane-mounted hydraulic equipment. The metallurgists said those low-grade rivets would have ripped apart more easily during the collision, causing the ship to sink more quickly that it would have if stronger rivets had been used.

The ship was going too fast: Many investigators have said that the ship's captain, Edward J. Smith, was aiming to better the crossing time of the Olympic, the TITANIC's older sibling in the White Star fleet. For some, the fact that the TITANIC was sailing full speed ahead despite concerns about icebergs was Smith's biggest misstep. "Simply put, TITANIC was traveling way too fast in an area known to contain ice, which was one of the major reason of the TITANIC disaster.

Iceberg warnings went unheeded: The TITANIC received multiple warnings about ice-fields in the North Atlantic over the wireless, but Corfield notes that the last and most specific warning was not passed along by senior radio operator Jack Phillips to Captain Smith, apparently because it didn't carry the prefix "MSG" (Masters' Service Gram). That would have required a personal acknowledgment from the captain. "Phillips interpreted it as non-urgent and returned to sending passenger messages to the receiver on shore at Cape Race, Newfoundland, before it went out of range," Corfield writes.

The binoculars were locked up: Corfield also says binoculars that could have been used by lookouts on the night of the collision were locked up aboard the ship – and the key was held by David Blair, an officer who was bumped from the crew before the ship's departure from Southampton. Some historians have speculated that the fatal iceberg might have been spotted earlier if the binoculars were in use, but others say it wouldn't have made a difference.

The steersman took a wrong turn: Did the TITANIC's steersman turn the ship toward the iceberg, dooming the ship? That's the claim made by Louise Patten, who said the story was passed down from her grandfather, the most senior ship officer to survive the disaster. After the iceberg was spotted, the command was issued to turn "hard a starboard," but as the command was passed down the line, it was misinterpreted as meaning "make the ship turn right" rather than "push the tiller right to make the ship head left," Patten said. She said the error was quickly discovered, but not quickly enough to avert the collision. She also speculated that if the ship had stopped where it was hit, seawater would not have pushed into one interior compartment after another as it did, and the ship might not have sunk as quickly.

Reverse thrust reduced the ship's maneuverability: Just before impact, first officer William McMaster Murdoch is said to have telegraphed the engine room to put the ship's engines into reverse. That would cause the left and right propeller to turn backward, but because of the configuration of the stern, the central propeller could only be halted, not reversed. Corfield said "the fact that the steering propeller was not rotating severely diminished the turning ability of the ship. It is one of the many bitter ironies of the Titanic tragedy that the ship might well have avoided the iceberg if Murdoch had not told the engine room to reduce and then reverse thrust."

There were too few lifeboats: Perhaps the biggest tragedy is that there were not enough lifeboats to accommodate all of the TITANIC's more than 2200 passengers and crew members. The lifeboats could accommodate only about 1200 people.

Do these nine causes cover everything, or are there still more factors I'm forgetting? Are there some lessons still unlearned from the TITANIC tragedy?

What Is Root Cause Analysis?

Looking at the TITANIC failure report, it shows that there is no single cause and no single train of events that lead to a failure. Rather, there are factors that combine at a particular time and place to allow a failure to occur. Sometimes the absence of any single one of the factors may have been enough to prevent the failure. Sometimes, though, it is impossible to determine, at least within the resources allotted for the analysis, whether any single factor was key. If failure analysts are to perform their jobs in a professional manner, they must look beyond the simplistic list of causes of failure that some people still believe. They must keep an open mind and always be willing to get help when beyond their own experience.

Different Levels of Causes

A failure is often the result of multiple causes at different levels. Some causes might affect other causes that, in turn, create the visible problem. Causes can be classified as one of the following:

- Symptoms. These are not regarded as actual causes, but rather as signs of existing problems.
- First-level causes. Causes that directly lead to a problem.
- Higher-level causes. Causes that lead to the first-level causes. They may not directly cause the problem, but form links in the chain of cause-and-effect relationships that ultimately create the problem.

Some failures often have compound reasons, where different factors combine to cause the problem. Examples of the levels of causes follow.

The highest-level cause of a problem is called the root cause:

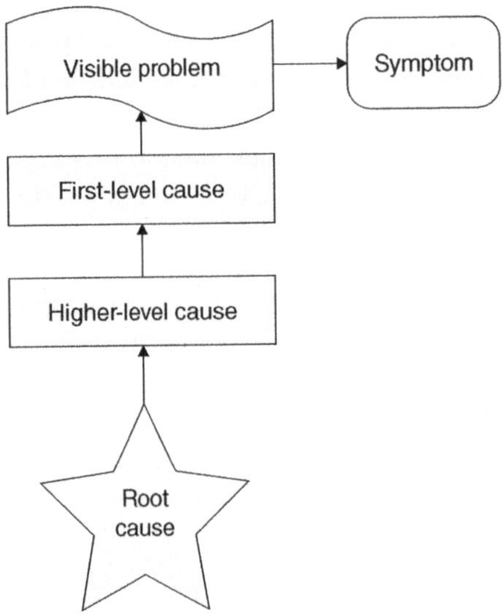

Hence, the root cause is "the evil at the bottom" that sets in motion the entire cause-and-effect chain causing the problem(s).

TrevoKletz said

> . . .root cause investigation is like peeling an onion. The outer layers deal with technical causes, while the inner layers are concerned with weaknesses in the management system. I am not suggesting that technical causes are less important. But putting technical causes right will prevent only the LAST event from happening again; attending to the underlying causes may prevent MANY SIMILAR INCIDENCES.

The difference between failure analysis and root cause analysis is that failure analysis is a discipline used for identifying the physical roots of failures, whereas the root cause analysis (RCA) techniques is a discipline used in exploring some of the other contributors to failures, such as the human and latent root causes. Root cause analysis is intended to identify the fundamental cause(s) that if corrected will prevent recurrence. The principles of RCA may be applied to ensure that the real root cause is identified to initiate appropriate corrective actions. RCA helps in correcting and preventing failures, achieving higher levels of quality and reliability, and ultimately enhancing customer satisfaction

Depending on the objectives of the RCA, one should decide how deeply one should analyze the case. These objectives are typically based on the risk associated with the failures and the complexity of the situation. The three levels of root cause analysis are physical roots, human roots, and latent roots. Physical roots, or the roots of equipment problems, are where many failure analyses stop. Physical root causes are derived from laboratory investigation or engineering analysis and are often component-level or materials-level findings. Human roots (i.e., people issues) involve human factors, where the error may be happened due to human judgment that may have caused the failure. Latent roots include roots that are organizational or procedural in nature, as well as environmental or other roots that are outside the realm of control.

Physical Roots

This is the physical mechanism that caused the failure, it may be fatigue, overload, wear, corrosion, or any combination of these. For example – corrosion damage of a pipeline, a bearing failed due to fatigue. Failure analysis must start with accurately determining the physical roots, for without that knowledge, the actual human and latent roots cannot be detected and corrected. The analysis may focus on physics of the incident. In the case of TITANIC, *the iron rivets were too weak.*

The steel plates of the TITANIC buckled as there were excessive stress applied to the hull *when* the ship hit the iceberg. The strength of steel and hull was not sufficient to prevent the hull from being breached by the steel plates buckling. The failure of the hull steel resulted from brittle fractures caused by the high sulfur content of the steel, the low temperature water on the night of the disaster, and the high impact loading of the collision with the iceberg. When the TITANIC hit the iceberg, the hull plates split open and continued cracking as the water flooded the ship.

Human Roots

The human roots are those human errors that result in the mechanisms that caused the physical failures. What is the error committed that lead to the physical cause?

Someone did the wrong thing knowingly or unknowingly. We asked what caused the person to commit this mistake. A good example is, the TITANIC was sailing full speed ahead despite concerns about icebergs was Smith's biggest misstep. the TITANIC was actually *speeding up* when it struck the iceberg as it was White Star chairman and managing director, Bruce Ismay's, intention to run the rest of the route to New York at full speed, arrive early, and prove the TITANIC's superior performance. Ismay survived the disaster and testified at the inquiries that this speed increase was approved by Captain Smith and the helmsman was operating under his Captain's direction.

Latent Roots

All physical failures are triggered by humans. But humans are negatively influenced by latent forces. The goal is to identify and remove these latent forces. Latent causes reveal themselves in layers. One after the other, the layers can be peeled back, similar to peeling the layers off an onion. It often seems as if there is no end. These forces within the organizations are causing people to make serious mistakes.

These are the management system weaknesses that include training, policies, procedures and specifications. People make decision based on these and if the system is flawed, the decision will be in error and will be the triggering mechanism that causes the mechanical failure to occur. These are the management system weaknesses. These include training, policies, procedures and specifications. The most proactive of all industrial action might be to identify and remove these latent traps. But all our attempts to identify and remove these latent causes of failure start at the human. Humans do things "inappropriately," for "latent" reasons. In order to understand these reasons, we must first understand what "errors" are being made. This puts people at risk – especially the "culprits." Once exposed. They are in danger of being inappropriately disciplined.

In the TITANIC case, the voyage had been so hastily pushed that the crew had no specific training or conducted any drills in lifesaving on the TITANIC, being unfamiliar with the lifeboats and their davit lowering mechanisms. Compounding this was a decision by White Star management to equip the TITANIC with only half the necessary lifeboats to handle the number of people onboard. The reasons are long established. White Star felt a full complement of lifeboats would give the ship an unattractive, cluttered look. They also clearly had a false confidence the lifeboats would never be needed.

To understand different level of root causes, let us take one industrial case.

Consider this example: During the overhauling of a large reciprocating compressor, the maintenance supervisor discovers a damaged compressor rod requiring replacement. So, he decides to have a rod made in a local shop by fabricating the rod with cut threads. But the OEM's design department has recommended the compressor rods for this frame size to have rolled threads. As a result of the improper fabrication, the rod fails due to fatigue in the thread area and causes extensive secondary damage inside the compressor.

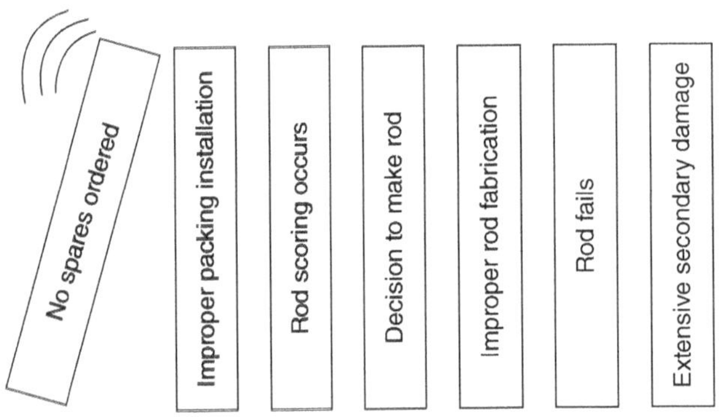

Figure 2.1 Events leading to compressor failure.

If you study this example, you can discern the following events leading to the costly failure:

- The warehouse did not stock spares for this rod because it was a new compressor installation.
- The maintenance supervisor decides to have a rod fabricated without drawings.
- Neither the user nor the local shop investigated the thread requirements.
- Because the compressor was not equipped with vibration shutdowns, it ran for a significant amount of time before it was shutdown.

There were several chances to break the chain of events leading to the catastrophic compressor failure. If the project engineer had ordered spare parts through the OEM, this failure probably would have been avoided. If either the maintenance supervisor or the local machine shop had talked to the OEM, or studied the failed rod, they would have been aware of the importance of rolled threads. Lastly, if a vibration shutdown had been in place, the compressor would have shutdown after only minimal damage. We see there were six major events leading to the secondary compressor damage. These events were as follows:

- No procedure in place to order spare parts for newly purchased equipment (latent root).
- The improper installation of the packing leads to rod scoring.

- Because a spare rod is not available and plant management wants the compressor back in operation as soon as possible, it was decided to have a replacement rod fabricated at a local machine shop.
- No one checks with the OEM about rod thread specifications (physical root).
- The rod fails after two days of operation.
- The broken rod causes extensive damage to the cylinder, packing box, distance piece, and cross-head.

After examining the vestiges of the failure, the rotating equipment (RE) engineer would discover a fatigue failure in the threaded portion of the rod. From this, he would conclude an improper thread design led to a stress riser and a shortened fatigue life. After talking to the OEM, he writes a report recommending that all compressor rods in the plant have rolled threads.

This recommendation will surely reduce rod failures, but the investigation did not uncover the latent root of failure. The stress riser, due to the improper thread design, is called the "physical root," because it did initiate the physical events leading to the secondary damage. However, there were significant events preceding the physical root that are of interest. If the RE engineer had the time and resources, he would have discovered that the absence of a procedure requiring new equipment to be purchased with adequate spares directly initiated the sequence of events. This basic event is called the "latent root."

By requiring spare parts be purchased from the OEM for all new equipment, the latent root is eliminated, not only for this scenario but, potentially, for many other similar events. This example demonstrates the importance of finding out the "latent root" of rotating equipment failures. Stopping at the "physical root," deprives the organization of a valuable opportunity for improvement. So, an RCFA is a detailed analysis of a complex, multi-event failure, such as the example above, in which the sequence of events is hoped to be found, along with the initiating event. The initiating event is called the root cause, and factors that contributed to the severity of the failure or perpetuated the events leading to the failure are called contributing events.

Industry personnel generally divides failure analysis into three categories in order of complexity and depth of investigation.

They are:

1) Component failure analysis (CFA) looks at the specific physical cause of failure such as fatigue, overload, or corrosion of the machine element that failed, for example, a bearing or a gear. This type of analysis mostly emphasizes to find the physical causes of the failure.
2) Root cause investigation (RCI) is conducted in greater depth than the CFA and goes substantially beyond the physical root of a problem. It investigates to find the human errors involved but doesn't involve management system deficiencies.
3) Root cause analyses (RCA) include everything the RCI covers plus the management system problems that allow the human errors and other system weaknesses to exist.

Although the cost increases as the analyses become more complex, the benefit is that there is a much more complete recognition of the true origins of the problem. Using a CFA to solve the causes of a component failure answers why that specific part or machine failed and can be used to prevent similar future failures. Progressing to an RCI, we find the cost is 5–10 times that of a CFA but the RCI adds a detailed understanding of the human errors contributing to the breakdown and can be used to eliminate groups of similar problems in the future. However, conducting an RCA may cost well into six figures and require several months. These costs may be intimidating to some, but the benefits obtained from correcting the major roots will eliminate huge classes of problems. The return will be many times the expenditure and will start to be realized within a few months of formal program implementation.

One thing that has to be recognized is that, because of the time, manpower, and costs involved, it is essentially impossible to conduct an RCA on every failure. The cost and possible benefits have to be recognized and judgments made to decide on the appropriate type of analysis.

When RCA Is Justified

Equipment Damage or Failure

RCFA are normally justified for those events associated with the partial or complete failure of critical production equipment, machinery, or systems. This type of incident can have a severe, negative impact on plant performance. Therefore, it often justifies the effort required to fully evaluate the event and to determine its root cause.

Operating Performance

Many a time deviations in operating performance occur without the physical failure of equipment or components. Chronic deviations may justify the use of RCFA as a means of resolving the recurring problem.

Product Quality

RCFA can be used to resolve most quality-related problems. However, the analysis should not be used for all quality problems.

Capacity Restrictions

Many of the problems or events that occur affect a plant's ability to consistently meet expected production or capacity rates. These problems may be suitable for RCFA, but further evaluation is recommended before beginning an analysis. After the initial investigation, if the event can be fully qualified and a cost-effective solution not found, then a full analysis should be considered. Note that an analysis normally is not performed on random, nonrecumng events or equipment failures.

Economic Performance

Deviations in economic performance, such as high production or maintenance costs, often warrant the use of RCFA. The decision tree and specific steps required to resolve these problems vary depending on the type of problem and its forcing functions or causes.

Safety

Any event that has a potential for causing personal injury should be investigated immediately. While events in this classification may not warrant a full RCFA, they must be resolved as quickly as possible. Isolating the root cause of injury-causing accidents or events generally is more difficult than for equipment failures and requires a different problem-solving approach. The primary reason for this increased difficulty is that the cause often is subjective.

Top Reasons Why We Need to Perform RCFA

1) Failures simply won't go away by fixing them all the time. We can only eliminate failures if we try to analyze them through Root Cause Failure Analysis. Then, only maintenance department can focus more on improving their asset performance.
2) To arrive at the correct solution to our equipment problems RCFA is not about addressing all the probable causes but rather failures being looked back in reverse to determine what really cause the problem. In performing RCFA, each hypothesis is verified until we have gathered enough evidence that these are the actual facts that lead to the failure itself. In completely eliminating the problem, it is important to address not only the physical cause but both the human and the latent cause.
3) Equipment failures might induce the possibility of secondary damage. Parts that are in the process of failing such as bearings will increase the vibration of equipment, this increase in vibration would be harmful to other parts that are directly coupled to the part that induce the vibration. Oftentimes secondary damage will be more costly than the parts that initially failed
4) Being proactive will give me a sense of security. Many maintenance personnel believes that a good backlog of maintenance work will ensure them of their job security. This is not the right mindset. Traditional maintenance people is confined to repairs and fixing failures but the scope of our job is beyond boundaries, our real job is to improve our equipment reliability and the scope of maintenance is beyond boundaries CBM, Oil Analysis, Lubrication, Tribology, Coaching their Operators on Basic Equipment Condition, Oil Contamination Control, Spare Parts Management, Maintenance Cost Reduction Team, just to name a few.
5) We all learn from the failure itself. For every failure that occurred and that had been thoroughly analyzed through RCFA, there is a learning that we can all can gained from these experience in order to prevent the recurrence of the failure itself. Sometimes failures speak to us in a different language.

Root Cause Analysis in a Larger Context

The roots of RCA method can be traced to the broader field of total quality management or TQM. TQM has developed in different directions more or less simultaneously. One of these directions is the development of a number of problem analysis, problem-solving, and improvement tools. Today, TQM possesses a large toolbox of such techniques. Further, problem-solving is an integral part of continuous improvement. Thus, root cause analysis is one of the core building blocks in an organization's continuous improvement efforts. However, it is important to keep in mind that root cause analysis must be made part of a larger problem-solving effort that embraces a relentless pursuit of improvement at every level and in every department or business process of the organization.

Conclusion

Root cause analysis (RCA) is a systematic process for identifying the root causes of problems or events and an approach for responding to them. By properly carrying out RCA, problems are best solved and root causes are eliminated. However, prevention of problem recurrence by one corrective action may not always possible by merely addressing the immediate obvious symptoms. Many organizations tend to focus on single factor when trying to identify a cause, which leads to an incomplete resolution. Root cause analysis helps avoid this tendency and looks at the event as a whole. It is also important not to focus on the symptoms rather than the actual underlying problems contributing to the issue, leading to recurrence. The advantage of RCA is that it provides a structured method to identify the root cause of known problems thus ensuring a complete understanding of problems under review. By directing corrective measures at root causes, it is more probable that problem recurrence will be prevented.

3

Root Cause Analysis Process

The key to a good root cause analysis is truly understanding it. Root cause analysis (RCA) is an analysis process that helps you and your team find the root cause of an issue. RCA can be used to investigate and correct the root causes of repetitive incidents, major accidents, human errors, quality problems, equipment failures, production issues, manufacturing mistakes, and can even be used proactively to identify potential issues.

The key to successful root cause analysis is understanding a process or sequence that works. The effect is the event – what occurred. A cause is defined as a set of circumstances or conditions that allows or facilitates the existence of a condition an event. Therefore, the best strategy would be to determine why the event happened. Simply put, eliminating the cause or causes will eliminate the effect.

What is root cause analysis

Root cause analysis is a logical sequence of steps that leads the investigator through the process of isolating the facts or the contributing factor surrounding an event or failure. Once the problem has been fully defined, the analysis systematically determines the best course of action that will resolve the event and assure that it is not repeated. A contributing factor is a condition that influences the effect by increasing the probability of occurrence, hastening the effect, and increasing the seriousness of the consequences. But a contributing factor will not cause the event. For example, a lack of routine inspections prevents an operator from seeing a hydraulic line leak, which, undetected, led to a more serious failure in the hydraulic system. Lack of inspection didn't cause the effect, but it certainly accelerated the impact.

There is a distinction between failure analysis, root cause failure analysis and root cause analisis.

Failure Analysis: Stopping an analysis at the Physical Root Causes. This is typically where most people stop, what they call their "Failure Analysis". The Physical Root is at a tangible level, usually a component level. We find that it has failed and we simply replace it. I call it a "parts changer" level because we did not learn HOW the "part failed."

Root Cause Failure Analysis: Indicates conducting a comprehensive analysis down to all of the root causes (physical, human and latent), but connotes analysis on mechanical items only. I have found that the word "Failure" has a mechanical connotation to most people. Root Cause Analysis is applicable to much more than just mechanical situations. It is an attempt on our part to change the prevailing paradigm about Root Cause and its applicability.

Root Cause Failure Analysis: A Guide to Improve Plant Reliability, First Edition. Trinath Sahoo.

Root Cause Analysis: Implies the conducting of a full-blown analysis that identifies the Physical, Human and Latent Root Causes of HOW any undesirable event occurred. The word "Failure" has been removed to broaden the definition to include such non-mechanical events like safety incidents, quality defects, customer complaints, administrative problems (i.e. – delayed shutdowns) and the similar events.

RCA can be done reactively (after the failure – RCFA) or proactively (RCA). Many organizations miss opportunities to further understand when and why things go well. Was it the project team involved? The change management methodology applied during implementation? The vendor used or the equipment selected? I would argue that performing RCA on successes is just as, if not more, important for overall success than performing RCFAs on failures

The objectives for conducting a RCA are to analyze problems or events to identify:

- What occurred
- How it occurred
- Why it occurred
- Actions for averting reoccurrence that can be developed and implemented

The root cause analysis process – RCA has five identifiable steps.

1) Define the problem
2) Collect data
3) Identify possible causal factors
4) Identify the root cause
5) Recommend and implement solution

Define the problem

One of the important steps in root cause failure analysis (RCFA) is to define a problem. Effective and event descriptions are helpful to ensure the execution of appropriate root cause analyses. The first step to define the problem is by asking the four questions:

- What is the problem?
- When did it happen?
- Where did it happen? and
- How did it impact the goals?

The investigator or the RCA analyst seldom present when an incident or failure occurs. Therefore, the first information report or FIR is the initial notification that an incident or failure has taken place. In most cases, the communication will not contain a complete description of the problem. Rather, it will be a very brief description of the perceived symptoms observed by the person reporting the problem.

It involves failure reporting regarding incident which includes details of failure time, place, nature of failure, and failure impacts on organization.

Consider a problem on a centrifugal pump AC Motor. A typical problem report could state "pump ABC motor has a problem". Even though this type of problem reporting could be worse, for example, "fan is bad" or "shrill noise from one of the pumps." "Pump ABC Motor has a problem" it is still not a very good definition.

A better definition may be "AC Motor of pump ABC" is hot. Can we do better with some basic Root Cause Analysis steps? Sure! Let's ask the traditional, WHAT, WHERE, WHEN, EXTENT. The problem is:

What: AC Motor of pump ABC (already answered)
Where: Motor is hot close to the front (belt drive side)
When: Don't know exactly, but 7 days ago a 138 F reading was recorded (normal)
Extent: Front of motor is running 210 F.

The above definition is usually enough to get a problem started. Is it ideal? Perhaps not, but it's pretty good for a problem statement. This level of problem reporting for craftspeople and operators would be a huge improvement for most plants in improving day-to-day Root Cause Analysis.

Collection of data

Data collection is the second and important phase of RCA process. Acquiring, gathering, or collecting the failure data regarding the incident are a key for getting the valuable results of RCA investigation. Comprehensive and relevant failure data are crucial to identify and understand the root causes of a failure accurately. Unavailability of correct, adequate, and sufficient data can lead to undesired results of RCA.

It is important to collect data immediately after occurrence of failure for accurate information and evidence collection before the data is lost. The information that should be collected consists of personnel involved; conditions before, during, and after the event; environmental factors; and other information required for root cause analysis process.

Every effort should be made to preserve physical evidence such as failed components, ruptured gaskets, burned leads, blown fuses, spilled fluids, partially completed work orders, and procedures. Event participants and other knowledgeable individuals should be identified. All work orders and procedures must be preserved and effort should be made to preserve physical evidence such as failed components and ruptured gaskets. After the data associated with the event have been collected, the data should be verified to ensure accuracy.

Data for any failure could include the previous failure reports, maintenance, and operations data, process data, drawings, design, physical evidences, failed part of equipment and any other necessary information related to the particular failure. It is not necessary that every failure required comprehensive data but sometimes data could be missing and gathered data is not sufficient to identify actual causes of the failure. So it is necessary that collected data must be accurate and relevant. Failure can't be investigated properly without availability of correct and related data. Usually, data collection consumes more time as compare to other steps of RCA process so data must be precise and meaningful for identifying the exact causes of failure. Information collected from gathered data is significant for making recommendation and conclusions.

When investigating an incident involving equipment failure, the first job is to preserve the physical evidence. The instrumentation and control settings and the actual reading before the failure happen should be fully documented for the investigating team. In addition, the operating and process data, approved standard operating (SOP) and standard maintenance procedure (SMP), Copies of log books, work packages, work orders, work permits, and maintenance records; eq should be preserved.

Some methods of gathering information include:

- Conducting interviews/collecting statements – Interviews must be fact finding and not fault finding. Preparing questions before the interview is essential to ensure that all necessary information is obtained.

- Interviews should be conducted, preferably in person, with those people who are most familiar with the problem. Although preparing for the interview is important, it should not delay prompt contact with participants and witnesses. The first interview may consist solely of hearing their narrative. A second, more-detailed interview can be arranged, if needed. The interviewer should always consider the interviewee's objectivity and frame of reference.
- Reviewing records: Review of relevant documents or portions of documents and reference their use in support of the root cause analysis.
- Acquiring related information: Some additional information that an evaluator should consider when analyzing the causes include:
  a) Evaluating the need for laboratory tests, such as destructive/nondestructive failure analysis.
  b) Viewing physical layout of system, component, or work area; developing layout sketches of the area; and taking photographs to better understand the condition.
  c) Determining if operating experience information exists for similar events at other facilities.
  d) Reviewing equipment supplier and manufacturer records to determine whether correspondence has been received addressing this problem.

Interviews

For critical incidents, all key personnel involved must be interviewed to get a complete picture of the incident. Individuals having direct or indirect knowledge that could help clarify the case should also be interviewed.

Questions to Ask

- What happened?
- Where did it happen?
- When did it happen?
- What changed?
- Who was involved?
- Why did it happen?
- What is the impact?
- How can recurrence be prevented?

Analyze Sequence of Events

The sequence of event helps in finding out which cause has first triggered the incident. This helps in organizing the information and establishes relationship between the event and incident.

Design Review

It is essential to clearly understand the design parameters and specifications of the systems associated with an event or equipment failure. Unless the investigator understands precisely what the machine or production system was designed to do and its inherent limitations, it is impossible to isolate the root cause of a problem or event. The data obtained from a design review provide a baseline or reference, which is needed to fully investigate and resolve plant problems.

The objective of the design review is to determine whether the machine is running within acceptable operating envelope. The condition of the machine and the process condition are being investigated. For example, a centrifugal pump may be designed to deliver 1000 m^3/h of water having a discharge Pressure of 20 kg/cm^2. If it is operated beyond this point, then the power will increase and due to running beyond design limit vibration may go up. The review should establish the acceptable operating envelope, or range, that the machine or system can tolerate without a measurable deviation from design performance. Evaluating variations in process parameters, such as pressures flow rate, and temperature, is an effective means of confirming their impact on the production system.

Operating and Maintenance Manuals

O&M manuals are one of the best sources of information. In most cases, these documents provide specific recommendations for proper operation and maintenance of the machine, equipment, or system. In addition, most of these manuals provide specific troubleshooting guides that point out many of the common problems that may occur. A thorough review of these documents is essential before beginning the RCA. The information provided in these manuals is essential to effective resolution of plant problems.

Operating Procedures and Practices

This part of the application and maintenance review consists of evaluating the standard operating procedures and the actual operating practices. Most production areas maintain some historical data that track its performance and practices. These records may consist of log books, reports, or computer data. These data should be reviewed to determine the actual production practices that are used to operate the machine or system being investigated.

This part of the evaluation should determine if the SOPs were understood and followed before and during the incident or event. The normal tendency of operators is to shortcut procedures, which is a common reason for many problems. In addition, unclear procedures lead to misunderstanding and misuse. Therefore, the investigation must fully evaluate the actual practices that the production team uses to operate the machine or system.

Maintenance History

A thorough review of the maintenance history associated with the machine or system is essential to the RCFA process. The primary details that are needed include frequency and types of repair, frequency and types of preventive maintenance, failure history, and any other facts that will help in the investigation.

Operating Envelope

Evaluating the actual operating envelope of the production system associated with the investigated event is more difficult. The best approach is to determine all variables and limits used in normal production. For example, define the full range of operating speeds, flow rates, incoming product variations, and the like normally associated with the system. In variable-speed applications, determine the minimum and maximum ramp rates used by the operators.

Maintenance Procedures and Practices

A complete evaluation of the standard maintenance procedures and actual practices should be conducted. The procedures should be compared with maintenance requirements defined by both the design review and the vendor's O&M manuals. Actual maintenance practices can be determined in the same manner as described earlier or by visual observation of similar repairs. This task should determine if the SMPs are followed consistently by all maintenance personnel assigned to or involved with the area being investigated. Special attention should be given to the routine tasks, such as lubrication, adjustments, and other preventive tasks. Determine if these procedures are being performed in a timely manner and if proper techniques are being used.

Misapplication

Misapplication of critical process equipment is one of the most common causes of equipment-related problems. In some cases, the reason for misapplication is poor design, but more often it results from uncontrolled modifications or changes in the operating requirements of the machine.

Management Systems

The common root causes of management system problems are policies and procedures, standards not used, and employee relations, inadequate training, inadequate supervision, wrong worker selection etc. Most of this potential root causes deal with plant culture and management philosophy. While hard to isolate, the categories that fall within this group of causes contribute to many of the problems that will be investigated. Many SOPS used to operate critical plant production systems are out of date or inadequate. This often is a major contributor to reliability and equipment-related problems. Training or inadequate employee skills commonlycontribute to problems that affect plant performance and equipment reliability. The reasons underlying inadequate skills vary depending on the plant culture, workforce, and a variety of other issues.

Identify Possible Causal Factors

What Is a Causal Factor?

A causal factor can be defined as any "major unplanned, unintended contributor to an incident (a negative event or undesirable condition), that if eliminated would have either prevented the occurrence of the incident or reduced its severity or frequency. Also known as a critical contributing cause."

What Is a Root Cause?

A root cause is "a fundamental reason for the occurrence of a problem or event." Analysts can look for the root cause of an event in order to prevent it from happening again in the future. The root cause is the primary driver of a process.

What Is the Difference Between a Causal Factor and a Root Cause?

The causal factor isn't the single factor that drove the event. Instead, a causal factor was one of a few influences. The event could still occur again or would have happened without the causal factor. In fact, during a root cause analysis, analysts often use techniques called the "5 whys," fish bone diagram, fault tree analysis etc to identify multiple causal factors until they find a root cause of an event. Put simply, the root cause is the primary driver of the event and causal factors are secondary or tertiary drivers.

During this stage, identify as many causal factors as possible. Too often, people identify one or two factors and then stop, but that's not sufficient. With RCA, you don't want to simply treat the most obvious causes – you want to dig deeper.

- What sequence of events leads to the problem?
- What conditions allow the problem to occur?
- What other problems surround the occurrence of the central problem?

The Five Whys

The Five Whys is a simple problem-solving technique that helps to get to the root of a problem quickly. The Five Whys strategy involves looking at any problem and drilling down by asking: "Why?" or "What caused this problem?" Invented in the 1930s by Toyota Founder Kiichiro Toyoda's father Sakichi and made popular in the 1970s by the Toyota Production System, the 5 Whys strategy involves looking at any problem and asking:

"Why?" and "What caused this problem?"

The idea is simple. By asking the question, "Why" you can separate the symptoms from the causes of a problem. This is critical as symptoms often mask the causes of problems. As with effective incident classification, basing actions on symptoms is worst possible practice. Using the technique effectively will define the root cause of any non-conformances and subsequently lead you to defining effective long-term corrective actions.

While you want clear and concise answers, you want to avoid answers that are too simple and overlook important details. Typically, the answer to the first "why" should prompt another "why" and the answer to the second "why" will prompt another and so on; hence the name Five Whys. This technique can help you to quickly determine the root cause of a problem. It's simple and easy to learn and apply.

The 5-Why analysis is the primary tool used to determine the root cause of any problem. It is documented in the Toyota Business Process manual and practiced by all associates.

When to Use 5 Why

When the problem and root cause is not immediately apparent When you want to prevent the problem from occurring in the future.

Ask yourself, "Will implementing the Systemic Corrective Action prevent the next failure?" If the answer is "NO," you must understand the deeper WHY.

If human error is identified, you must understand why the human committed the error. What management controlled factor impacted performance? What system must change to eliminate (or significantly reduce) the chance for error? "Training the Operator" is rarely the best response.

Why was the operator not trained properly? Why was the training not effective? What environmental factors caused the operator to not do his/her best work? Did he/she have to go around the system due to other issues or pressures? Can the system be error-proofed? All root cause analysis must include a look at the associated Management Systems For virtually every incident, some improvement(s) in the Management Systems could have prevented most (or all) of the contributing events – ASQ estimates 82–86% Correct the process that created the problems.

During the 5 Why analysis, you should ask yourself if there are similar situations that need to be evaluated perform a "Look Across" the organization. If this situation could apply to multiple funds, then the corrective action must address all funds.

How to Use the 5 Whys

1) Develops the problem statement. Be clear and specific.
2) Assemble a team of people knowledgeable about the processes and systems involved in the problem being discussed. They should have personal knowledge about the non-conformance of the system.
3) On a flip chart, presentation board, or even paper; write out a description of what you know about the problem. Try to document the Problem and describe it as completely as possible. Refine the definition with the team. Come to an agreement on the definition of the Problem at hand.
4) The team facilitator asks why the problem happened and records the team response. To determine if the response is the root cause of the problem, the facilitator asks the team to consider "If the most recent response were corrected, is it likely the problem would recur?" If the answer is yes, it is likely this is a contributing factor, not a root cause.
   - If the answer provided is a contributing factor to the problem, the team keeps asking "Why?" until there is agreement from the team that the root cause has been identified.
   - It often takes three to Five Whys, but it can take more than five! So keep going until the team agrees the root cause has been identified.

The 5 Whys can help you uncover root causes quickly. However, making a single mistake in any question or answer can produce false or misleading results. You may find that there is more than one root cause for each non-conformance; corrective actions should be implemented for each of these.

Fishbone Diagram

One of the more popular tools used in root cause analysis is the *fishbone diagram,* otherwise known as the *Ishikawa diagram,* named after Kaoru Ishikawa, who developed it in the 1960s. A fishbone diagram is perhaps the easiest tool in the family of cause and effect diagrams that engineers and scientists use in unearthing factors that lead to an undesirable outcome.

A fishbone diagram is a visual way to look at cause and effect. It is a more structured approach than some other tools available for brainstorming causes of a problem (e.g., the Five Whys tool). The problem or effect is displayed at the head or mouth of the fish. Possible contributing causes are listed on the smaller "bones" under various cause categories. A fishbone diagram can be helpful in identifying possible causes for a problem that might not otherwise be considered by directing the team to look at the categories and think of alternative causes. Include team members who have personal knowledge of the processes and systems involved in the problem or event to be investigated.

Fishbone Diagram Structure

The left side of the diagram is where the **causes** are listed. The causes are broken out into major cause categories. The causes you identify will be placed in the appropriate cause categories as you build the diagram.

The right side of the diagram lists the **effect**. The effect is written as the **problem statement** for which you are trying to identify the causes.

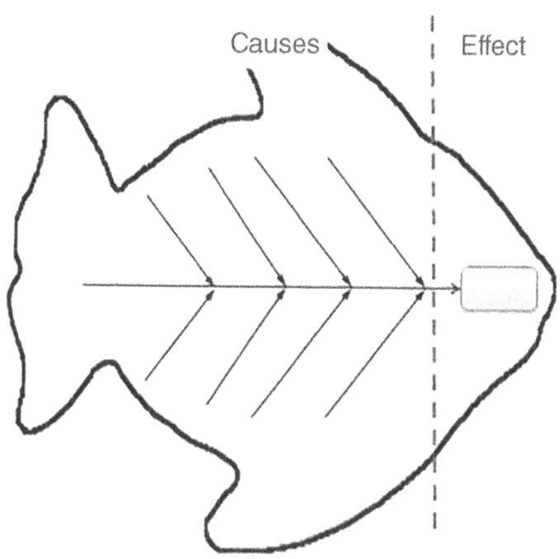

Ishikawa Fish Bone Diagram

The diagram looks like the skeleton of a fish, which is where the *fishbone* name comes from.

How to Create a Cause and Effect Diagram

A cause and effect diagram can be created in six steps.

1) Draw Problem Statement
2) Draw Major Cause Categories
3) Brainstorm Causes
4) Categorize Causes
5) Determine Deeper Causes
6) Identify Root Causes

1) Draw Problem Statement

   The first step of any problem-solving activity is to define the problem. You want to make sure that you define the problem correctly and that everyone agrees on the problem statement.

   Once your problem statement is ready, write it in the box on the right-hand side of the diagram.

2) Draw Major Cause Categories
After the problem statement has been placed on the diagram, draw the *major cause categories* on the left-hand side and connect them to the "backbone" of the **fishbone chart**. In a manufacturing environment, the traditional categories are

- Machines/Equipment
- Methods
- Materials
- People

In a service organization, the traditional categories are. . .

- Policies
- Procedures
- Plant
- People

You can start with those categories or use a different set that is more applicable for your problem. There isn't a perfect set or specified number of categories. Use what makes sense for your problem.

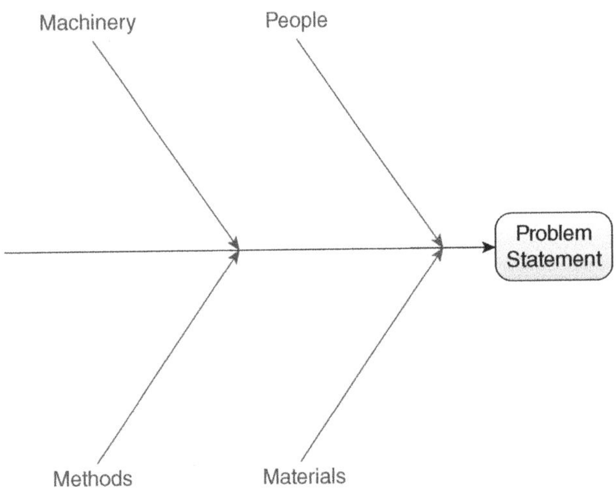

Cause and Effect Diagram - Major Cause Categories

3) Brainstorm Causes
Brainstorming the causes of the problem is where most of the effort in creating your Ishikawa diagram takes place.
Some people prefer to generate a list of causes before the previous steps in order to allow ideas to flow without being constrained by the major cause categories.
However, sometimes the major cause categories can be used as catalysts to generate ideas. This is especially helpful when the flow of ideas starts to slow down.

4) Categorize Causes
   Once your list of causes has been generated, you can start to place them in the appropriate category on the diagram.

   - Draw a box around each category label and use a diagonal line to form a branch connecting the box to the spine.
   - Write the main categories your team has selected to the left of the effect box, some above the spine and some below it.
   - Ideally, each cause should only be placed in one category. However, some of the "People" causes may belong in multiple categories. For example, Lack of Training may be a legitimate cause for incorrect usage of Machinery as well as ignorance about a specific Method.
   - Establish the major causes, or categories, under which other possible causes will be listed. You should use category labels that make sense for the diagram you are creating.

   Identify as many causes or factors as possible and attach them as subbranches of the major branches

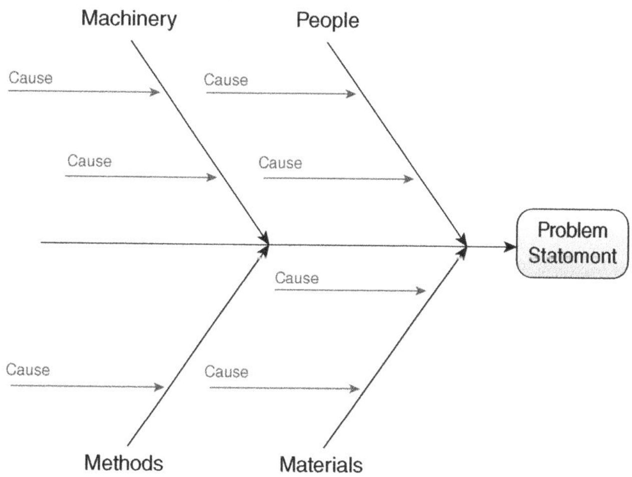

Ishikawa Diagram - Categorize Causes

5) Determine Deeper Causes
   Each cause on the chart is then analyzed further to determine if there is a more fundamental cause for that aspect. This can be done by asking the question, "Why does it happen?"
   This step can also be done for the deeper causes that are identified. Generally, you can stop going deeper when a cause is controlled one level of management removed from your group. Use your judgment to decide when to stop.

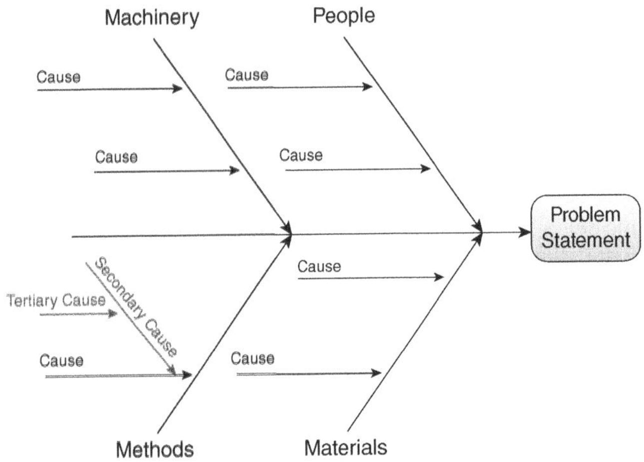

Fishbone Chart - Deeper Causes

6) Identify Root Causes
   The final step for creating a fishbone diagram is to identify the root causes of the problem. This can be done in several ways. . .

   - Look for causes that appear repeatedly
   - Select using group consensus methods
   - Select based on frequency of occurrence

   Fishbone diagrams are an excellent way to explore and visually depict the causes of a problem. They enable the root causes of a problem to be determined. This will help you be more effective by focusing your actions on the true causes of a problem and not on its symptoms. It Encourages group participation, Uses an orderly, easy-to-read format to diagram cause and effect relationships.

Fault Tree Analysis

Fault tree analysis helps determine the root cause of failure of a system using Boolean logic to combine a series of lower level events. FTA is a deductive analysis depicting a visual path of failure. It is a top-down analysis that helps determine the probability of occurrence for an undesirable event. The analysis creates a visual record showing the logical relationships between events and failures that lead to the undesirable event. It easily presents the results of your analysis and pinpoints weaknesses in the system.

The fault tree analysis (FTA) was first introduced by Bell Laboratories and is one of the most widely used methods in system reliability, maintainability and safety analysis. It is a deductive procedure used to determine the various combinations of hardware and software failures and human errors that could cause undesired events (referred to as top events) at the system level.

To do a comprehensive FTA, follow these steps:

1) Define the fault condition, and write down the top-level failure.
2) Using technical information and professional judgments, determine the possible reasons for the failure to occur. Remember, these are level two elements because they fall just below the top-level failure in the tree.
3) Continue to break down each element with additional gates to lower levels. Consider the relationships between the elements to help you decide whether to use an "and" or an "or" logic gate.
4) Finalize and review the complete diagram. The chain can only be terminated in a basic fault: human, hardware, or software.
5) If possible, evaluate the probability of occurrence for each of the lowest level elements and calculate the statistical probabilities from the bottom up.

Drawing Fault Trees: Gates and Events

Gate symbols represent results of interactions among contributing failure events and can vary among tools. Basic gates used to construct the Fault Tree can be seen below:

Gate Symbol	Name	Causal Relation
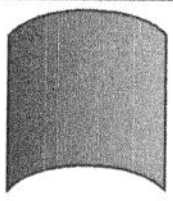	OR	Output event occurs if any one of the input events occurs
	AND	Output event occurs if all input events occur
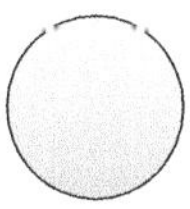	BASIC	Basic event for which failure data is available.
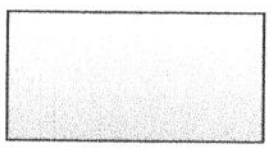	INTERMEDIATE EVENT	System or component event description
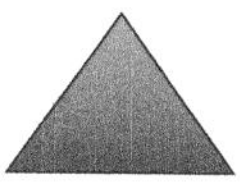	TRANSFER	Indicates that this part of the fault tree is developed in a different part of the diagram or on a different page.

FTA Gate Notes

Fault Tree probabilities can be computed by simple arithmetic only if basic events (component failures without lower level contributors) are independent. Independence is determined by ensuring the failure of one basic event has no effect on any other and groups of basic events cannot fail from common causes such as shock. For independent basic events with very small failure rates, typically found in electronic components, an AND gate output probability can be computed as the product of its input failure probabilities, and an OR gate output probability can be computed as the sum of its input failure probabilities.

Below is a Basic Fault Tree Analysis Example Structure:

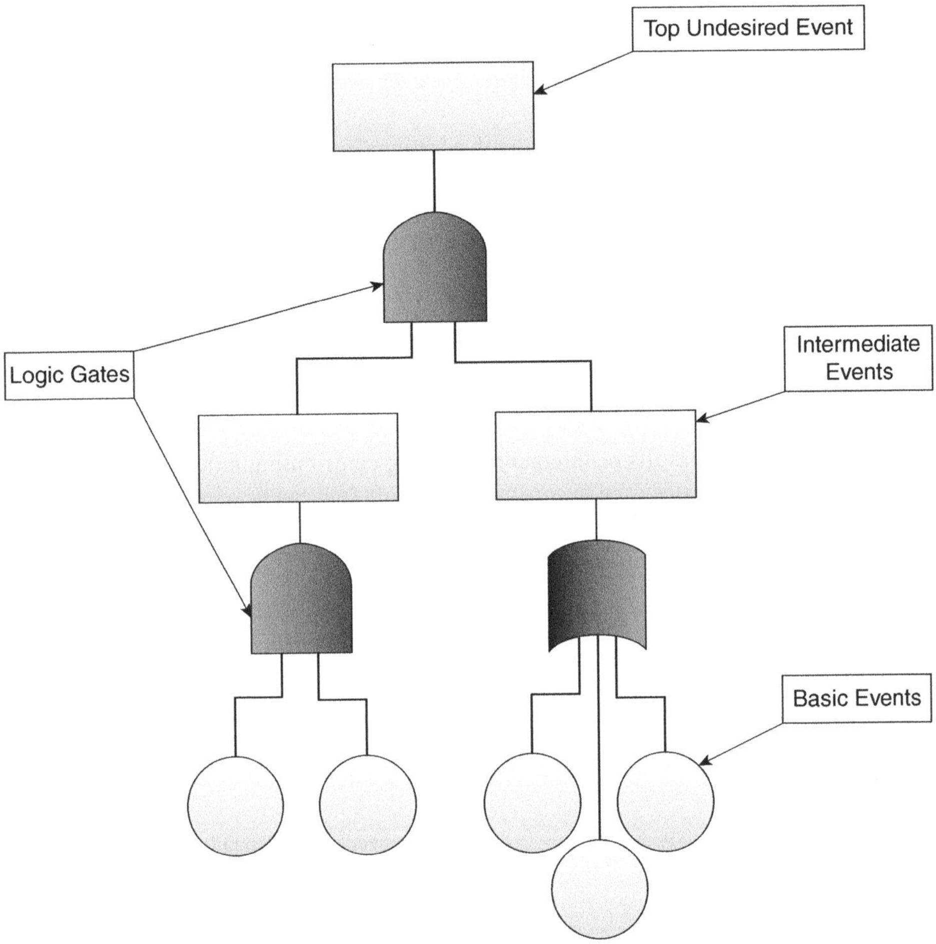

The five basic steps to perform a Fault Tree Analysis are as follows:

1) Identify the Hazard
2) Obtain Understanding of the System Being Analyzed
3) Create the Fault Tree
4) Identify the Cut Sets
5) Mitigate the Risk

Top-level event is called a Cut Set. There are many cut sets within the FTA. Each has an individual probability assigned to it. The paths related to the highest severity / highest probability combinations are identified and will require mitigation.

How to Undertake a Fault Tree Analysis?

Although the nature of the undesired event may be quite different, fault tree analysis has the same procedure for any types of undesired event. To do a comprehensive fault tree analysis, simply follow the process below:

1) Define and identify the fault condition (hazard) as precisely as possible based on the aspects such as the amount, duration, and related impacts.
2) Using technical skills and existing facility details to list and decide all the possible reasons for the failure occurrence.
3) Break down the tree from the top level according to the relationship between different components until you work down to the potential root cause. The structure of your fault tree analysis diagram should be based on the top, middle (subsystems), and the bottom (basic events, component failures) levels.
4) If your analysis involves the quantitative part, evaluate the probability of occurrence for each of the components and calculate the statistical probabilities for the whole tree.
5) Double-check your overall fault tree analysis diagram and implement modifications to the process if necessary.
6) Collect data, evaluate your results in full details by using risk management, qualitative, and quantitative analysis to improve your system.

Benefits of Fault Trees

A fault tree creates a visual record of a system that shows the logical relationships between events and causes that lead to failure. It helps others quickly understand the results of your analysis and pinpoint weaknesses in the design and identify errors. A fault tree diagram will help prioritize issues to fix that contribute to a failure. In many ways, the fault tree diagram creates the foundation for any further analysis and evaluation. For example, when changes or upgrades are made to the system, you already have a set of steps to evaluate for possible effects and changes. You can use a fault tree diagram to help design quality tests and maintenance procedures.

Identify the Root Cause

Look over your list of potential causal factors and determine the real reason this problem or issue occurred in the first place. These data should have provided enough insight into the failure for the investigator to develop a list of potential or probable reasons for the failure. Dig deep to examine each level of cause and effect and the events that led to the unfavorable outcomes. The problem is that in the real world it is never possible to prove a single event that solely initiates a whole chain of other events. That is because there are always other events before the so-called "root cause event." This may seem like semantics, but for problem-solvers, it is important to keep in mind that there never is a silver-bullet answer.

Analyzing the short list of potential root causes is to verify each of the suspect causes is essential. In almost all cases, a relatively simple, inexpensive test series can be developed to confirm or eliminate the suspected cause of equipment failure.

Most equipment problems can be traced to misapplication, operating or maintenance practices and procedures. Some of the other causes that are discussed include training, supervision, communications, human engineering, management systems, and quality control. These causes are the most common reasons for poor plant performance and equipment reliability. However, human error may contribute to, or be the sole reason for, the problem.

Recommend and Implement Solution

When working on solutions, keep your Root Cause Analysis aim in view. You don't just want to solve the immediate problem. You want to prevent the same problem from recurring.

Ask the following questions for finding a solution,

- What can you do to prevent the problem from happening again?
- How will the solution be implemented?
- Who will be responsible for it?
- What are the risks of implementing the solution?

A short list of potential corrective action are generated. Each potential corrective action should be carefully scrutinized to determine if it actually will correct the problem. Because many time the analyst Try to fix the symptoms of problems rather than the true root cause. Therefore, care should be taken to evaluate each potential corrective action so that the right one can be implemented to eliminates the real problem. Many a time all corrective actions are not financially justifiable. In some cases, the impact of the incident or event is lower than the cost of the corrective action. In these cases, the RCA should document the incident for future reference, but recommend that no corrective action be taken on some occasions, implementing a temporary solution is the only financially justifiable course of action which can only correct the symptoms. In these instances, the recommendation should clearly define the reason the limitations why this decision was taken and what impact it will have on plant performance.

Also, consider whether the changes you plan to make will impact other areas of your business. Changes to processes can have knock-on effects. Be sure you aren't setting yourself up for a new set of problems when you implement the solution. To do this, you need to look at your process flows and how they relate to one another.

The final part of the solution design process is to decide on checks and balances that will tell you whether your business is implementing the solution you've devised and whether it works as planned.

Implementation means change, and change must be carefully managed. Everyone concerned needs to know about your solution and the reasoning that led you to believe that you can solve the problem.

So, explain the root cause analysis process and how you arrived at your conclusion. Explain your solution and how you want it to be implemented. Ensure that everyone involved has the knowledge and resources they need to follow through and devise method for testing your new system.

Keep in mind, though, that it's always better to first apply the solution on a small scale. You can never know what could go wrong. Once you're certain that the new solution brings results, you can start applying it company-wide.

Conclusion

When you designed the solution, you decided on key indicators that would allow you to see whether the solution works. Use these indicators to follow up. In this instance, you're going to see whether the symptoms are gone. The presence or absence of the issues that launched you on your root cause analysis and problem-solving initiative will tell you whether you have successfully solved the problem. Remember to watch out for new issues that may arise elsewhere as a result of the changes you made.

4

Managing Human Error and Latent Error to Overcome Failure

Everyone can make errors no matter what their level of skill, experience or how well trained and motivated they are. Commonly cited statistics claim that human error is responsible for anywhere between 70 and 100% of failure. Many major failures, e.g. Texas City, Piper Alpha, Chernobyl were contributed by human failure. To enhance reliability, companies need to manage human failure as robustly as they manage technical and engineering failures. It is important to be aware that human failure is not random; understanding why errors occur and the different factors which make them worse will help you develop more effective controls.

Human error was a factor in many highly publicized accidents in recent memory. The costs in terms of human life and money are high. Placing emphasis on reducing human error may help to reduce these costs. This chapter provides an insight view about the causes of human errors and suggests the way to reduce the errors.

Review of Some of the Accidents

Over the last few decades, we have learnt much more about the origins of human failures. The industries/organizations must consider human factor as a distinct element to be assessed and managed effectively in order to control risks. Some of the following accidents of Table 4.1 in different sectors provide clues to understand failures.

Table 4.1 illustrates how the failure of people at many levels within an organization can contribute to a major disaster. For many of these major accidents, the human failure was not the sole cause but one of a number of causes, including technical and organizational failures, which led to the final outcome. Remember that many "everyday" minor accidents and near misses also involve human failures. All major disasters lead to huge human, property, and environmental losses.

All this evidence shows that human error is a major cause of unreliability or causation of accidents.

Types of Human Failure:

What Types of Errors Do Humans Make?

The consequences of human failures can be immediate or delayed and the failures can be grouped into the following categories:

Root Cause Failure Analysis: A Guide to Improve Plant Reliability, First Edition. Trinath Sahoo.

Table 4.1 Industrial accidents caused by human error.

Accident, industry and date	consequences	Human contribution and other cause
Union Carbide Bhopal, 1984 (Chemical Unit)	The plant released a cloud of toxic methyl isocyanate. Death toll was 2500 and over one quarter of the city's population was affected by the gas.	The leak was caused by a discharge of water into a storage tank. This was the result of a combination of operator error, poor maintenance, failed safety systems, and poor safety management.
Space Shuttle Challenger 1986 (Aerospace)	An explosion shortly after lift-off killed all seven astronauts on board	An O-ring seal on one of the solid rocket boosters split after take-off releasing a jet of ignited fuel. Inadequate response to internal warnings about the faulty seal design. Decision taken to go for launch in very cold temperature despite faulty seal. Decision-making result of conflicting scheduling/safety goals, mindset, and effects of fatigue.
Piper Alpha 1988 (Offshore)	167 workers died in the North Sea after a major explosion and fire on an offshore platform	Formal inquiry found a number of technical and organizational failures. Maintenance error that eventually led to the leak was the result of inexperience, poor maintenance procedures, and poor learning by the organization. There was a breakdown in communications and the permit-to work system at shift changeover, and safety procedures were not practiced sufficiently
Texaco Refinery, 1994 (Petroleum Industry	An explosion on the site was followed by a major hydrocarbon fire and a number of secondary fires. There was severe damage to process plant, buildings and storage tanks. 26 people sustained injuries, none serious.	The incident was caused by inflammable hydrocarbon liquid being continuously pumped into a process vessel that had its outlet closed. This was the result of a combination of: an erroneous control system reading of a valve state, modifications which had not been fully assessed, failure to provide operators with the necessary process overviews and attempts to keep the unit running when it should have been shut down.

Active failures- Active failures are the acts or conditions precipitating the incident situation. Active failures have an immediate consequence and are usually made by front-line people such as drivers, control room staff or machine operators. In a situation where there is no room for error, these active failures have an immediate impact on failure.

Latent failures- Though active failures are the acts or conditions precipitating the incident situation. Latent human error is made due to systems or routines that are formed in such a way that humans are disposed to making these errors.

Active Failures

There are 3 types of active human error:

- Slips and lapses – made inadvertently by experienced operators during routine tasks
- Mistakes – decisions subsequently found to be wrong, though the maker believed them to be correct at the time
- Violations – deliberate deviations from rules for safe operation of equipment

Familiar tasks carried out without much conscious attention are vulnerable to slips and lapses if the worker's attention is diverted: for example, missing a step in a sequence because of an interruption.

Mistakes occur where a worker is doing too many or complex tasks at the same time or is under time pressure: for example, misjudging the time and space needed to complete an overtaking maneuvre.

Violations, though deliberate, usually stem from a desire to perform work satisfactorily given particular constraints and expectations.

Factors that are most closely tied to the failure and can be described as active failures or actions committed by the operator that result in human error. We have identified these active failures or actions as Errors and Violations.

i) **Errors:** Errors are factors in a mishap when mental or physical activities of the operator fail to achieve their intended outcome as a result of skill-based, perceptual, or judgment and decision-making errors, leading to an unsafe situation. Errors are unintended. We classified Errors into two types:
   a) **Skill-based Errors**: When people are performing familiar work under normal conditions, they know by heart what to do. They react almost automatically to the situation and do not really have to think about what to do next. For instance, when a skilled automobile driver is proceeding along a road, little conscious effort is required to stay in the lane and control the car. The driver is able to perform other tasks such as adjusting the radio or engaging in conversation without sacrificing control. Errors committed at this level of performance are called slips or lapses.
   b) **System based**: are a more complex type of human error where we do the wrong thing believing it to be right. The failure involves our mental processes which control how we plan, assess information, make intentions and judge consequences. These errors are judgment and decision-making errors. Misperception of an object, threat or situation (such as visual, auditory, proprioceptive, or vestibular illusions, cognitive or attention failures).

ii) **Violations:** Violations are any deliberate deviations from rules, procedures, instructions, and regulations. The breaching or violating of rules or maintenance procedures is a significant cause of many failures. Removing the guard on dangerous machinery or driving too fast will clearly increase the risk. Our knowledge of why people break rules can help us to assess the potential risks from violations and to develop control strategies to manage these risks effectively.

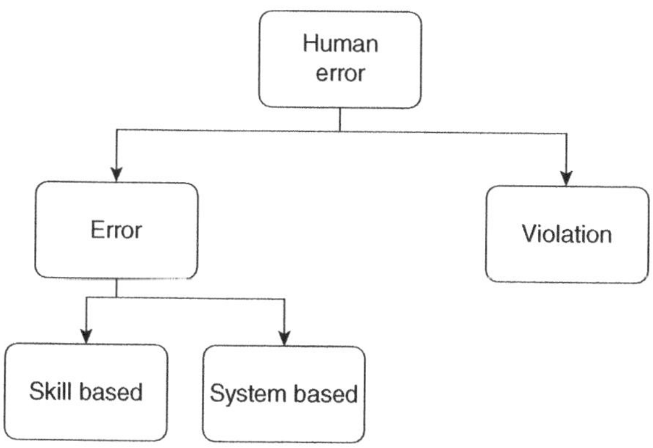

Figure 4.1 Contributing factors to human error.

Latent Failures

Latent failures are normally present in the system well before an failure occur and are most likely bred by decision-makers, regulators, and other people far removed in time and space from the event. These are the managerial influences and social pressures that make up the culture ("the way we do things around here"), influence the design of equipment or system, and define supervisory inadequacies. They tend to be hidden until triggered by an event. Latent failures may occur when several latent conditions combine in an unforeseen way. Efforts should be directed at discovering and solving these latent failures rather than by localizing efforts to minimize active failures by the technician. Also, there are organizational influences such as communications, actions, omissions, or policies of upper-level management directly or indirectly affect supervisory practices, conditions, or actions of the operator(s) and result in system failure or human error.

A distinction between active failures and latent conditions rests on two differences. The first difference is the time taken to have an adverse impact. Active failures usually have immediate and relatively short-lived effects. Latent conditions can lie dormant, doing no particular harm, until they interact with local circumstances to defeat the systems' defenses. The second difference is the location within the organization of the human instigators. Active failures are committed by those at the human–system interface, the front-line activities. Latent conditions, on the other hand, are spawned in the upper echelons of the organization and within related manufacturing, contracting, regulatory, and governmental agencies that are not directly interfacing with the system failures

The consequences of these latent conditions permeate throughout the organization to local workplaces – control rooms, work areas, maintenance facilities etc. – These local workplace factors include undue time pressure, inadequate tools and equipment, poor human–machine interfaces, insufficient training, under-manning, poor supervisor–worker ratios, low pay, low morale, low status, macho culture, unworkable or ambiguous procedures, and poor communications.

Within the workplace, these local workplace factors can combine with natural human performance tendencies such as l limited attention, habit patterns, assumptions, co complacency, or mental shortcuts. These combinations produce unintentional errors a and intentional violation committed by individuals and teams at the "sharp end," or the direct t human-system interface (active error).

Latent failures are those aspects of an organization which influence human behavior and make active failures more likely. Factors include:

- Ineffective training;
- Inadequate supervision;
- Ineffective communications;
- Inadequate resources (e.g. people and equipment); and
- Uncertainties in roles and responsibilities;
- Poor SOPs.
- poor equipment design or workplace layout
- work pressure, long hours, or insufficient supervision
- distractions, lack of time, inadequate procedures, poor lighting, or extremes of temperature

Latent failures provide great, potential danger to active failures. Latent failures are usually hidden within an organization until they are triggered by an event likely to have serious consequences.

What Factors Influence Human Reliability?

For practical application, it is important to understand the errors personnel are likely to make. Steps can be taken to eliminate them or, if this is not possible, to minimize the consequences.

The main causes of human error are as follows:

- unfamiliarity: a job or situation is important but occurs infrequently or is novel,
- time shortage: not enough time is available to complete the job by following procedure or for error detection and correction,
- understanding: People do not understand the job properly or no means available to convey information such that it is easy to understand,
- "Mental models": the way the operator imagines a system to work is different to how the designer imagined it,
- information overload: simultaneous presentation of information goes beyond a persons capacity to understand,
- new techniques: the need to learn new techniques which may follow philosophies opposing those that have been used previously,
- feedback system: feedback is poor, ambiguous, or inappropriate,
- conformation: no clear conformation is available from the system of the action that is required to control it,
- inexperience: the circumstance present requires experience, to understand and control the situation, beyond that of the person involved,
- information quality: specified procedures, or instructions from other humans, are of poor quality such that they are inappropriate to the situation present when followed,
- diversity: the system has no diversity to allow checking of information presented,
- physical ability: the person does not have the physical ability to perform the required tasks,
- mental stimulation: the person is required to spend a lot of time either inactive or involved in highly repetitive, menial tasks,
- disruption: work patterns cause disruption to normal sleep and rest cycles,
- pacing: other people influence the pace at which tasks can be performed,
- over manning: more people present than required to do the job satisfactorily.

What Factors Influence Human Variability

Human performance depends on a lot of factors which means they perform differently in different situations. Below are list of factors that affects people performance.

- reaction to stress,
- fatigue,
- supervisor's expectations,
- social interaction,
- social pressure,
- group interaction and identification,
- crew efficiency,
- morale,
- time at work,
- idle time, and
- repetition of work.

These factors should be taken into account for considering reliable system. What is the possibility that good engineering practice will not be followed? It is also important to consider these factors when selecting personnel or setting up teams.

The Prevention of Human Error

If we eliminate human error or at least reduce the consequences, we will have gone a long way to preventing failures. To improve human reliability, we need to understand what affects it. Human's will always make errors, the reason that failure happen because of latent errors that are present in the systems which do not give the operator a chance. These latent errors are the root cause of most failures. The way to reduce human error through effective management such that reliability is considered important by everyone involved.

Ways to Reduce Human Error

The Procedures

Procedures are very important and should be well thought out.

- They should be clear, precise, and easy to read.
- They should be easily understood by those who will use them. The best way to achieve this is to write procedures with close participation with those people.
- They should be easily accessible.
- Checklists should be provided, to be followed for unusual circumstances. The system should be checked for any shortcuts that may be easy to take and that will cause risk.

Audits

These will generally be regular checks made by independent assessors covering a wide range of features like whether maintenance procedures are followed or not, whether equipments are running within operating window or not etc. They provide a good indication of the obvious problem areas.

Increased Discipline

Here people are punished for the failures they are involved in. Once again, this is more likely to reduce reporting of errors rather than necessarily the number of failures. It also requires placing the blame on certain people. Although an operator might have been directly involved, the failure is more likely to have been caused by latent human errors which may not be so obvious and are out of their control. Of course, it is usually the management of a company that would decide who to punish, blaming an operator is a lot easier than blaming a manager who is really more responsible by allowing latent errors to be present in the system.

Increased Automation

Here human manual control is replaced by automatic controls, generally electronic devices. These devices will do as instructed without the problems of human variability and unpredictabilty. With a high amount of automation, the human has different tasks to perform. They are basically there to deal with unforeseen circumstances or to perform tasks that the designer cannot automate. This generally means that during normal operation there is less to be done. Automation can be useful but must be designed correctly. The information should match the operators own mental idea of what is happening. The information should be useful. Important information should be given priority, cross checking should be possible to validate information, alarm analysis and decision aids should be included. Operators probably need more training to operate automated plant although it would appear they have less to do than on a manually operated plant.

Automation can be useful but must be designed correctly. The information should match the operators own mental idea of what is happening. The information should be useful. Important information should be given priority, cross-checking should be possible to validate information, alarm analysis and decision aids should be included.

Improved Training

Training is very important in the effort to reduce human errors and hence accidents. Safety training is vital for everybody involved in the system. Operator training will not, however, improve reliability when the root cause is bad design or poor management. Training given should be well planned and appropriate to the job. Realistic simulation and role-play exercises are some of the best ways to train people. Everybody must be familiar with the system and made aware of the risks involved and how their actions effect reliability. Training should cover the use of all job aids including procedures, and other ancillary and emergency equipment. Recovery procedures should be explained for use after errors have been made. Personnel performance checks and evaluations should be used and good, constructive feedback given at regular intervals. Refresher training should also be used to prevent behavior patterns building up such that variations of equipment and procedures can not be handled.

Motivational Campaigns

This is a system where some sort of reward is offered for operating in a reliable manner. This usually involves analyzing failure rates. There's little doubt that advertising and campaigns can significantly affect how people think and act. It is not a set offixed rules; it is purely a collection of recommendations. Pick and choose what you want, adapt them to your individual circumstances. People are in undated with information every day. To grab their attention, your message must be short, simple and relevant to your target audience. They must immediately understand its importance and what you are asking them to do and why. Try to distil your main message down to a jargon-free statement, ideally no longer than two sentences. One way to do this is to think in terms of 'problem and solution'. Other method for campaign are

- Raise awareness of reliability problems, including their causes, amongst managers, employees and people who advise them on these issues.
- Provide practical solutions through 'good practice models'.
- Alert people to new risks and possible solutions.

Conclusion

Human error does account for a large number of accidents, however, it is the latent errors that are the real root cause. It is far too easy to blame operators for causing accidents but it must be appreciated that all humans will make errors. It is the job of the management to ensure that system and procedures are in place to avoid such incidents. The responsibility starts at the very top, with the managing director, and must work its way to all levels in the company. That way the company culture will improve to support reliability first.

5

Metallurgical Failure

Many a times unanticipated equipment failures do occur for a variety of reasons. These events often become too costly as well as disruptive to plant operations and may also have safety implications. To minimize the frequency and severity of such failures, it is necessary for personnel who have equipment responsibility to understand the failures and to confront their causes. Potential causes of the failure of the components and their mechanism are numerous. Therefore, procedure of the failure analysis of each failed component should be different and the same must be developed after giving proper thought on possible sequence of events before failure along with proper evaluation of the situation and consideration of material, manufacturing process, service history and actual working condition, etc. Since the failure analysis involves lot of efforts, time and use of resources therefore at the end of analysis failure analyst should be in a position to come out with few most potential causes of the failure so that suitable recommendations can be made to avoid reoccurrence of the similar failure.

It may sound a little far-fetched, but experts say that the causes for more than 90% of all plant failures can be detected with a careful physical examination using low power magnification and some basic physical testing. Inspection of the failure component will show the forces involved, whether the load applied cyclically or was single overload, the direction of the critical load, and the influence of outside forces such as residual stresses or corrosion. Then, accurately knowing the physical roots of the failure, you can pursue both the human errors and the latent causes of these physical roots.

In this chapter, an overview of the processes involved in a typical metallurgical failure analysis is provided. The discussion describes various failure mechanisms in metals that can be examined, some of the tests and processes that are used in an analysis, for failed components.

The metallurgical failure analysis can be defined as a scientifically based systematic laboratory examination of metallurgical evidence and the gathering of background information related to an equipment failure. This analysis helps in establishing the cause of the failure. Because the approach to the failure analysis is usually determined by the nature of the failure, all analyses do not require the same procedure. Laboratory procedures focus on the failed equipment itself and most commonly consist of general and detailed macrophotography, metallo-graphic examination, chemical analysis of the failed part and of any extraneous or foreign materials present, mechanical property determinations, fractographic examination, and others.

Root Cause Failure Analysis: A Guide to Improve Plant Reliability, First Edition. Trinath Sahoo.

Understanding the Basics

Before explaining how to diagnose a failure, we should review the effects of stress on a component. When a load is put on a part, it distorts. In a sound design the load isn't excessive, the stress doesn't exceed the "yield point," and the part deforms elastically, i.e., when the load is released the part returns to its original shape. This is shown in Figure 5.1, a "stress–strain" diagram that shows the relationship between loads and deformation. In a good design, the part operates in the elastic range, the area between the origin and the yield strength. Beyond this point, the part will be permanently deformed, even greater increases in load will cause the part to break.

Elastic Limit

The elastic limit is the limit beyond which the material will no longer go back to its original shape when the load is removed, or it is the maximum stress that may be developed such that there is no permanent or residual deformation when the load is entirely removed.

Elastic and Plastic Ranges

The region in stress–strain diagram from O to P is called the elastic range. The region from P to R is called the plastic range.

Yield Point

Yield point is the point at which the material will have an appreciable elongation or yielding without any increase in load.

Ultimate Strength

The maximum ordinate in the stress–strain diagram is the ultimate strength or tensile strength.

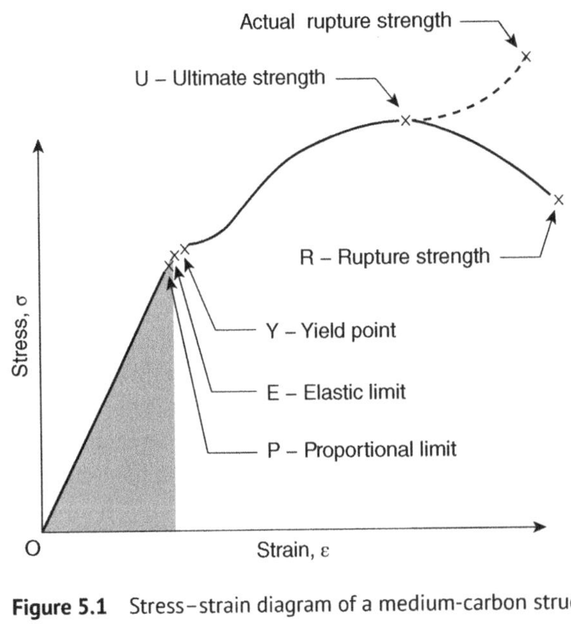

Figure 5.1 Stress–strain diagram of a medium-carbon structural steel.

Rapture Strength

Rapture strength is the strength of the material at rupture. This is also known as the breaking strength.

Elastic Deformation

Elastic deformation occurs when stiffness of the component is less and the same is primarily determined by modulus of elasticity and cross section. Elastic deformation can lead to the failure of mechanical components especially in high precision assemblies and machinery where even small elastic deformation under operating conditions is not acceptable.

Plastic Deformation

Excessive plastic deformation of the mechanical components can lead to the failure in two conditions (i) externally applied stress is beyond the yield strength limit and (ii) component is subjected to applied stress lower than yield stress but exposed to high temperature conditions enough to cause creep.

To avoid the failure by plastic deformation owing to externally applied stress more than yield strength, the cross section should be designed after taking proper factor of safety and considering the yield strength of materials of which component is to be made. For mechanical components that are expected to be exposed in high temperature, creep resistant materials should be selected so that under identical load condition, low steady-state creep rate of creep-resistant materials can allow desired longer creep life.

Identification of Types of Failures

Failure analysis is separated into two distinct parts, the first being the mode of failure, and second, the cause of failure. The mode is the failure process, and the cause is the part that can be altered or changed to prevent future occurrence. Some commonly recognized failure modes are as follows:

- Fracture (Ductile and Brittle)
- Fatigue (mechanical and thermal)
- Stress Corrosion
- Hydrogen Damage
- Corrosion
- Wear and Erosion

Fracture – Ductile Overload vs. Brittle Overload Failures "Ductile failure" is one where there is a great deal of distortion of the failed part. Commonly, a ductile part fails when it distorts and can no longer carry the needed load, like an overloaded steel coat hanger. However, some ductile parts break into two pieces and can be identified because there is a great deal of distortion around the fracture face, similar to what would happen if you tried to put too much load on a low carbon steel bolt.

The term "brittle fracture" is used when a part is overloaded and breaks with no visible distortion. This can happen because the material is very brittle, such as gray cast iron or hardened steel, or when a load is applied extremely rapidly to a normally ductile part. A severe shock load on the most ductile piece can cause it to fracture like glass.

An important point about failures is that the way the load is applied, i.e., the direction and the type, can be diagnosed by looking at the failure face. A crack will always grow perpendicular to the plane of maximum stress. Below we show examples of the difference in appearance between ductile overload and brittle overload failures.

We know we can look at an overload failure and knowing the type of material, tell the direction of the forces that caused the failure. Common industrial materials that are ductile include most aluminum and copper alloys, steels and stainless steels that are not hardened, most nonferrous metals, and many plastics. Brittle materials include cast irons, hardened steel parts, high strength alloyed nonferrous metals, ceramics, and glass.

One note of caution is that the type of fracture, ductile, or brittle should be compared with the nature of the material. There are some instances where brittle fractures appear in normally ductile materials. This indicates that either the load was applied very rapidly or some change has occurred in the material, such as low temperature embrittlement, and the material is no longer ductile. An example of this was a low carbon steel clip used to hold a conduit in position in a refrigerated (−50 °F) warehouse. The clip was made from a very ductile material, yet it failed in a brittle manner. The investigation showed it had been hit by a hammer, a blow that would have deformed it at normal temperatures.

In a brittle overload failure, separation of the two halves isn't quite instantaneous, but proceeds at a tremendous rate, nearly at the speed of sound in the material. The crack begins at the point of maximum stress, then grows across by cleavage of the individual material grains. One of the results of this is that the direction of the fracture path is frequently indicated by chevron marks that point toward the origin of the failure.

Brittle vs. Ductile Fracture characteristics

Ductile materials- extensive plastic deformation and energy absorption ("toughness") before fracture.

Brittle materials- little plastic deformation and low energy absorption before fracture.

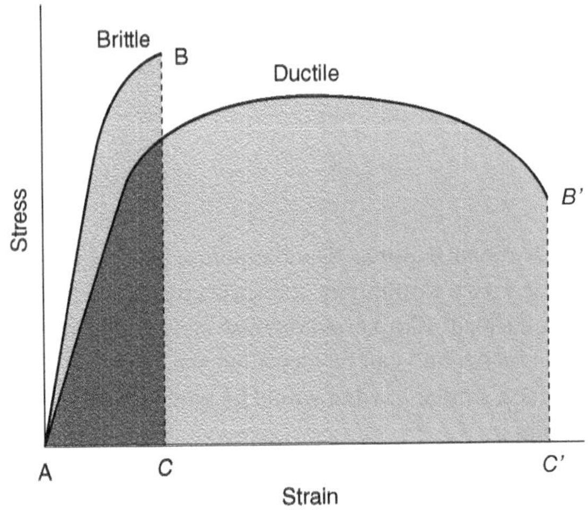

Figure 5.2 Stress–strain curve of brittle and ductile material.

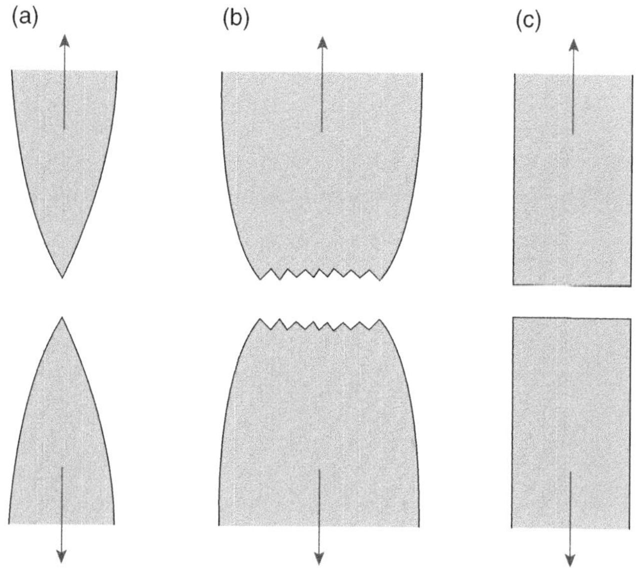

Figure 5.3 Ductile vs brittle fracture. (a) Very ductile, soft metals (e.g., Pb, Au) at room temperature, other metals, polymers, and glasses at high temperature. (b) Moderately ductile fracture, typical for ductile metals. (c) Brittle fracture, cold metals, ceramics.

Ductile Fracture

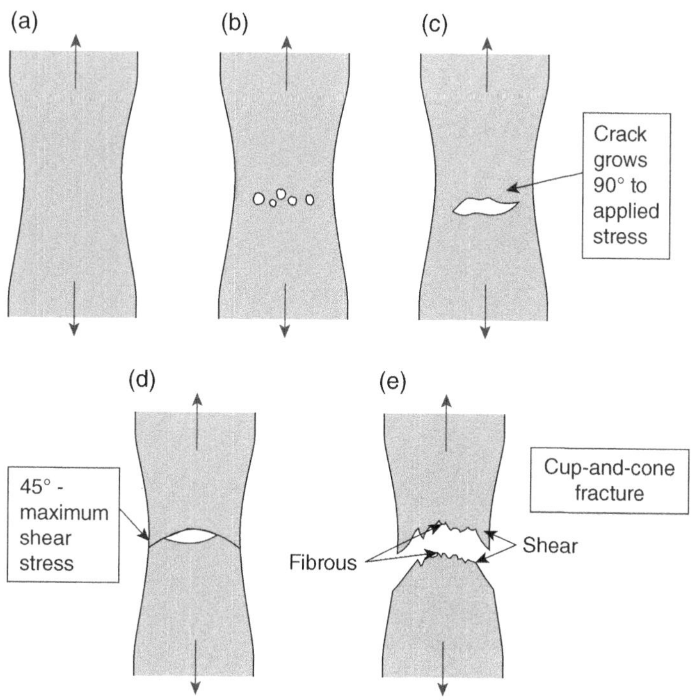

Figure 5.4 Different stages before ductile fracture. (a) Necking (b) formation of microvoids (c) coalescence of microvoids to form a crack (d) crack propagation by shear deformation (e) fracture.

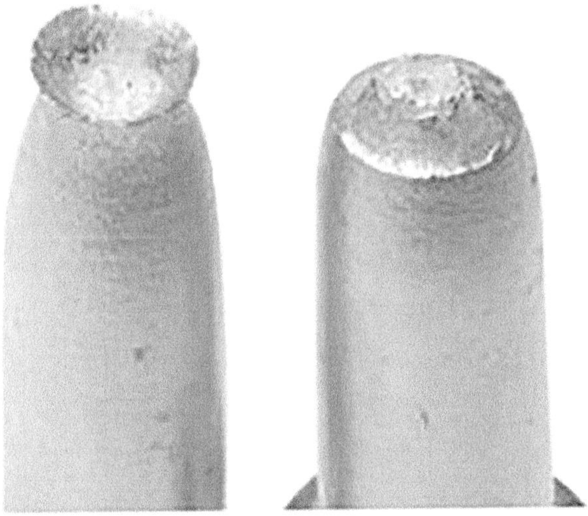

Figure 5.5 Cup and cone fracture in Al. *Source*: Callister, William D. and Rethwisch, David G. (2009). Materials Science and Engineering: An Introduction, 8e. Wiley.

Brittle Fracture

- No appreciable plastic deformation
- Crack propagation is very fast
- Crack propagates nearly perpendicular to the direction of the applied stress
- Crack often propagates by cleavage – breaking of atomic bonds along specific crystallographic planes (cleavage planes).

Figure 5.6 Brittle fracture in a mild steel. *Source*: Callister, William D. and Rethwisch, David G. (2009). Materials Science and Engineering: An Introduction, 8e. Wiley.

Example: Failure of a Pipe

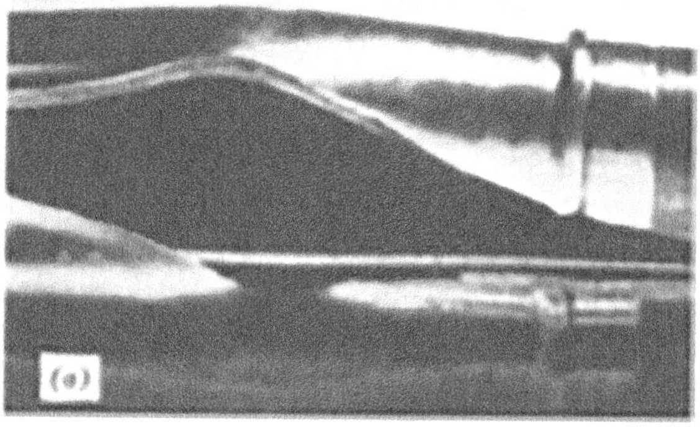

Figure 5.7 Ductile failure: -one piece -large deformation (after some amount of plastic deformation). *Source*: Colangelo, Vito J. and Heiser, Francis A. (1987). Analysis of Metallurgical Failures. 2e. Wiley.

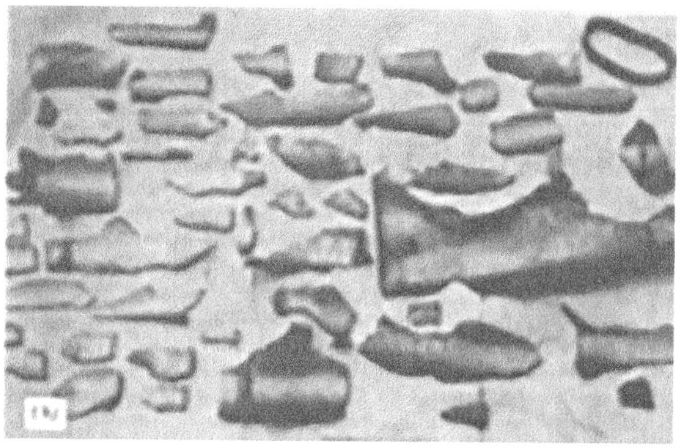

Figure 5.8 Brittle failure: -many pieces -small deformation, (even when the stress is within the elastic range).

Depending on the ability of material to undergo plastic deformation before the fracture two fracture modes can be defined – ductile or brittle. Ductile fracture is characterized by large amounts of plastic deformation. Proportionally large amounts of energy will therefore be required to induce this fracture. When designing a structure, it is usually preferable for the material to fail in a ductile manner as there will be sufficient warning (evidence of deformation) before the final failure occurs. In a brittle fracture, little plastic deformation and low energy absorption before fracture take place.

Ductile fracture- most metals (not too cold):

- Extensive plastic deformation ahead of crack.
- Crack is "stable".
- Resists further extension unless applied stress is increased.

Stages in Ductile Fracture

- Necking
- Cavity formation (micro cracks)
- Crack formation
- Crack propagation
- Fracture

Brittle Fracture Characteristics

- Takes place without any appreciable plastic deformation and by rapid crack propagation.
- Direction of crack motion is nearly perpendicular to the direction of applied tensile stress.
- Flat surface.

Brittle fracture- ceramics, ice, cold metals:

- Relatively little plastic deformation.
- Crack is "unstable."
- propagates rapidly without increase in applied stress.

Origin of Fractures (Ductile and Brittle)

Every structure has a load limit beyond which it is considered unsafe. An applied load that exceeds this limit is known as overload. When a component fails because due to a single application of a load greater than the strength of the component, it is termed as overload failure. The nature of fracture arising due to overload failure could either be ductile or brittle or a combination of the two.

In general, ductile fractures are associated with metal flow at failure zone due to plastic deformation and fibrous-surface appearance. In brittle fractures, plastic deformation is almost absent and the surface shows irregular bright facets of a cleavage type. Establishing the origin of a fracture is essential in failure analysis, and the location of the origin determines which measures should be taken to prevent a repetition of the fracture. The fracture-surface characteristics that show the direction of crack propagation (and conversely, the direction toward the origin) include features such as chevron marks, crack branching, and river patterns. Features that help identify the crack origin include concentric fibrous marks, radial marks, and beach marks. By a study of these features, crack progress can be traced back to the point of origin, and then, it can be ascertained whether the crack was initiated by an inclusion, a porous region, a segregated phase, a corrosion pit, a machined notch, a forging lap, a nick, a mar, or another type of discontinuity, or was simply the result of overloading.

Some of the questions that should be raised concerning the nature, **history**, functions, and properties of the fractured part, and the manner in which it interacts with other parts, to find out root cause of failure are-

- *Loading.* Where the nature, rate, and magnitude of the applied load correctly anticipated in the design of the part? Were repeated or cyclic loadings involved? What was the direction of the principal stress relative to the shape of the part? Where residual stresses present to an undesirable degree?

- *Material.* Was the recommended alloy used? Where its mechanical properties at the level expected? Where surface or internal discontinuities present that could have contributed to failure? Did the microstructure conform to that prescribed?
- *Shape.* Did the part comply with all pertinent dimensional requirements of the specification? Did the part have sufficient section thickness to prevent local overloading? Where fillets formed with sufficiently large radii? Where there adequate clearances between interacting parts? Where any of the contours deformed during service? Was there evidence of mechanical surface damage?
- *Environment.* Was the part exposed to a corrosive environment or to excessively high or low temperatures? Was the surface of the part suitably protected? Where the properties of the part altered by the exposure? Was there interaction (for example, galvanic) between the material of the part and that of adjacent components?

Fatigue Failures

So far we've talked about the gross overloads that can result in immediate, almost instantaneous, catastrophic failures. Another type of failure occurs by means of progressive brittle cracking under repeated alternating or cyclic stresses of an intensity considerably below the normal strength. Although the fracture is of brittle type, it may take some time to propagate, depending on both intensity and frequency of the stress cycles. The number of cycles required to cause fatigue failure at a particular peak stress is quite large but it decreases as stress increases. A very important distinction is that fatigue cracks take time to grow across a part. In a fatigue failure, an incident of a problem can exceed the material's fatigue strength and initiate a crack that will not result in a catastrophic failure for millions of cycles. There were cases where fatigue failures in 1200 rpm motor shafts that took less than 12 hours from installation to final fracture, about 830 000 cycles. On the other hand, there are cases the crack growth in slowly rotating process equipment shafts has taken many months and more than 10 000 000 cycles to fail.

Fatigue fracture results from the simultaneous action of repeated or fluctuating cyclic stress, tensile stress, and plastic strain. No fatigue crack starts or grows in the absence of any of these three active components. Cyclic stress initiates the crack and tensile stress produces the crack growth. There are many variables in service that influence the fatigue behavior or characteristic pattern. These include the magnitude and frequency of application of the fluctuating stress, the presence of a mean stress, temperature, environment, part size and shape, state of stress and residual stresses, surface finish, surface damages, and microstructure. The occurrence of fatigue may be considered as a three-stage process.

1) Initiation of the surface or sub-surface fatigue crack under a fluctuating load.
2) Crack propagation under tensile stress until the section thickness critically bears the imposed load.
3) Sudden fracture under overload

In the presence of a severe stress concentration, multiple crack origins are formed which eventually unite to form a single crack front. When the stress concentration on the surface is absent, cracks propagate more readily near the center of a section than at the surface. While

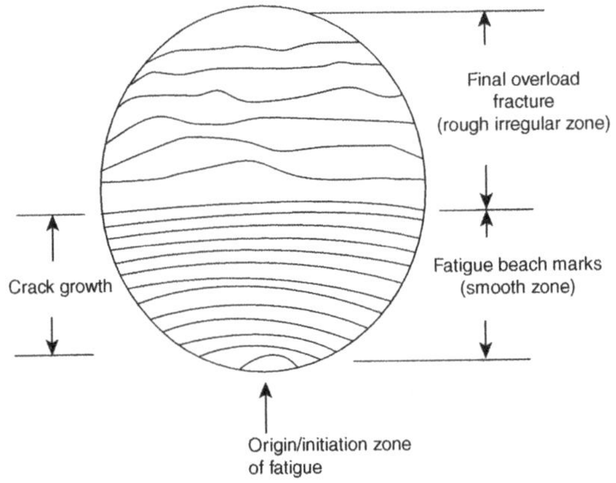

Figure 5.9 Schematic representation of the fatigue crack (three stage) phenomenon.

there is a stress concentrating 'notch, W-shaped crack fronts are-observed. For a given material, zone of the final fracture increases with increasing applied load. On rotating or bending the final fracture', region is often rotated or offset toward the origin in the direction opposite to the direction of rotation. Also, with increasing nominal stress, the final fracture moves toward the center.

Stress Concentration

A stress concentration is a physical or metallurgical condition that increases the local stress in the part by some factor. A good example is the shaft shown in Figure 5.10. We see that the stress in the area of the radius varies depending on the size of the radius. A small radius can increase the stress dramatically

Thermal Fatigue Failure

Thermal fatigue failure is the result of repeated thermal cycling which generates repeated stress gradients within a free component or stress cycling within a component constrained to fixed dimensions. Thermal fatigue fractures are characterized by surface cracking formed by multiple initiation sites that join randomly by edge sliding to form the main crack. Other features to identify thermal fatigue are as follows:

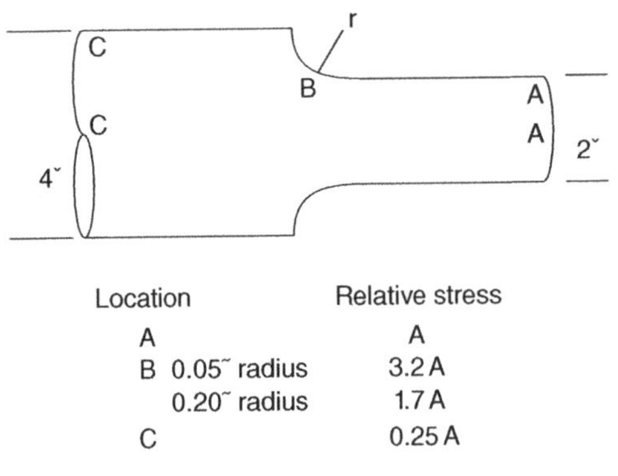

Figure 5.10 Stress concentration at corners.

- Fractures are planner and transverse with no visible plastic deformation.
- Fracture is mostly transgranular.
- Oxidized fracture surfaces and oxide wedge filled cracks further characterize thermal fatigue failures

Fatigue resistance is affected by a number of controllable factors:

- The chemistry of the material and its resultant microstructure have a profound effect on fatigue strength. In fact, they can equally influence on mechanical strength (tensile and yield). Alloying elements, such as chromium, nickel, and moly, have the greatest effect on the iron base system. Solid solution alloys show the maximum increase in fatigue strength.
- Grain size appears to be a strong determining factor in inhibiting the plastic deformation process that occurs with crack propagation.
- Environmental factors such as cyclic temperature, temperature gradient, and corrosion pitting that result in stress concentrations. The thermal fatigue failure shown in was due to the temperature gradient across the thick wall section.
- Reduction of localized surface stress concentrations by such techniques as case hardening, shot peening, auto frottage, and thread rolling.
- Proper heat treatment can markedly improve fatigue resistance. As an example, for steels, a tempered martensitic.

Stress Corrosion Cracking

Stress corrosion cracking (SCC) is a fracture process that involves the combined and simultaneous action of a tensile stress and a corrosive environment. SCC occurs when the tensile stress and a specific environment are able to cause failure by their combined action, but are insufficient to cause failure by either one acting alone. In fact, the tensile stresses are usually below the metal's yield strength. Furthermore, the metal would suffer only minimal corrosion in the absence of the applied stress. The problem itself can be quite complex. The situation with buried pipelines is a good example of such complexity.

There are three requirements for SCC to occur:

1) A susceptible metal.
2) Tensile stresses applied to the metal.
3) A specific environment containing an aggressive species that promotes SCC.

This form of corrosion is particularly dangerous because it may not occur under a particular set of conditions until there is an applied stress. The corrosion is not clearly visible prior to fracture and can result in catastrophic failure. Many alloys can experience stress corrosion, and the applied stress may also be due to a residual stress in the material. An example of a residual stress could be a stress remaining in a material after forming, or a stress due to welding. Stress corrosion cracking will usually cause the material to fail in a brittle manner, which can have grave consequences as there is usually little or no warning before the failure occurs.

The impact of SCC on a material usually falls between dry cracking and the fatigue threshold of that material. The required tensile stresses may be in the form of directly applied stresses or in the form of residual stresses. Cold deformation and forming, welding, heat treatment, machining, and grinding can introduce residual stresses. The magnitude and importance of such stresses is often underestimated. The residual stresses set up as a result of welding operations tend to approach the yield strength. The buildup of corrosion products in confined spaces can also generate significant stresses and should not be overlooked.

How to Prevent SCC

SCC can be controlled by any of the following three approaches:

1) Design, which includes selection of the mechanical and materials aspects of components,
2) Controlling the materials, and
3) Controlling the environment.

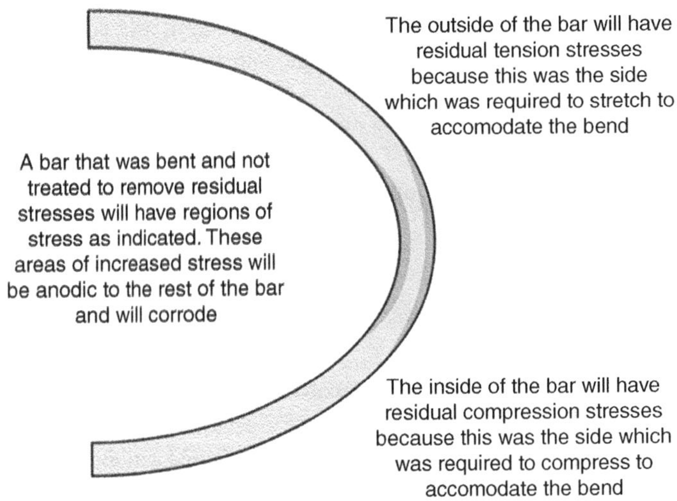

Figure 5.11 Stress corrosion on a bar.

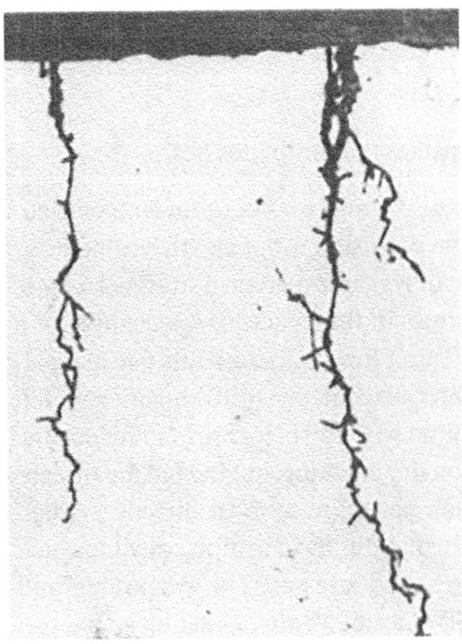

Figure 5.12 Stress corrosion cracking. *Source*: Stress Corrosion Cracking/Industrial Metallurgists, LLC.

The source of the tensile stress which causes SCC may be externally applied stress or residual stresses. Externally applied stresses arise from applied mechanical loads such as tensile or bending loads. Residual stress is an internal stress that exists in a metal without an external load being applied. Residual stresses can result from cold working, heat treating, or welding. Increasing the yield strength of a metal is one way to improve its resistance to SCC because the threshold stress for SCC increases as the yield strength increases. The yield strength can be increased through alloying, heat treating, cold-working, and combination of these approaches. There is one very important consideration when increasing the yield strength. The increase in strength must not be accompanied by a significant reduction of the metal's toughness, because decreasing the toughness will have a detrimental effect on a metal's resistance to SCC and on its fracture toughness.

The environmental factors, such as pH and temperature, also influence the severity of SCC. By controlling the environmental factors, SCC can be controlled.

Chloride SCC

One of the most important forms of stress corrosion that concerns the nuclear industry is chloride stress corrosion. Chloride stress corrosion is a type of intergranular corrosion and occurs in austenitic stainless steel under tensile stress in the presence of oxygen, chloride ions, and high temperature. It is thought to start with chromium carbide deposits along grain boundaries that leave the metal open to corrosion. This form of corrosion is controlled by maintaining low chloride ion and oxygen content in the environment and use of low carbon steels.

Hydrogen Damage

Hydrogen can diffuse into metals and alloys from several sources during processing and subsequent service. These sources include the dissociation of moisture during casting and welding, thermal decomposition of gases, and pickling and plating operations. Hydrogen can also be generated from cathodic reactions during corrosion in service and from cathodic protection measures by sacrificial anodes and impressed current.

The effects of hydrogen are well known in ferritic and martensitic steels, where it can diffuse to suitable sites in the microstructure and develop local internal pressure resulting in the characteristic form of hydrogen embrittlement. In low carbon steels, which have inherent ductility, hydrogen may not give rise to cracking but will cause blisters to develop at inclusions. This can lead to delamination in-plate due to the directional nature of the inclusions. Steels for sour gas service, where the environment contains wet hydrogen sulfide, must have very low sulfur levels or have been treated with additions to control the shape of the inclusions during deoxidation to minimize the danger of hydrogen embrittlement and blistering.

Failure Due to Hydrogen Damage

Failure is time-dependent and occurs at low rates of strain as the load-bearing cross section is reduced during slow crack growth in the embrittled region. Susceptibility for embrittlement is higher in alloys with higher yield strengths, i.e. those that are cold-worked, age-hardened or in their martensitic form. The sites at which hydrogen is trapped include the original austenite grain boundaries and the interfaces between the matrix and non-metallic inclusions, for example, manganese sulfides. These then result in both intergranular cracking (with separation at the prior austenite boundaries) and transgranular cracking (flaking or quasi-cleavage) which is associated with the inclusions. Hydrogen can assist in the propagation of corrosion fatigue cracks and can also cause sulfide stress corrosion cracking in ferritic and martensitic steels, including the stainless grades.

Addressing Hydrogen Damage

The first and foremost method for preventing hydrogen damage is the obvious option of preventing direct contact between a metal and the hydrogen-containing agent. Controlling the environment during operations such as casting and melting will allow for the exposure to hydrogen to be moderated. Other than preventing exposure, it is also possible to give the metal or alloy a metallurgical treatment, which would serve to reduce the susceptibility of the material to damage caused by hydrogen, chemical means, or otherwise.

Corrosion Damage

Corrosion damage can be apparent in many different ways, including loss of material, surface pitting, and the buildup of corrosion deposits, but it is convenient to classify corrosion by visual observation of the corroded material before any cleaning is conducted. There are generally considered to be eight basic forms of corrosion.

- General attack (uniform corrosion)
- Galvanic corrosion
- Crevice corrosion
- Pitting
- Intergranular corrosion
- Selective leaching
- Stress corrosion
- Erosion-corrosion

Although the distinctions between the eight basic categories of corrosive attack have become blurred, particularly when fundamental mechanisms are considered, this classification may help (at least in the first instance) to simplify the analysis. The identification of the factors associated with the forms of corrosion can guide failure investigators. A listing of the most important factors would ensure that engineers with little or no corrosion training are made aware of the complexity and multitude of variables involved.

Temperature can affect the corrosion behavior of materials in different ways. If the corrosion rate is only controlled by the metal oxidation process, the corrosion rate will increase exponentially with an increase in temperature. The higher the fluid temperature the faster the rate of oxidation. Experience shows that corrosion is more pronounced in hot water lines. Galvanic corrosion, also known as electrolysis, occurs when different metals come into contact with each other. Chemical composition of the fluid may have differing effects on the corrosive forces at play. When water velocities exceed 4 ft/s in oversized circulation pumps, installation of undersized distribution lines, multiple or abrupt changes in the direction of the pipe, corrosion may take place. The pH of a solution is also an important factor in the corrosion of materials.

How to Control Corrosion

There are many ways to organize and operate successful corrosion management systems, each of which is asset specific depending on factors such as Design, Stage in life cycle, Process conditions, and Operational history. The corrosion policy provides a structured framework for identification of risks associated with corrosion, and the development and operation of suitable risk control measures.

Corrosion is caused by a chemical reaction between the metal and gases in the surrounding environment. By taking measures to control the environment, these unwanted reactions can be minimized. Sacrificial coating involves coating the metal with an additional metal type that is more likely to oxidize, hence the term "sacrificial coating." There are two main techniques for achieving sacrificial coating: cathodic protection and anodic protection. The most common example of cathodic protection is the coating of iron alloy steel with zinc, a process known as galvanizing. Anodic protection involves coating the iron alloy steel with a less active metal, such as tin. Tin will not corrode, so the steel will be protected as long as the tin coating is in place.

Another simple way to prevent corrosion is to use a corrosion-resistant metal such as aluminum or stainless steel. Depending on the application, these metals can be used to reduce the need for additional corrosion protection. Though the application of a paint coating is a cost-effective way of preventing corrosion. Paint coatings act as a barrier to prevent the transfer of electrochemical charge from the corrosive solution to the metal underneath. Corrosion inhibitors can be applied as a solution or as a protective coating using dispersion techniques. Corrosion inhibitors are commonly applied via a process known as passivation. Corrosion inhibitors are chemicals that react with the surface of the metal or the surrounding gases to suppress the electrochemical reactions leading to corrosion. They work by being applied to the surface of a metal where they form a protective film. Cathodic protection (CP) is by far the best way to stop corrosion on pipelines. It uses impressed currents from a fixed anode to interfere with the electrical circuit in the corrosion cell. It is 100% effective against most forms of external pipe corrosion.

Wear and Erosion Damage

Wear and erosion involve loss of material. This may be, for example, because of the absence of adequate lubrication, the rubbing together of components that are supposed to have clearance between them, or from the handling of abrasive materials that impinge on the component that continually removes surface material. In many cases, the presence of debris buildup may provide critical understanding of the specific wear/erosion mechanism involved in the failure. A common wear problem that is encountered relates to bearings, either roller-element bearings or plain bearings. Features to be looked for include the nature of the damage, for example, overall wear or scores in the bearing surfaces. In the latter case, it is particularly important to identify any hard particles that may be embedded and trapped at the ends of the score marks. Where unexpected wear between components in relative motion is encountered, and lubrication is observed to have been adequate, it is important to look for possible contaminants. These may arise from some other component in the system (for example, through the loss of hard chromium plating particles or through the development of corrosion products that are then transmitted through the system in circulating lubricant).

Failure Investigation

Two of the critical goals in a failure investigation are to identify both the failure mechanism(s) and failure mode(s) that are present. Failure mechanisms are a key component in categorizing damage and failures.

Stages of a Failure Analysis

Depending on the nature of the failure and the availability of physical evidence or background information, there are stages that are common to all successful failure analysis. Steps may be followed by assessment of the damage and conditions leading to failure. These stages may differ depending on whether fracture, corrosion, and/or wear conditions are being investigated.

Generally, a failure analyst will start with a broad range of possible explanations but over time will narrow and refine the existing possibilities. Often, a likely theory develops during the course of the investigation. This can be helpful, but only if the investigator does not let the theory influence his or her objectivity.

The Principal Task of a failure analyst during a physical cause investigation is to identify the sequence of events involved in the failure. Technical skills and tools are required for such identification, but the analyst also needs a mental organizational framework that helps evaluate the significance of observations.

The basic steps are:

1) Collect data
2) Identify damage modes present
3) Identify possible damage mechanisms
4) Test to identify actual mechanisms that occurred
5) Identify which mechanism is primary and which is/are secondary
6) Identify possible root causes
7) Test to determine actual root cause
8) Evaluate and implement corrective actions

Collection of Background Data and Samples

The first step in a failure investigation involves gathering background information. This includes material, manufacturing process, circumstances surrounding the failure, engineering drawings, associated specifications and other background data. In addition to the failed component, it is also helpful to have an intact, unused, "exemplar" for examination. An experienced failure analyst can assist in the discovery process to obtain relevant documentation

- Information about the failure:
- Date and time of failure, temperature, and environment
- Extent of damage, sequence of failure, and injuries
- Stage of operation when failure occurred
- Blueprints, photographs, or sketches of the failure and adjacent areas
- Any service deviations that might have contributed to the failure
- Opinions of operating personnel regarding the failure

But for the analysis of some components, it may be impractical or impossible for the failure analyst to visit the failure site. Under these circumstances, data and samples may be collected at the site by field engineers or by other personnel under the direction of the failure analyst. A field failure report sheet or checklist can be used to ensure that all pertinent information regarding the failure is recorded.

Visual Examination

The failed part, including all its fragments, should be subjected to a thorough visual examination before any cleaning is undertaken. For example, traces of paint or corrosion found on a portion of a fracture surface may provide evidence that the crack was present in the surface for some time before complete fracture occurred. The preliminary examination should begin with unaided visual inspection. The unaided eye has exceptional depth of focus, the ability to examine large areas rapidly and to detect changes of color and texture. If required, macroscopic examination is performed to document the main features by using low-power magnification. It is important at this stage to fully document the "as-received" condition and photograph overall fracture and position. The temptation to put fracture surfaces back together should be avoided as it can damage fracture features. Because metallic parts are prone to oxidation, a reaction between the metal and the oxygen in the air, failed components should ideally be examined as soon as possible. Documentation of damage and cracking surrounding a fracture, including damage patterns and crack origins, scores, scuffing, dents, distortions, evidence of plastic deformation, and fractures is an important step in assessing relevance to final failure.

Testing

First and foremost, a test protocol must be developed and agreed to by all parties. The protocol is usually separated into non-destructive and destructive evaluation. Destructive evaluation, in the forensic sense, includes any process that alters the evidence. Testing also includes cleaning and some types of "non-destructive" testing. Significant care must be taken prior to any destructive testing. Again, details of all testing, destructive and non-destructive must be shared and agreed to by all involved parties.

Non-Destructive Evaluation/Non-Destructive Testing (NDE/NDT)

Depending on the metal alloy, various types of non-destructive inspection can be performed. Non-destructive inspection can reveal discontinuities or additional cracking in the component. The most common types of NDE/NDT are Liquid Penetrant Testing (PT), Fluorescent Penetrant Inspection (FPI), Magnetic Particle Inspection (MPI), Acoustic Emission Testing (AE), Radiographic Testing (RT), and Ultrasonic Testing (UT). All these tests are used to detect surface cracks and discontinuities. Radiography is used mainly for internal examination. A photographic record of the results of non-destructive inspection is a necessary part of record keeping in the investigation.

Mechanical Testing

Mechanical testing determines properties of a material when force is applied, therefore indicating its appropriate use in mechanical applications. The mechanical properties of metals can be expressed in numerous ways: strength, ductility, hardness, toughness, etc. The hardness and the strength of a material are closely related. Hardness is useful for estimating wear resistance and approximate strength; and is defined as the resistance of a material to surface indentation. Rockwell hardness is the most widely used method for determining hardness and several different Rockwell scales are utilized for materials of a variety of hardness ranges. Aluminum alloys, brass, and soft steels are often measured on the Rockwell B scale or HRB, whereas harder steels and titanium are measured on a Rockwell C scale or HRC. Other ways to obtain mechanical properties include tensile testing, compression testing, impact testing, fatigue testing, and fracture toughness testing, etc., depending on the application and performance requirements of the component.

Macroscopic Examination

Macroscopic examination is an extension of the visual examination and evaluates quality and homogeneity of the part. It is used to determine the origin of the failure and the type of fracture such as ductile, brittle, torsion or fatigue. Microstructural features can be used to assess internal quality, presence of hydrogen flakes, chemical segregation, hard cases, flow lines, and welds. A stereomicroscope (1–50×) is often utilized for the macroscopic examination. It is during this step that the fracture surface is evaluated. The first piece of information often observed is where the fracture initiated, that is, locating the fracture "origin" or "origins" and determining whether it/they are located at the surface or below the surface. It is also important to note the fracture direction in relation to the normal or expected loading. Markings on the fracture surface formed during the initiation and propagation of the crack can be used to evaluate the fracture and determine origin of cracking. Examination of the origin by the experienced failure analyst may reveal the cause and determine any contributing factors toward failure.

Microscopic Examination

Microscopic examination is usually performed in a Scanning Electron Microscope or SEM. A Scanning Electron Microscope is a high-power magnifying and imaging instrument that uses an accelerated electron beam as a light source. The SEM allows magnifications up to 50 000× and improves depth-of-field resolution. A Metallurgical Expert can examine the fracture surface in the SEM and determine fracture topography while inspecting the origin area for anomalies. It is in the SEM that the Metallurgist can classify the fracture and determine the fracture type. Semi-quantitative chemical information can also be gathered while in the SEM by Energy Dispersive X-Ray Analysis.

Metallographic Examination

Metallography is defined as the science of the constitution and structure (or microstructure) of a metal. During metallographic inspection, the failure analyst or technician, sections the area of interest usually through abrasive-wheel cutting and mounts the specimen without deformation. The metallographic cross section is then polished to a mirror-like finish. The sample is subsequently examined by the Metallurgist in a metallurgical microscope.

Many important macroscopic properties of metallic materials are highly sensitive to the microstructure. Critical mechanical properties, like tensile strength or elongation, as well as other thermal or electrical properties, are directly related to the microstructure.

Metallographic examination involves the use of high magnification microscopy. These examinations may include optical and scanning electron microscopy. Optical microscopic examination is used to determine grain size, microstructure, and inclusion type and content. On the other hand, scanning electron microscopy is used to determine abnormalities, such as inclusions, segregation, and surface layers, as well as fracture features.

Chemical Analysis

A chemical analysis is sometimes performed by the Metallurgical Expert to determine bulk chemistry, local elemental concentration, surface corrosion products, and coating chemistry. A failed component chemically analyzed to determine whether the grade is indeed as claimed, because mixes occasionally occur at the mill, in the warehouse, or at the fabrication or manufacturing shop. A small percentage of all failures are caused by grade mixes. Chemical analysis can verify conformance to a standard or specification, detect impurities, identify alloys, and analyze trace elements. Analytical chemistry can be performed by a variety of techniques including optical emission spectroscopy, atomic emission spectroscopy, and inductively coupled plasma analysis, to name a few.

Simulations

Occasionally, the investigator must simulate the environmental conditions encountered during service to ascertain suitability of the material to environmental conditions and to determine the effect of prior heat treatment or other processing on the service performance of the material. For example, improper heat treatment may render the material susceptible to certain types of attack. Simulation of the heat treatment as reported in the case history is valuable for both confirmation purposes and for further testing, particularly on a comparison basis (e.g., comparison of the toughness of the failed material as received and after various experimental heat treatments). Certain types of simulation tests require accelerated testing to obtain the desired information in a reasonable length of time. Interpretation of accelerated test data must be done with care.

Data Analysis, Conclusions and Report

After the completion of the outlined steps, the investigator should be ready to interpret and summarize the data that has been collected. Some of the work performed may not contribute in determining the root cause of the failure, yet it may be helpful in eliminating some possible causes. In combination, the steps that have been outlined will, in most cases, enable the investigator to conclude the root cause of the failure. The report should provide the following:

- Description of the failed component
- Conditions at the time of failure
- Background service history
- Mechanical and metallurgical data about the failed part
- Evaluation of the material quality
- Discussion of mechanisms that explain the root cause of the failure
- Recommendations for prevention of future failures or for action to be taken with similar parts.

Impact of working condition on metallurgical failure

The failure of an engineering component in actual working conditions can occur due to very large of factors related with design, materials, manufacturing, service conditions, etc. To have systematic understanding on various factors which can lead to metallurgical failure of engineering components, these can be groups under following headings.

- Improper design
- Improper selection of materials
- Defects and discontinuities in metal itself
- Improper processing of materials
- Poor service conditions
- Poor assembling
- Poor maintenance

Improper Design

The deficiency in design of a component can be in various forms such as presence of stress raisers owing to sharp change in cross section, changing the design without proper consideration of its influence on stress distribution especially in high stress areas of the component. Many a time duplicating a successful design for more severe loading conditions or the design is developed owing to lack of knowledge to use proper criteria for designing the engineering components may lead to failure. Designers frequently also come across the situation when accurate calculations and clear analysis of stress (under prevailing technological understanding and capabilities) is not practicable due to complexity in geometry of the component.

Improper Selection of the Materials

For each type of expected failure mechanism, a combination of the mechanical, physical, and chemical properties should be possessed by the material to be selected for developing a design. For example, if failure of a component is expected to occur by excessive plastic deformation at room temperature and high temperature conditions, then yield strength and creep respectively will be important criterion for design. Similarly, if failure of a component is expected to occur by fracture under overloads, fluctuating loads, and impact loads then ultimate strength, endurance strength, and impact strength respectively should be considered for design purpose. Deficient material selection can occur due to reliance on tensile data for selection of materials, and inability to select of metal in light of the expected failure mechanism and so as to develop suitable criteria for the design purpose. The problem of the materials selection is further complicated when the performance of materials varies as function of time, e.g. creep, corrosion, embrittlement, etc. The criteria for the selection of metal for designing a component for a particular service conditions must be based on the expected failure mechanism.

UnFavorable Manufacturing Processing Conditions

A wide range of manufacturing processes are used for obtaining the desired size, shape, and properties in stock material which includes primary and secondary shaping processes such as castings, forming, machining, and welding apart from the processes like heat treatment. The selection of inappropriate combination of the process parameters for each of above mentioned manufacturing processes can lead to development of discontinuities, defects, unfavorable transformation, and metallurgical changes and so deterioration in the performance of final product during the service. These imperfections and discontinuities are mostly process specific and can exist in variety of forms due to improper selection of manufacturing process and their parameters. Therefore, due care must be taken by failure analyst to investigate the presence of any defect, discontinuity or unfavorable features in end produced by manufacturing processes and failed prematurely during the service.

Poor Assembling

Error in assembly can result due to various reasons such as ambiguous, insufficient or inappropriate assembly procedure, misalignment, poor workmanship. Sometimes, failures are also caused by the inadvertent error performed by the workers during the assembly. For example, failure of nut and stud assembly (used for holding the car wheel) by fatigue can occur owing to lack of information regarding sequence of tightening the nuts and torque to be used for tightening purpose; under such conditions any sort of loosening of nut which is subjected to external load will lead to fatigue failure.

Poor Service Conditions

Failure of an engineering component can occur due to abnormal service condition experienced by them for which they are not designed. These abnormal service conditions may appear in the form of exposure of component to excessively high rate of loading, unfavorable oxidative, corrosive, erosive environment at high or low temperature conditions for which it has not been designed. The contribution of any abnormality in service conditions on the failure can only be established after thorough investigation regarding compatibility of the design, manufacturing (such as heat treatment) and material of the failed components with condition experienced by them during the service.

Stress overloads, excessive temperatures(hot or cold), inadequate abnormal corrosive environments, excessive speed, shock, thermal and mechanical, and excessive wear are all common misuse or misoperations that can promote premature and material failures. Loss of metal from corrosion including chemical attack, liquid metal corrosion, stress accelerated corrosion, and corrosion fatigue are common. Inadequate or maintenance such as poor lubrication, weld repairs, contamination, and cold straightening without thermal stress relieving and wear compensation can lead to accelerated wear resulting in deformation that promote fatigue, of rotating or static components.

Poor Maintenance Strategy

The failure of many moving mechanical components takes place due to poor maintenance plan. A well-developed maintenance plan indicating each and every important step to be used for maintenance such as what, when, where, who, and how for maintenance, is specified explicitly. Lack of information on proper schedule of maintenance, procedure of the maintenance, frequently causes premature failure of moving components.

Conclusion

In failure analysis, investigation takes place to determine the cause of failure, usually with the aim of taking corrective action to fix the problem and mitigate against further failures. A metallurgical failure analysis takes in to account as much of the information as possible during analysis. The end goal of failure analysis is to provide a determination of the root cause. Analysis of a failed part can be done using destructive testing or non-destructive testing (NDT). Failure analysis is undertaken across all branches of manufacturing industry to prevent future asset and product fails as well as protecting against potentially dangerous risks to people and the environment.

6

Pipe Failure

Process piping systems are essential to the vast majority of the oil and gas, chemical, petrochemical and process industries. The catastrophic failure of a pipe or component in a process piping system may endanger plant personnel. If an explosion or fire occurs, or a hazardous chemical is released, widespread fatalities and or undesirable environmental consequences may result. To improve the availability of process piping systems, it is essential for operating and management personnel to have an understanding of the most common causes of failures in process piping systems.

Classification of Failure Mechanisms

The cause of many past failures in industrial piping systems have been studied. A review reveals that failures of pipes are associated with one -and sometimes more than one – of the following mechanisms.

1) Low-temperature brittle cleavage
2) Ductile fracture
3) Fatigue (mechanical and/or thermal)
4) Embrittlement
5) Creep stress-rupture
6) Stress-corrosion cracking

Low-Temperature Brittle Cleavage

The failure will be abrupt, taking the form of a simultaneous tensile separation between the faces of adjoining planes in the structure, and indicating a complete lack of cohesion or homogeneity. The failure is trans-crystalline. Low-temperature brittle cleavage failures are often catastrophic in nature. A classic example occurred during World War II with the failure of over 1000 welded Liberty merchant ships, many of which broke completely in two, some even before they had been put in service. Failure usually initiates in a notch defect or sharp stress concentration – in the Liberty ship failures, many were reported to have initiated at the sharp corners of a deckhatch opening.

Root Cause Failure Analysis: A Guide to Improve Plant Reliability, First Edition. Trinath Sahoo.

Ductile Rupture

While generally less spectacular than brittle cleavage failures, ductile rupture failures are numerically far more common. The mechanism differs from brittle behavior only in the manner in which "tearing" action (plastic slippage) takes place, the ensuing crack propagating linearly at a relatively slow pace. In piping, such a crack will often first be evidenced by a minor leak of the contents, making it possible to repair the break before complete failure ensues. Since ductile rupture is virtually independent of temperature, its prediction is very difficult.

Fatigue (Mechanical and/or Thermal)

Fatigue in piping may emanate from a variety of causes. Although generally resulting from cyclic conditions, it may also follow from sustained primary loads not necessarily of a cyclic nature. Fatigue failures generally display the characteristic of ductile rupture. In the low-cycle category are failures resulting from the relatively low frequency of operating cycles from one shutdown to another. Fatigue due to cyclic thermal expansion and contraction normally happen. In the high-cycle category are failures associated with severe mechanical vibration, and to extremely rapid fluctuations in fluid temperatures at points local to mixing tees due to turbulent flow conditions.

Embrittlement

Embrittlement can result from many causes ranging from phase changes within the alloy to interaction between the alloy and its environment. Sigma phase formation in stainless steels at elevated temperatures is one example of embrittlement. The effect of absorbed hydrogen on high strength steel is an example of environmental interaction. The only way to avoid embrittlement is to select alloys that are stable under the operating conditions and are not adversely affected by the environment.

Creep Stress-Rupture

Under high stress at normal temperatures, most materials exhibit hardening and strengthening as a result of elastic-plastic deformation. As the stress is increased, plastic flow will increase until instability occurs resulting in failure. The essential parameters of the creep-resistant properties of steels are stress, temperature, and time.

Stress-Corrosion Cracking

The mechanism of stress-corrosion failure is a complex one in which mechanical and chemical actions combine to produce effects more severe than either one alone. Small cracks on the surface of the material, produced by cold-working or by residual welding stresses, will tend to propagate rapidly when subjected to prolonged static stresses in the presence of a corrosive medium. Under cyclic loading conditions, stress-corrosion fatigue is also possible.

Causes of Premature Failures in Piping

Premature failures in process piping can occur as a result of a variety of different causes in actual operating environment. These may include design deficiencies or material defect in the pipe or fittings introduced by the manufacturing process. These may include defect conditions introduced during shop fabrication of pipe spools, or the erection or welding or heat treatment of the process piping system. They may also include deterioration associated with normal or upset operating conditions. We will discuss each of the points in detail in the following paragraphs.

Design Deficiencies

Deficiencies in design can lead to failures in process piping systems. Important aspects of the design process are selecting the most suitable pipe diameter and ensuring that the pipe has adequate wall thickness for the intended service conditions. It also includes the selection of components such as branch connection, tees, nozzle branches, etc., and the specification of forgings, as specific piping products are determined by the design of the piping system. It also include the selection of materials on the basis of the expected and possible maximum exposures of the piping to temperatures (high or low). Finally it includes the recognition of potential upset conditions and an allowance for those upset conditions in the design of the piping system.

Material Defect

Material not furnished to specification can create failures in many ways, whether the material is the wrong specification, or fails to meet one of the many chemical, physical, or geometrical requirements of the specification. It may cover the full range of metallurgical problems from those related to material mis-application to heat treatment and corrosion related types. If the wrong material is initially selected and goes unchallenged, no subsequent inspections will correct the error. Indeed, they will only serve to perpetuate it. Rigid procedures are therefore necessary to ensure that the initial selection is correct. In large industrial complexes where many different materials are being used, errors of this type are perhaps understandable, although still unexcusable. Nevertheless, emphasis must be placed in the Piping Design Guide on ensuring proper initial material selection for all piping components. There are cases when a fabricator in-advertently substitutes a carbon steel fitting for a low alloy steel fitting, then the substitution is classified as a fabrication defect.

Base Material Defects

The base metal is subject to a wide range of manufacturing defects. Forged or wrought items can have slag or impurity inclusions on their surface or internally, resulting in slivers, scabs, laminations, or seams. The forming operations can leave cracks, gouges, tool marks, and surface laps. Cast items can also suffer from slag or sand inclusions and are more susceptible than wrought materials to internal voids due to gas pockets. In addition castings are subject to shrinkage voids, segregation, cracking, and occasionally a chaplet or core support will not fuse into the casting.

Manufacturing Defects

Manufacturing defects can and have produced failures in process piping systems. These defects, if sufficiently severe, have the potential to produce failures including shrinkage, blowholes, cracks, and cold shut in castings such as valves, fittings and other components. Defects also include laminations, hot tears, seams and laps in forged or extruded components. Electric resistance welded pipe has also been utilized in process piping, lead to failures as a result of poorly fuse seam welds. Failures have occurred also as a result of accelerated preferential corrosion of the seam welds. Some of the leaks that have occurred have resulted in major fires or explosions from the leakage of hydrocarbon fluids or gases.

Some of the other manufacturing defects are as below.

Welding defect- There is extensive literature on weld defects and failures, covering both historical and theoretical considerations. Poor fit-up and poor end preparation are the two main causes. Poor fit-up can occur with or without backing rings and often takes the forms of wall mismatch, improper or uneven root gap or mis alignment. These often result in notch type stress risers and cracks or poor penetration into the pipe or backing ring. So many of these defects are associated with backing rings that their use should not be permitted. The angle and shape of the bevels on butt-weld ends is also quite important and if improperly done can result in poor penetration, burn through, or other fusion zone defects. The establishment of suitable weld profile geometries, together with rigorous inspection and checking, will be the main prevention of this type of defect.

Many of the other defects are arc strikes, burn through, centerline crevices, cracks of many types, crater pits, mismatch, incomplete fusion, lack of penetration, overlap, oxidation, porosity, sink or concavity, slag inclusions, slugging, undercut, weld reinforcement, and others. It is sufficient to say that the use of the best qualified welders, welding equipment, weld procedures, and an extremely thorough testing and inspection procedure will be necessary to keep these problems to a minimum.

Heat Treatment Defects

Heat treatment if not done correctly or if it is simply not done can cause problems for the material or application. Consideration of the proper heat treatment, in a pressure vessel or piping component includes preheat as well as post heat treatments. Time versus temperature curves must be obtained and checked for all heat treatments, and the heating method verified for its appropriateness to avoid these failures. Among the failures related to improper heat treatment are:

1) Excessive hardness differences
2) Variations in metallurgical structures
3) Localized stresses
4) Surface defects
5) High transition temperature
6) Reduced creep strength

Dimensional and Geometrical Defects

Defects such as lack of straightness, out-of-roundness or ovality, corrugations, and wall thinning can result from manufacturing processes. As they are relatively easily identified, it is unusual for them to be the source of failures. However, these defects may assume much greater significance in piping during fabrication which may create defect afterwards.

Storage- (Environmental Effects)

During storage and often after installation, steps must be taken to protect piping from hostile environments. This will be a problem if it is located near an ocean, as many will be for a source of cooling water or a heat sink. The large amounts of stainless steel used at such locations must be carefully protected against chloride corrosion from the marine atmosphere.

Fabrication and Erection Defect

All the weld defects that can occur in the manufacture of piping materials can also occur during the fabrication and erection of piping systems. However, good erection welding can be much more difficult to accomplish due to field conditions and to the fact that welds are being made on large, continuous systems. Difficulties are inherent in the immobility of the large sections to be joined, coupled with the necessity for maintaining close dimensional tolerances. Space and other limitations may make it impossible to rotate the work during welding, causing the welder often to work in awkward and cramped positions. Inert gas welding is considered mandatory for some piping, and purge gas control may prove to be a problem due to environmental factors. In large pipe sizes it may be undesirable or impractical to purge the entire system internally. Not only will defective welding be more likely to occur in the field, but they may be harder to detect due to limited or difficult access to the welds when using portable inspection equipment. All of the foregoing considerations point up the necessity for formulating stringent welding procedures and quality control methods, enforceable with equal assurance both in the fabrication shop and in the field. Also, the most common type of fabrication defect involves weld defects are incomplete penetration, lack of fusion, or cracking. Fabrication and erection of piping can create the additional problems of overlooked cold springing, excessive loading by misplaced or ineffective supports, and excessive stresses due to incorrect ineffective, or nonexistent thermal expansion anchors, guides, and other engineered restraints.

Service-Related Deterioration

Process piping system, depending upon the fluid which they transport, are subject to a variety of different service condition. The piping system may be subject to sub zero operating temperatures where embrittlement is a real possibility. The piping system may also be subject to elevated operating temperatures where failure by creep becomes a concern. Embrittlement and creep are two of the several characteristics of metals associated with service related deterioration. A variety of other service conditions may result in different type of deterioration including hydrogen damage, erosion, corrosion, stress corrosion, fatigue, stress relief cracking, etc. account for the vast majority of failures in process piping system.

A considerable number of process piping systems are subject to operation at temperatures of 800 °F and higher. Piping systems that operate at these temperatures for prolonged periods of time are subject to creep. Creep reprints the slow flow of metal subjected to high temperatures, or great pressures. Metallurgically, creep is defined as the time dependent strain occurring under stress. The creep strain occurring at a diminishing rate is called primary creep, that occurring at a minimum and almost constant rate, secondary creep, that occurring at an accelerating rate. Creep is not the only adverse side effect of elevated operating temperatures. Elevated operating temperatures will also produce graphitization in carbon steels and carbon ½ molybdenum steels. Conditions of hydrogen embrittlement and hydrogen damage can also occur in carbon steels, low alloy steels and even stainless steel, depending upon the temperatures and exposure times.

Erosion is another mechanism which produces failures in process piping system. Typically, it occurs in those systems where two phase flow exists at locations where the direction of flow suddenly changes. Erosion also occurs in systems carrying abrasive materials or slurries. In general terms, failures produced strictly by erosion are less common then failures produced by erosion in conjunction with corrosion. Typically, in failures proceed by erosion, the erosion removes the protective oxide coating, exposing a bare metal surface. The process fluid then corrodes the base metal. As the erosion endures that the protective oxide coating never has a chance to reform, the corrosion can progress unabated. Corrosion also accounts for failures in process piping system. Corrosion can take many forms including general corrosion and pitting corrosion. In certain environments, corrosion, in conjunction with tensile stresses, may also produce stress corrosion cracking.

Specific Operational and Service Problems

Overpressure

A common source of over pressure failure is the blocking in of an unrelieved liquid-filled line when the fluid is subject to expansion. Valve mal-operation and faulty equipment operation are other sources of over pressure. Water hammer can result from the too-rapid opening or closing of block valves in the system.

Thermal Strain Cycling

In the low-cycle fatigue, failures occur from the relatively low frequency of operating cycles from one shutdown to another. Fatigue due to cycling thermal expansion and contraction is typical. The thermal spot causes membrane stress and enhances bending stress in the pipe. Increased stress shortens the fatigue life and accelerates the crack propagation rate. The mechanism to increase stress was found to be the structural constraint of thermal strain by the thermal spot.

Mechanical Strain Cycling

Fatigue failures have occurred in systems subject to repeated mechanical loads such as vibration. Sources of mechanical strain cycling are as,

1) Pressure cycling
2) Variations in flow
3) System change

External Factors

External factors include vibrations induced by reciprocating machinery and other similar excitations. Mechanical fatigue is basically a system design problem and has to be minimized by the isolation or reduction of equipment vibrations, careful pressure reduction valve design, designing the piping layout for optimum flow conditions, and in designing for the expected thermal expansions.

Excessive loading- Failures due to excessive primary or secondary loads – i.e. in excess of those for which the system was initially designed – are relatively rare, and are caused by oversight in plant layout and piping design, or by system mis-operation or malfunction. Inadequate pipe support design will result in high primary stresses; and the omission of measures to take care of thermal expansion will cause excessive secondary stresses.

Overheating- Although many failures have arisen from internal overheating, most have been due to external heating in boilers, heaters, and heat exchanger tubes. Failures due to overheating are usually due to equipment failure or a higher process temperature than was originally specified, resulting in a material breakdown. Proper material selection, and careful operation, will do much to avoid this type of failures.

Embrittlement- Embrittlement from service conditions has resulted in many severe failures. It can arise from many sources, the most common being the graphitization of steel from excessively high temperature operation, the use of material below its transition temperature, and the use of materials in an atmosphere which results in hydrogen embrittlement. Other sources are the high temperature formation of sigma phase in some stainless.

Corrosion and/or Erosion

Corrosion and/or erosion was the most frequent cause of pipe failure found arising from a great variety of causes. General corrosion, which involves reductions in thickness over large area may occur from the inside diameter of the pipe as a result of a corrosive process fluid, or some contaminant inadvertently introduced into the process fluid. It also may occur from the outside diameter of the pipe. Corrosion problems have been increasing in frequency and severity over the past three decades, reducing your pipe system's life expectancy down to only a few years.

There are multiple reasons why corrosion can occur in a pipe system. It could be the years of wear and tear and conditions that were not addressed and often ignored from the beginning. The worse thing you can do is ignore minor issues that occur within a pipe system. It is convenient to classify corrosion by the forms in which it manifests itself, the basis for this classification being the appearance of the corroded metal. Each form can be identified by mere visual observation. In most cases the naked eye is sufficient, but sometimes magnification is helpful or required. Valuable information for the solution of a corrosion problem can often be obtained through careful observation of the corroded test specimens or failed equipment. Examination before cleaning is particularly desirable. Some of the eight forms of corrosion are unique, but all of them are more or less interrelated. The eight forms are: (i) uniform, or general attack, (ii) galvanic, or two-metal corrosion, (iii) crevice corrosion, (iv) pitting, (v) inter granular corrosion, (vi) selective leaching, or parting, (vii) Erosion corrosion, (viii) stress corrosion. This covers practically all corrosion failures and problems. Below, the eight forms of corrosion are discussed in terms of their characteristics, mechanisms, and preventive measures. Hydrogen damage, although not a form of corrosion, often occurs indirectly as a result of corrosive attack, and is therefore included in this discussion.

Uniform Attack

Uniform attack is the most common form of corrosion. It is normally characterized by a chemical or electrochemical reaction which proceeds uniformly over the entire exposed surface or over a large area. The pipe becomes thinner and eventually fails. A carbon steel pipe will show rusting over its entire outside surface.

Uniform attack, or general overall corrosion, represents the greatest destruction of metal on a tonnage basis. This form of corrosion, however, is not of too great concern from the technical standpoint, because the life of equipment can be accurately estimated on the basis of comparatively simple tests. Merely immersing specimens in the fluid involved is often sufficient. Uniform attack can be prevented or reduced by (i) proper materials, including coatings, (ii) inhibitors, or (iii) cathodic protection.

External Corrosion

More pipe damage occur due to external (outer surface) corrosion than internal causes. Most is due to insulation failure and the fact that it is hidden from view until a leak, maintenance or some other event prompts a visual investigation. For uninsulated pipe, such as roof level condenser water lines, a surprisingly high volume of pipe is left to deteriorate and could have been avoided through simple maintenance. In most examples, surface rust is minor, and can be easily addressed by the maintenance staff by mechanical wire wheel and the application of an effective rust reverser and outer protective coating. Left to continue, surface rust develops into stratified layers under which deep pitting accelerates, and against which only sand blasting will prove effective at its removal.

Galvanic or Two-Metal Corrosion

A potential difference usually exists between two dissimilar metals when they are immersed in a corrosive or conductive solution. If these metals are placed in contact (or otherwise electrically connected), this potential difference produces electron flow between them. Corrosion of the less corrosion-resistant metal is usually increased and attack of the more resistant material is decreased, as compared with the behaviour of these metals when they are not in contact. The less resistant metal becomes anodic and the more resistant metal cathodic. Usually the cathodic metal corrodes very little or not at all in this type of couple. Because of the electric currents and dissimilar metals involved, this form of corrosion is called galvanic, or two-metal, corrosion.

Crevice Corrosion

Intense localized corrosion frequently occurs within crevices and other shielded areas on metal surfaces exposed to corrosives. This type of attack is usually associated with small volumes of stagnant solution caused by holes, gasket surfaces, lap joints, surface deposits, and crevices under bolt and rivet heads. As a result, this form of corrosion is called crevice corrosion.

Pitting

Pitting is a form of extremely localized attack that results in holes in the metal. These holes may be small or large in diameter, but in most cases they are relatively small. Pits are sometimes isolated or so close together that they look like a rough surface. Generally a pit may be described as a cavity or hole with the surface diameter about the same as or less than the depth.

Pitting is one of the most destructive and insidious forms of corrosion. It causes equipment to fail because of perforation with only a small percent weight loss of the entire structure. It is often difficult to detect pits because of their small size and because the pits are often covered with corrosion products. In addition, it is difficult to measure quantitatively and compare the extent of pitting because of the varying depths and numbers of pits that may occur under identical conditions. Pitting is also difficult to predict by laboratory tests. Sometimes the pits require a long time (several months or a year) to show up in actual service. Pitting is particularly vicious because it is a localized and intense form of corrosion, and failures often occur with extreme suddenness.

Inter-Granular Corrosion

Grain boundary effects are of little or no consequence in most applications or uses of metals. If a metal corrodes, uniform attack results since grain boundaries are usually only slightly more reactive than the matrix. However, under certain conditions, grain interfaces are very reactive and inter granular corrosion results. Localized attack at and adjacent to grain boundaries, with relatively little corrosion of the grains, is inter-granular corrosion. As a result the alloy disintegrates (grains fall out) and loses its strength.

Inter-granular corrosion can be caused by impurities at the grain boundaries, enrichment of one of the alloying elements, or depletion of one of these elements in the grain-boundary areas. Small amounts of iron in aluminum, wherein the solubility of iron is low, have been shown to segregate in the grain boundaries and cause inter-granular corrosion. It has been shown that based on surface tension considerations the zinc content of a brass is higher at the grain boundaries. Depletion of chromium in the grain-boundary regions results in inter-granular corrosion of stainless steels.

Selective Leaching

Selective leaching is the removal of one element from a solid alloy by corrosion processes. The most common example is selective leaching of zinc from brass alloys containing more than 15% zinc (dezincification) in the presence of oxygen and moisture, e.g. from brass taps in chlorine-containing water. It is believed that both copper and zinc gradually dissolve out simultaneously, and copper precipitates back from the solution. The material remaining is a copper-rich sponge with poor mechanical properties, and a color changed from yellow to red. Dezincification can be caused by water containing sulfur, carbon dioxide, and oxygen. Stagnant or low velocity waters tend to promote dezincification. To combat this, arsenic or tin can be added to brass, or gunmetal can be used instead. Dezincification resistant brass (DZR), also known as Brass C352 is an alloy used to make pipe fittings for use with potable water. Similar processes occur in other alloy systems in which aluminum; iron, cobalt, chromium, and other elements are removed. Selective leaching is the general term to describe these processes.

Erosion Corrosion

Erosion corrosion is the acceleration or increase in rate of deterioration or attack on a metal because of relative movement between a corrosive fluid and the metal surface. Generally, this movement is quite rapid, and mechanical wear effects or abrasion are involved. Metal is removed from the surface as dissolved ions, or it forms solid corrosion products which are mechanically swept from the metal surface. Erosion corrosion is characterized in appearance by grooves, gullies, waves, rounded holes, and valleys and usually exhibits a directional pattern. In many cases, failures because of erosion corrosion occur in a relatively short time, and they are unexpected largely because evaluation corrosion tests were run under static conditions or because the erosion effects were not considered.

Stress-Corrosion Cracking

Stress corrosion cracking, occur due to a combination of stress, a corrosive environment, and metallurgical conditions. In carbon steel piping system, stress corrosion cracking relates to applied or residual stresses in conjunction with a caustic environment, in stainless steel systems, cracking is generally associated with chloride contamination. The tensile stresses necessary to produce stress corrosion cracking in piping system are frequently associated with welding. Tensile stresses can also result from end sizing for fit up or cutting, surface grinding etc. Fatigue may also initiate or contribute to stress corrosion cracking. The rate of attack depends on the concentration of the specific corrosive environment, the temperature and the stress level. The higher the concentration of corrosive, the higher the operating temperature, and/or the higher the stress levels, the more severe the corrosion will be.

Microbiologically Influenced Corrosion

Microbiologically influenced corrosion (MIC) is, by far, the most severe and threatening form of corrosion to piping and fire protection systems. It is caused by the presence of various microbiological agents under specific environmental conditions and can, in some cases, result in an advanced and widespread failure of entire piping systems within only a few years. Microbiologically influenced corrosion produces large and deep pits in the steel pipe as an energy source (often as an alternative to oxygen), as well as through the production of strongly corrosive metabolic by-products, such as sulfuric acid, which further assist the microorganism in dissolving pipe metal. MIC exists to varying degrees of severity, and is not exclusive to carbon steel piping systems or open condenser water systems. MIC is commonly found in closed chill water piping, especially those winterizing with glycol, and has been documented to destroy copper, brass and stainless steel pipe.

Other Causes for Pipe Failure

Interior Pipe Deposits

Internal rust deposits, commonly termed tuberculation, are prone to damage most piping systems. They are the lighter and less dense end product of steel pipe corrosion. Once established by high and uncontrolled corrosion conditions, internal deposits initiate much greater deep pitting. The highest corrosion loss is more likely at horizontal lines and in low flow or dead end areas where rust and other deposits will settle, but can affect vertical lines and main risers as well. Random areas of high tuberculation are actually the result of a high pitting or cell corrosion condition directly beneath it, with the volume or height of the rust deposit directly proportionate to the depth and volume of pipe wall loss.

Insulation Failure

Standard fiberglass insulation provides an ineffective moisture barrier for cold water pipes. Condensed area humidity then produces a secondary, and typically hidden, corrosion condition at the pipe exterior. Most common at chill water and dual temperature systems, external corrosion losses will exceed internal corrosion rates by up to 10 times or more. Over decades and hidden from view, insulation failure can destroy entire piping systems. Missing, damaged, and broken insulation defines a likely problem worth investigating. The threat of insulation failure is greatest in smaller diameter piping due to its inherently lesser wall thickness combined with generally thinner insulation applied.

Weathering Damage

Of all the forms of corrosion caused to piping systems, weathering damage due to rain, snow, atmospheric conditions or cooling tower overspray is the easiest to prevent. Most weathering damage requires decades to produce a failure, and is simply due to a lack of maintenance. Smaller diameter piping is always most vulnerable due to its inherently lesser wall thickness.

Wet Pipe Insulation

Enough water can be hidden inside pipe insulation to develop piping failure. This is due to the fact that fiberglass insulation offers very little true moisture barrier and allows humidity to condense at the cold pipe surface. With fiberglass insulation; water will eventually penetrate to the outside to produce wetness, discoloration, and crystallization, thereby providing telltale evidence of a problem. If allowed to continue, substantial piping damage is likely. If acted upon, corrosion under insulation (CUI) problems can be avoided and the often mild deterioration present at early stages of such a problem can be corrected.

Failure Due to Vibration

Vibration of process plant piping occur due to flow induced vibration (FIV), acoustic induced vibration (AIV) and Flow induced pulsation and these are related to the flow of the main process fluid through the piping system.

Flow Induced Vibration

Flow induced vibration is the result of turbulence in the process fluid, which occurs due to major flow discontinuities such as bends, tees, partially closed valves, and small bore connections. The high levels of broadband kinetic energy created downstream of these sources is concentrated at low frequencies, generally less than 100 Hz, and can lead to excitation of vibration modes of the piping and connected equipment. The extent of this problem depends on the piping design, support configuration and stiffness, valve operation, and other related factors which determine the severity of the resulting vibration. Management of the risks related to FIV can be tackled through screening activities to pinpoint piping sections of concern, vibration monitoring and investigation of identified problems, and development of targeted modifications to operational parameters, the pipework configuration, and supports.

Acoustic Induced Vibration

A relief or control valve on piping systems in gas service, or other pressure reducing devices, can generate high levels of high frequency acoustic energy, an effect commonly referred to as acoustic induced vibration. In addition to high noise levels arising external to the piping, this excitation can result in high frequency vibration of the pipe wall, with the potential for high dynamic stresses at welded features such as supports and small bore connections. This in turn can lead to the possibility of fatigue cracking within a relatively short period of time.

Assessment of the risks related to AIV can be carried out for pressure reducing devices on gas systems, such as control valves, pressure relief valves, and blow down valves, to screen for this failure mechanism, through prediction of the acoustic noise generated downstream, and estimation of the risk of failure of the piping. Detailed investigation and the development of solutions would then be conducted for any sections considered to be at high risk.

Flow Induced Pulsation

Flow induced pulsation (FIP) can be caused by dead leg branches in pipework, which can be excited as acoustic resonances with discrete frequencies. These resonances can induce large shaking forces in the pipework, leading to integrity and safety risks. Pulsation analysis to industry design criteria is highly recommended for both design and engineering of complex pipework and manifolds (e.g. compressor stations), and to assess the operation of existing gas plants. Mitigation actions can therefore be focussed on curing the source of the pulsation and vibration instead of limiting the effects.

Pipe Failure Due to Water Hammer in Steam Systems

Water hammer occurs when water, accelerated by steam pressure or a low-pressure void, is suddenly stopped by impact on a valve or fitting, such as bend or tee, or on a pipe surface. Water velocities can be much higher than the normal steam velocity in the pipe, especially when the water hammer is occurring at start up. When these velocities are destroyed by impact, the kinetic energy in the water is converted into pressure energy, and a pressure shock is applied to the obstruction.

In mild cases, there is noise and perhaps movement of the pipe. More severe cases lead to fracture of the pipe or fittings with almost explosive effect and consequent escape of live steam at the fracture. Fracturing of pipes or steam system components can propel fragments that can cause injury or loss of life.

There are two types of water hammer: a steam-flow-driven water hammer and condensate-induced water hammer. A steam-flow-driven water hammer is an impact event, where a slug of rapidly moving water strikes a stationary object. The exchange of momentum creates a pressure of perhaps a few hundred psi in the impact area.

A condensate-induced water hammer is the more powerful of the two types. It is a rapid condensation event that occurs when a steam pocket, being totally surrounded by cooler condensate, collapses into a liquid state. Depending on the pressures and temperatures involved, the reduction in volume may be by a factor of several hundred to well over a thousand, and the resulting low-pressure void allows the pressurized surrounding condensate to rush in, resulting in a tremendous collision. This in turn generates a severe over-pressurization that can easily exceed 1000 psi. Gaskets, fittings and valves – virtually any piping component – are susceptible to failure, often with tragic consequences. Common places to look for both types of water hammer are steam mains, steam tracing lines and air heating coils.

Failure Due to Pipe Support

The integrity of a piping system is not only dependent on the soundness of the "pressure envelope" but also on the pipe supports. A proper support system is required to take the weight of the pipe (including its contents, fittings, valves, etc.), to control movement of the piping, and protect sensitive equipment, through all its operational modes. If supports are not properly designed, installed and maintained, excessive stress can be created in the piping system, and excessive loads can be transmitted to support structures and connected equipment. This may exhibit as pipe deformation, leaking joints, damage to connected equipment, and failure at welds and other "high stress" points. If the excessive stress coincides with deterioration mechanisms (e.g., internal and external corrosion and stress corrosion cracking), premature pipe failure may occur.

Experience across the process industries indicates that corrosion under pipe supports is a major area of concern for plant maintenance engineers and inspectors. API RP 574 (supplements the API 570 piping inspection code), provides guidance how to inspect piping supports.

How to Overcome Piping System Failure

A process piping system failure or leakage can have a significant impact on the business due to interruptions in production, or a catastrophic effect if an explosion or fire occurs or a hazardous fluid is released. Piping integrity is everyone's responsibility. Many piping incidents are caused by operational excursions outside of the design limits or ineffective management of change, not simply "deterioration" or "ineffective maintenance."

Implementing an effective operational integrity program (e.g., Integrity Operating Windows) helps focus the inspection plan on the key vulnerabilities and locations where potential problems may occur.

But unexpected conditions cause the material to degrade and experience unforeseen wear. To enhance reliability of the piping system it must start from design stage have control on operating environment and manage a good maintenance system.

Design, Construction and Installation

Piping systems need to be designed, manufactured, fabricated, installed, inspected, and tested in compliance with their specifications. However, experience shows that specific areas of weakness exist during this early stage. For example, it is not uncommon for the piping to be incorrectly installed (e.g., pipe is over-stressed during assembling process, wrong welding procedures, bad practices during the welding process, including fit-up and joint preparation, bad practices during support and insulation installation, etc.). This is generally due to lack of technical understanding of the installation requirements of equipment by the construction group, and of the consequences of such shortcomings – the potential to introduce or accelerate damage mechanisms.

Operation and Maintenance (Useful Life)

This stage of the life cycle is where operation and maintenance should be aligned in order to extend as much of the piping's useful life as possible.

Operating Window

In order to maintain the integrity and reliability of pressure equipment during operation of any process unit, a set of operating ranges and limits needs to be established for key process variables in order to achieve the desired results (i.e., product within specification, safe operation, reliability, etc.). These limits are generally called operating limits or operating envelopes. It is important that operating boundaries are established, and documented, including allowable excursions during start-up and shut-down of facilities. It is equally important that systems are in place to communicate when piping has been operated outside of the agreed limits, so that appropriate action can be taken. A good practice for piping is the establishment, implementation, and maintenance of integrity operating windows (IOWs). IOWs are a specific subset of these key operating limits that focus on maintaining the integrity or reliability of process equipment.

Factors That Affect Rate of Corrosion on a Piping System

The pH of the Fluid

Acidity of a fluid sample is measured on a pH scale. The pH scale ranges from zero (maximum acidity) to 14 (maximum alkalinity); the middle of the scale, 7, represents the neutral point. Acidity increases from neutral toward 0. In copper pipe systems when the pH is more than 8, a copper oxide film is usually formed on the pipe walls. This film acts as a barrier that slows the effects of corrosion. However, when the pH in the water supply is lower than 8 in the water supply the copper oxide film (barrier) is dissolved leaving no protective barrier and the pipe is subject to the corrosive action of the water.

Amount of Oxygen in the Fluid

Dissolved air in consists of about 30% oxygen and the rest is mostly nitrogen, which is non-corrosive. Oxygen degrades metals through an electro-chemical process of internal oxidation. The result is that metal gradually gets converted to an oxide (rust), becoming thinner and weaker in the process. As the pipe corrodes the impurities are deposited in the water lines. Encrusted build up is the direct result of the oxidation process.

While oxygen content decreases under higher fluid temperature and higher pressure it is these higher temperatures and higher pressure conditions that speed up the oxidation process. Experience shows that corrosion is more pronounced in hot lines.

Chemical Makeup of Fluid

Chemical composition of the fluid may have differing effects on the corrosive forces at play. For example a moderate to high level of calcium would help form a protective coating on the pipe which could slow down the corrosive effects. However, high levels of calcium may cause a calcium build up in the pipe.

Galvanic Corrosion

Galvanic corrosion, also known as electrolysis, occurs when different metals come into contact with each other. When conflicting metals are touching each other one of the metals has a tendency to give up electrons. Basically at the point of contact the metal giving up the electrons dissolves over time. The most frequent cases of this happening are when galvanized pipe and copper are connected; copper pipes touch steel studs, or steel pipe hangers.

Contrary to common belief, the effects of galvanic corrosion are limited to the immediate area of contact. The use of dielectric fittings helps stop the problem but does not repair the resulting thin walled and damaged pipe.

Temperature

The higher the fluid temperature the faster the rate of oxidation. Experience shows that corrosion is more pronounced in hot water lines.

Water Velocity

Water velocity problems are usually associated with a "closed" loop piping system where the need to pump or circulate the water is required. Erosion corrosion occurs at locations where water turbulence develops. Turbulence can be caused by excessive velocity, sudden changes in direction (sharp turns, elbows) and through "flow" obstacles such as burrs and solder excess.

The major contributing factors to this type of erosion corrosion occur:

When water velocities exceed 4 ft/sec, oversized circulation pumps, installation of undersized distribution lines, multiple or abrupt changes in the direction of the pipe, burrs on the inside of the pipe, improper soldered joints.

How to Mitigate Corrosion

Coatings- Coatings can be used to stop or mitigate corrosion on both inside diameters (IDs) and outside diameters (ODs). Thousands of coating systems are available. To select the proper coating system for the job, the engineer needs to consider the pipe's design, environment, contents, pressures, external influences, design life, and costs. Conventional external coatings such as mastic, and many newer types such as epoxies and polymers, have proven to be successful corrosion inhibitors. Internal coatings are available for contents corrosion control, cable pulling friction reduction, and erosion (impingement) corrosion control.

Inhibitors- Inhibitors generally are used to control corrosion on the inside of pipelines. The specific inhibitor depends on the material to be protected and the solutions transported inside the pipe. For external protection, several types are available. Some absorb moisture so it can be collected in strainers. Others are developed to react with specific corrosion-causing agents to render them inert. Still another type is used to augment a concrete coating, greatly increasing the concrete's protective capability.

Inhibitors are so numerous and have such a variety of capabilities that they can solve many problems that formerly seemed unsolvable.

Cathodic Protection (CP)

Because corrosion develops in an electrical circuit, stopping corrosion is a matter of disrupting the electrical circuit. Cathodic protection (CP) is by far the best way to stop corrosion on pipelines. It uses impressed currents from a fixed anode to interfere with the electrical circuit in the corrosion cell. It is 100% effective against most forms of external pipe corrosion.

Galvanic CP connects a high-energy metal, such as zinc or magnesium, to the pipe (anode). The zinc or magnesium acts as a sacrificial anode, protecting the pipe. The sacrificial anode operates as a galvanic anode with impressed voltages as high as −1.4 to −2.1 V (and in some cases, even more negative). If a counter-impressed current from a sacrificial anode is more negative than −0.85 V, corrosion in steel stops.

How to Reduce the Risk of Water Hammer

Operators can reduce their risk of water hammer by preventing or resolving steam system design issues.

1) Avoid water hammer completely by taking steps to ensure that water (condensate) is drained away before it accumulates in sufficient quantity to be picked up by the steam.
2) Improve steam quality, keeping steam as dry as possible at all times. Install steam-conditioning stations upstream of meters and any other critical steam system components.
3) Do not allow steam velocities to become excessive as a result of system modifications. The higher the velocity, the higher the force of impact during a steam flow-driven event.
4) Steam traps: Make sure the steam traps used are of correct type and capacity.
5) Piping: Correct any occurrences of pipe sagging and missing, wet or damaged insulation that could cause condensate accumulation and exceed steam traps' capacities.

Inspection and Maintenance Plan to Avoid Failure

To ensure a catastrophic failure does not catch a plant off guard, ongoing maintenance and inspection is imperative. Regularly executed inspections can help plants remain more efficient. So, how does one determine the optimum levels of inspection and maintenance for piping? Risk based Piping Inspection Analysis can be used to improve the current condition of piping inspection program where piping system assessment, predicted inspection interval, and database management can be simulated.

The piping inspection plan could consume a large portion of the maintenance budget if it is not designed, planned, and implemented properly. Generally, process plants (depending of the plant size) can have between 1000 and 5000 piping systems. For that reason, inspection resources for piping should be appropriately focused using systematic, consistent techniques, such as RBI and/or criticality analysis, to identify which system to inspect and what inspection methods to employ. The inspection plan should identify what, where, when, and how piping should be inspected.

Understand the Ways in Which the Piping Can Fail

An effective risk-based assessment can only be carried out with a sound knowledge of the ways in which in the piping can fail. This includes an understanding of its vulnerabilities, such as:

- Damage mechanisms (e.g., internal corrosion, external corrosion).
- Design limitations (e.g., number of cycles to avoid fatigue failure, locations of high stress).
- Local effects (e.g., injection points, dead legs).
- Vulnerable components (e.g., bellows expansion joints, pipe supports).
- Current condition of the piping.

Such a deterioration assessment should be carried out by a team with the necessary range of expertise – design, materials/corrosion, operations, inspection, maintenance, and process safety. As many companies do not have such resources internally, external specialists should be brought in, as appropriate. From these vulnerabilities, the team should determine how the piping is likely to fail (e.g., catastrophic sudden failure, initial small leak that might gradually develop into a structural failure). There can often be more than one "credible failure scenario."

Determine the Likelihood of Piping Failure

The assessment team should determine the likelihood of occurrence for each of the failure scenarios. As with the consequence assessment, a series of word models can be very effective for this assessment. This is particularly the case in the early stages of the development of a piping strategy, when perhaps knowledge of the piping systems, their condition, and inspection/maintenance/operational history may be limited.

Carefully Determine the Scope and Frequency of Inspection

The extent and frequency of inspection of piping depend on the forms of deterioration that can affect the piping and consequence of a piping failure. API 571 provides a substantial list and details of damage mechanisms affecting oil and gas facilities.

A widely used method for determining piping inspection intervals is based on the "half of the remaining life" concept. However, this type of fixed-time approach may not be most appropriate for all piping. This is particularly the case where the fluid composition changes over time, such as in offshore installations, or refineries which are likely to experience varying crude oil composition, or waste treatment facilities that are required to handle a wide range of chemicals. This approach may also not be appropriate when deterioration mechanisms are localized. In such cases, a simple statistical approach is unlikely to give adequate focus to specific vulnerabilities.

It is not enough to base future inspection plans only on prior reported history of equipment condition. A fundamental understanding of the process/operating conditions and resulting deterioration and failure mechanisms is required in order to establish and maintain an inspection program, along with an IOW (integrity operating window) program. RBI is commonly regarded as good practice and the most appropriate methodology to define the scope and the inspection interval of piping systems, especially for those piping systems with high and medium criticality. RBI is a condition with consequence-based approach and provides a rational basis for focusing inspection efforts. The risk analysis that supports the RBI program may be qualitative, quantitative, or a combination of the two (semi-quantitative). In each case, the risk analysis approach should be used to systematically screen for risk, identify areas of potential concern, and develop a prioritized list for more in-depth inspection or analysis.

Specify the Inspection

How does one determine the type of NDT to use, select appropriate locations to inspect, and decide how much piping to inspect? The type of NDT technique to be applied will depend on the type of damage mechanism likely to occur and if it is internal or external. ASME PCC-3 table C-1 provides a useful reference to help to identify the inspection method. The challenge for piping inspection is that many of these examination methods depend upon proper access and surface preparation and thus will not be appropriate for all situations. The single most frequent damage mechanism leading to pipe replacement is corrosion. A key to the effective monitoring of piping corrosion is identifying and establishing condition monitoring locations

(CML). CMLs are designated areas in the piping system where measurements are periodically taken. Ultrasonic (UT) thickness measurements are obtained within examination points on the pipe. "A good practice is to develop a piping circuit layout and associated CMLs identified on inspection isometric to aid the inspector in performing inspection tasks."

Assess the Inspection Results

The piping inspection plan should be designed with the following philosophy: plan to inspect, inspect to detect, detect to assess, and correct or repair to keep the piping operating in a safe and reliable way. The piping's inspection tasks normally generate a large quantity of data that if not collected, recorded and analyzed systematically, critical information about the piping integrity condition could be overlooked.

Assessing the Piping Integrity

Piping integrity should be assessed by a "Competent Person" (CP) who has sufficient training, knowledge and experience to carry out the required assessment. The level of competence required will depend on the complexity of the situation. Different organizational arrangements and practices exist indifferent countries. For example in many countries, API 570 authorized inspectors create the inspection strategies, assess the deterioration rate, the remaining life and time to next inspection.

Report and Monitor Trends

A record of thickness data obtained during periodic inspections provides a means of determining corrosion or erosion rates and expected equipment life. This task can be greatly assisted by an appropriate computer system whereby the data can be shown on graphs and diagrams or presented as tabulated information. In addition to wall thickness measurements, means of corrosion monitoring (such as corrosion coupons or corrosion probes) may be useful in supplementing the wall thickness measurements and in establishing the timing of these thickness measurements. Permanent records should be maintained throughout the service life of each piping system. As a part of these records, progressive inspection and maintenance records should be regularly updated to include new information pertinent to the operation, inspection, and maintenance history of the piping system.

Use the Integrity Outcomes to Revise the Inspection Approach

Investigation of asset-related failures, incidents and nonconformities should be performed for all piping. The investigation should determine not only the direct cause of the failure or "near miss," but also the underlying and contributory factors. This will enable a thorough analysis of the implications of the incident for your systems, procedures and practices, as well as the equipment. Only then can the outcome of the investigation be effectively used to update the piping inspection approach, and corrective actions taken to address all the causes of identified non-conformances in order to prevent, or reduce the likelihood, of future similar incidents.

A structured review of the integrity process, including key performance indicators, should be regularly carried out to monitor integrity performance so that positive trends can be reinforced and unfavorable trends can be corrected.

Conclusion

The mechanical integrity of process piping is critical to effectively manage process safety, environmental hazards, and business risks in the oil and gas, chemical, petrochemical and process industries. Failure (loss of containment) of piping systems is more likely to occur than the failure of pressure vessels, and the consequences are significantly greater, when measured across the process industry as a whole. In this chapter some of the good practices are highlighted that can be applied to help organizations develop an optimal piping integrity strategy. Doing this in a cost-effective way is not an easy task. It is important to understand that people, processes, and systems should be brought together within a clear framework (Asset Integrity Management System) in order to achieve the organization's business goals. All piping should receive a certain level of attention through-out its life cycle. Using the approaches described in this chapter can help develop a proactive and risk-based piping integrity strategy.

7

Failure of Flanged Joint

A bolted flange connection is a complex mechanical system whose components must be selected and assembled properly to provide reliable sealing over a wide range of operating conditions. All of the various components of the assembled bolted flange connection are important to the proper operation of the joint. The components consist of the piping or vessels, the flange(s), the gasket(s) and bolts. In addition to the components themselves, the joint design and assembly are critical to the long-term operation of the joint.

Creating the Seal

The seal is created by the clamping forces acting upon the gasket surface, compressing the gasket and causing the gasket to conform to flange imperfections. The conformance of the gasket material to the flange surface under the compressive load (contact pressure) fills any leak paths and prevents the escape of the contained fluid from the bolted flange connection while maintaining a specified leakage rate. To achieve a successful seal, the gasket must be resilient enough to conform to any irregularities in the mating surfaces. The gasket must also be sufficiently tough (rugged) enough to resist extrusion, creep and blowout under the operating conditions and unexpected pressure/temperature excursions. Sufficiently high bolt load must also remain on the gasket to minimize leakage.

Forces Acting on a Gasket Joint

When a flanged joint is pressurized it is subjected to following stresses.

1) Compressive stress produced by bolt load (assembly load) that flows the gasket material into surface imperfections to form a seal.
2) Hydrostatic end force (working pressure multiplied by area of aperture in gasket) that tends to separate flanges when the system is Pressurized.
3) Gasket itself subject to side load and due to internal pressure tending to extrude it through the flange clearance space. There are other shock forces that may be created due to sudden changes in temperature and pressure. Creep relaxation is another factor that may come into the picture.

Root Cause Failure Analysis: A Guide to Improve Plant Reliability, First Edition. Trinath Sahoo.

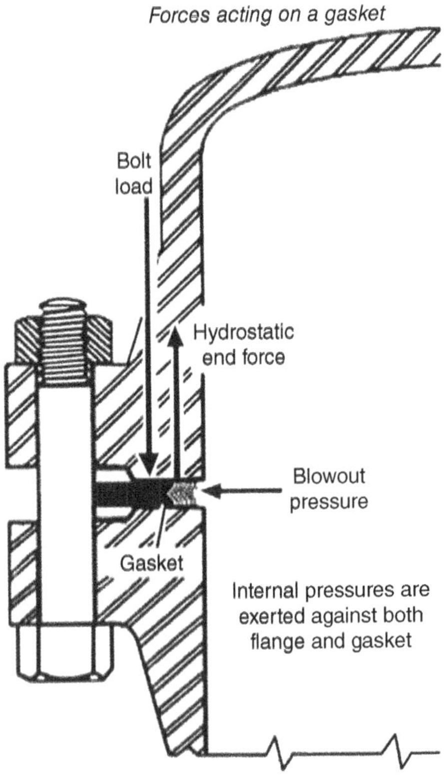

Figure 7.1 Forces acting on a gasket.

The equation below correlates the above three loads.

(Bolt Load − hydrostatic end thrust) = Residual gasket load.

Residual gasket stress is the stress remaining on the gasket at operating condition and is determined as a function of the internal pressure and gasket factor.

(Gasket factor × internal pressure) = Residual gasket stress.

So residual gasket stress must be higher than the fluid pressure if leakage is to be avoided. Codes traditionally classify gasket materials by two property values.

Y factor = the initial gasket stress or Minimum design seating stress

(This is due to preload prior to pressurization).

m = Ratio of residual gasket stress to fluid pressure

= **bolt load − hydraulic end load**

gasket area × internal pressure

Most of the pressure vessel codes have the value of Y and m.

How to Achieve a Successful Seal

When the bolted flange connection is assembled, the gasket is subject to compressive load between the faces of the flanges. The compressive load on the gasket must be high enough to compress the gasket into the surface finish of the flanges to fill any potential leak paths. The initial compression force applied to a joint must serve several purposes.

1) It must be sufficient to initially seat the gasket and flow the gasket into the imperfections on the gasket seating surfaces regardless of operating conditions.
2) Initial compression force must be great enough to compensate for the total hydrostatic end force that would be present during operating conditions.
3) It must be sufficient to maintain a residual load on the gasket/flange interface.

The residual gasket load which is the compressive load minus the hydrostatic end load must be sufficiently high to minimize leakage. It is also influenced by other factors, such as thermal expansion behavior of flanges and bolts. Once seated, a gasket must be capable of overcoming minor alignment issues, flange sealing face imperfections and operating variations such as, non-parallel flange faces, thermal Cycling, pressure Cycling, startup and shutdown process variations, hydro test pressure during leak testing etc.

Integrity of the Bolted Flange Connections

There are four major factors for consideration for integrity of the bolted flange connections: fastener related, gasket related, flange related, and the gasket-flange-fastener system as a whole. Considerations related to the fasteners include material properties of bolts/studs, nuts and washers. Specific gasket related considerations include the temperature, pressure and fluid interactions against the gasket; as well as the gaskets thickness required, stress to seal and storage of the actual gaskets. Flange related considerations include the type of flange in use, its surface finish, flange compatibility with semi-metallic and metallic gaskets due to electrochemical (or "galvanic") corrosion and flange strength and deformation. When you begin to put these three targeted categories together you must have more holistic considerations such as gasket size, overall joint "tightness," gasket factors (design) and torque to control joint integrity.

Fastener Related

In a flange system when the bolt is tightened, it acts like springs, pulling the flanges together. As the system is pressurized (during startup) these bolts need to be stretched enough to keep the load on the gasket.

When a standard threaded fastener fails in pure tension, it typically fractures through the threaded portion (as this is characteristically its smallest and therefore the weakest area). To avoid this risk, most carbon or alloy steel bolts have a defined proof load, which represents the usable strength range for that particular fastener. By definition, the proof load is an applied tensile load that the fastener must support without permanent deformation. In other words, the bolt returns to its original shape once the load is removed.

The relationship between tension and bolt stretch can be observed on a Tensile Stress-Strain diagram. To the left is the stress elongation curve. Steel possesses a certain amount of elasticity as it is stretched. Thus, a bolt that is properly tensioned should be functioning in the elastic range (as viewed on the Diagram). If the load is removed and the fastener is still within the elastic range, the fastener will always return to its original shape. However, if the

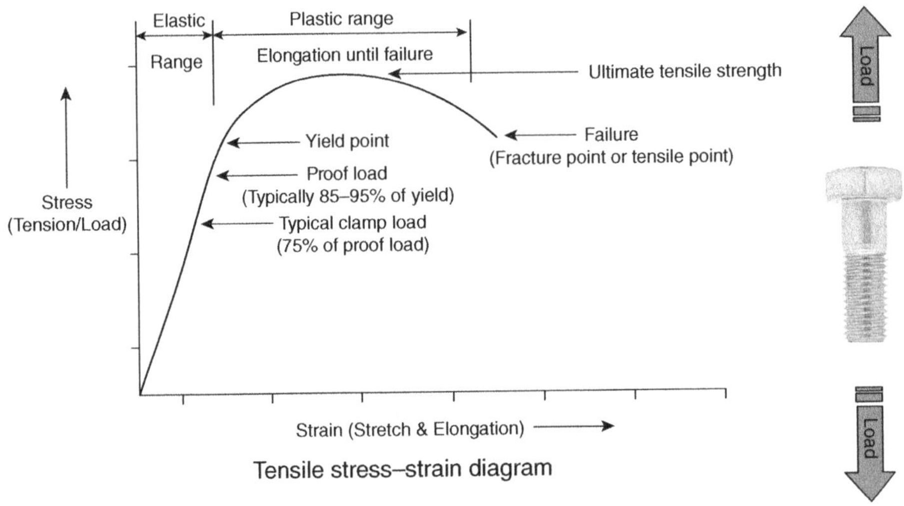

Figure 7.2 Tensile stress strain diagram of fastener. *Source*: Bolted Joint Design/Fastenal Company.

load applied causes the fastener to exceed its yield point, it enters the plastic range. At this point, the steel is no longer able to return to its original shape if the load is removed. The yield strength is the point beyond which a specified amount of permanent deformation occurs. If we would continue to apply a load, we would reach a point of maximum stress known as the ultimate tensile strength. Past this point, the fastener continues to neck down and elongate further with a reduction in stress. Additional stretching will ultimately cause the fastener to break at the tensile point.

Proper amount of tensioning of the bolts is vital. With too little clamping force, the joint may loosen. If the joint is exposed to cyclical loads, too little clamping force can shorten the bolt's fatigue life. Too much clamping force can also cause severe problems. By over-tightening the bolt, one may exceed the proof load of the bolt. Even if the bolt does not fail during assembly, it may later break under the external tensile load. Over-tightening of the bolt can also encourage the advancement of hydrogen embrittlement or stress corrosion cracking. The joint members can also be damaged or warp from too much clamp force.

After the initial assembly of a bolted joint, there are many other factors acting on the joint that work to compromise the initial seal (Cycling, Gasket Material, Bending moments). Most applications undergo pressure and thermal cycles which work to decrease the compressive load on the gasket. Pressure cycles change the hydrostatic end load. Thermal cycles can change the stretch in the bolts through various mechanisms, including bolt material modulus changes due to temperature, differential thermal expansion of the flange component materials and thermal transients, in heat-up and cool-down. These pressure and thermal cycles cause fluctuations on the compressive load of the gasket and may increase the potential for a leak. A fastener subjected to repeated cyclic loads can break suddenly and unexpectedly, even if the loads are well below the strength of the material. The fastener fails in fatigue. Fatigue strength is the maximum stress a fastener can withstand for a specified number of repeated cycles prior to its failure.

If a constant load is applied to a fastener and the service temperature is increased, the temperature places the bolt in its creep range; the bolt will begin to stretch even if the load is well within the fastener's mechanical limits. Eventually, the bolt may stretch to a point where it cannot support the load and will fail.

The minimum bolt load should be at fifty percent (50%) of bolt yield strength to ensure the "spring" is stretched enough. The bolt load typically used is at eighty percent (80%) or even one hundred percent (100%) of bolt yield; if the calculation method and assembly is very accurate. The flange connection will lose compressive load due to system relaxation. If the bolt is not stretched enough, the gasket residual load may drop below the load required to maintain a seal, thereby, causing a leak. Friction between the bolt, nut and the flange surface is a force often overlooked during gasketed joint assembly. When determining the torque required to properly tighten the bolts, a nut factor or friction factor needs to be taken into consideration. Using well lubricated nuts, bolts, and nut and bolt faces can help control the frictional load losses during assembly. The use of hardened steel washers between the nut and bolt heads and the flange is recommended to help control frictional losses from embedment of the bolt/nut face into the flange. Bolt tightening should be controlled to insure proper bolted flanged connection assembly.

Factors to Be Considered During Selection of Bolts

Tensile Area: The threaded portion of a bolt can be characterised by two dimensions, the major diameter and the pitch diameter. The major diameter is the diameter measured across the outmost parts of the threads. The pitch diameter is the diameter measured half way between the highest and the lowest point on the threads. A bolt subjected to a tensile load behaves as if its cross section has a diameter of effectively the average of the major and pitch diameter.

Strength and Grade of bolt: Bolt grades depend on the bolt materials, heat treatment etc., which lead to different strengths. Different grades of bolts have specific head markings on them to allow for easier sorting/identification of the bolt's estimated strength. A measure of a bolt's strength is its proof strength. The proof strength of a bolt is the maximum tensile load a bolt can with stand without incurring permanent deformation. Proof strength is slightly less than the yield stress (about 96%) of the material due to stress concentration in the threads. The strength of many standard metric threaded fasteners is specified according to an ISO standard. The strength of a bolt is specified using a two digit grading system. The grade is called a property class. A property class designation for bolts consists of two numbers separated by a decimal point. The number before the decimal point is one-hundredth of the nominal tensile strength of the bolt in N/mm^2. The number after the decimal point is 10 times that of the ratio between the nominal yield stress and nominal tensile stress. Consider a bolt having property class – 5.8.

Nominal tensile strength $= 5 \times 100 = 500\,N/mm^2$ Yield stress $= 5 \times 8 \times 10 = 400\,N/mm^2$

Threads per inch: This is the number of threads per inch on the bolt. This figure is important when calculating +how many turns of the nut on the bolt will be required to clamp the member or when determining how much of the thread will extend beyond the nut when clamped. Nuts and bolts are not perfectly rigid but stretch slightly under load. The distribution of stress on threads is not uniform. In fact the first thread takes a third of the load, the first three threads, takes three quarters of the load and the first six threads takes essentially the whole load. It is important to note that three full threads are all that are required to develop the full bolt strength. There is therefore little point in having more than six threads. Nuts are usually stronger than the bolts they are on; which is to say that the bolt will usually break before the nut strips. It is often said that two full threads must be exposed above a nut, the reason being the first two threads of a bolt are often poorly formed, and may not engage the nut properly. If they are not doing their share, the other threads in the nut will be overloaded and the nut may strip.

Use of Washers: It is preferable that the threads of a bolted joint do not fail before the bolt itself fails in tension. Thus heat treated nuts and hardened washers are used if the parts being connected are not harder than are required for high strength bolts. The reasons for using washers are

Bearing stress: Using a washer reduces the bearing stress on the member surface by distributing the load over the washer area instead of just the bolt head area. For softer materials this is critical because the bolt head or nut might start to sink into the member surface before an adequate preload is reached.

Vibration: Washers help prevent nuts and bolts from coming loose during operation of the joint, especially when vibration is present.

Tightening: Washers make it easier to tighten a bolt and nut joint by preventing the nut, or bolt, from turning when the other part is being turned.

Strength: Although really only an issue when high strength joints are being used, washers help prevent damage to the nut and bolt from any burrs that might exist from the drilling of the holes. Burrs on the surface can cause failure when high stresses are imposed on the joint.

No of Bolts: A joint is illustrated in Figure 1. There are three principal forces acting on any gasketed joint. They are: Bolt load and/or other means of applying the initial compressive load that flows the gasket material into surface imperfections to form a seal. The hydrostatic end force, that tends to separate flanges when the system is pressurized. Internal pressure acting on the portion of the gasket exposed to internal pressure tending to blow the gasket out of the joint. The effect of the preload is to place the parts in compression for better resistance to the external tensile load and to create friction between the parts to resist any shear load. The shear load does not affect the final bolt tension, and we shall ignore this load for now.

This implies that thinner and longer bolts are better. However the load carrying capacity of a bolt is proportional to the square of the bolt diameter and therefore a reduction in diameter will have some negative effects. Nonetheless, a larger number of small diameter bolts are better than a small number of larger diameter bolts. The benefit of high preload is leakage in a pipe joint is reduced and fastener loosening is minimised. As a rule of thumb, bolt preload should be 70–80% of the proof load of the bolt. Also it should exceed the maximum load in the system by 15%.

Torque: The most important term used when installing a bolted joint is the torque. Approximately 50% of the applied torque is wasted in overcoming the mating friction under the head, 40% friction wasted in overcoming resistance to threads and only 10% of the total torque exerted is used to produce tension in the bolt. As tightening proceeds a maximum torque value will be attained, followed by a sharp decrease in torque as additional turning is attempted. The decrease in torque is an indication of loss of tightness and the maximum bolt strength is exceeded. It is extremely important to properly torque the bolted joint to the correct level. To under-torque a fastener is just as bad as to over-torqueit. If the fastener is under-torqued, that will cause it to cycle many more times than normal and eventually shorten the fatigue life drastically.

Over-torquing runs the risk of shearing the fastener head off or even actually exceeding the yield point of the fastener. There is a relationship between torque and tension, but it is highly variable and must be used with caution.

$$T = K \times P \times D$$

where
T = torque
K = coefficient of friction (nut factor)
D = bolt nominal diameter
P = bolt desired tensile load (usually70% of yield stress)

The variables in the torque- tension relationship include lubrication, thread fit, tension in the bolt, bolt diameter and bolt grade.

$K = 0.2$ for plain finished bolts.

$K = 0.1$ for waxed or highly lubricated bolts.

Several factors affect how much tension occurs when a given amount of tightening torque is applied. It takes more force to tighten a 3/4″ bolt than to tighten a 3/8″ bolt because it is larger in diameter. It takes more force to stretch an SAE grade 8 bolt than to stretch an SAE grade 5 bolt because of greater material strength. The nut factor or coefficient of friction also plays an important role. The harder and smoother bolting surfaces such as threads and bearing surfaces require less torque to stretch a bolt than do softer, rougher and stickier surfaces. Surface finish of a bolt plays a determinant role on the clamping load. Lubricated and plated bolts should be torqued to a lower value. If you start lubricating the bolts to make assembly easier but maintain the same torque as was used when they were plain finished, then the bolt will break

Protection of Bolts

Plating

Plating on fasteners is done by an electrolytic process. After plating, the fasteners are baked at a temperature well below the decomposition temperature of the plating material to prevent hydrogen embrittlement. However, heating the plating to its decomposition temperature can generate free hydrogen again. Thus exceeding the safe operating temperature of the plating can cause premature fastener failure due to hydrogen embrittlement as well as loss of corrosion protection. So a bolt's plating material is usually the limiting factor on maximum service temperature.

Cadmium plating: Cadmium plating is done by electro-deposition and easy to accomplish. However, cadmium plated parts must be baked at 375 °F for 23 h, within 2 h after plating, to prevent hydrogen embrittlement. Since cadmium melts at 600 °F, its useful service temperature limit is 450 °F.

Zinc plating: Zinc can be electrode posited or can be hot dip galvanised. Zinc has a dull finish compared to cadmium. Zinc is a sacrificial material. It will migrate to uncoated areas that have had their plating scratched off, thus continuing to prevent corrosion resistance.

Passivation

Stainless steel fasteners will create galvanic corrosion or oxidation in a joint unless they are passivated or pre-oxidised prior to assembly. Passivation is the formation of a protective oxide coating on the steel by treating it briefly with an acid. The oxide coating is almost inert. Pre-oxidation is the formation of an oxide coating by exposing the fasteners to approximately 1300 °F temperature in an air furnace. The surface formed is inert enough to prevent galling due to galvanic corrosion.

Thread lubricants

Lubricants are used to reduce the coefficient of friction. Generally 0.2 is used for an un-lubricated coefficient of friction and 0.09 for a lubricated co-efficient of friction. It is observed that torque required for a lubricated fastener is less than half that of an un-lubricated fastener. The most common thread lubricants are:

Oil and Grease: Oil and grease are the most common type of thread lubricants, but they are limited to an operating range of 250 °F. Above this temperature the oil or grease will melt or boil off. Oil can not be used in a vacuum environment. However, oil and grease are good for both lubrication and for corrosion prevention.

Graphite: A fine carbon powder that needs moisture or oil to become a lubricant. Therefore, its maximum temperature is limited to the boiling point of oil or water. It also cannot be used in a vacuum environment without losing its moisture. As dry graphite is abrasive its use is detrimental to a bolted joint.

Molybdenum disulphide: One of the more popular dry lubricants. It can be approximately 750 °F. Molybdenumtri-sulphide is an abrasive rather than a lubricant.

Elevated Temperatures

In case of an elevated temperature joint design, three things are important – preload, temperature and time. As a result of the effects of temperature such considerations as selection and strength of materials, bolt relaxation and accelerated corrosion and oxidation come into play. The factor of time is critical for a designer, to know how long the structure will be exposed to the elevated temperature. With a fluctuating load there will be continued loss of preload and the possibility of rapid fatigue failure will occur. To eliminate the fatigue problem, an initial high preload is essential. At elevated temperatures the induced bolt load will change with time even if the elastic limits are not exceeded, and this can adversely affect fastener performance. Hence it is necessary to compensate for strength at high temperature conditions while selecting the fastener materials, as at high temperature the strength of the material reduces.

Gasket

Selecting the right gasket for a leak-free seal requires much more thought and consideration.

The primary selection of a gasket material is influenced by three factors.

1) The temperature of the fluid or gas to be contained.
2) The pressure of the fluid or gas to be contained.
3) The corrosive characteristics of the fluid or gas to be contained.

The mechanical factors are also important in the design of the joint also.

Gasket Seating Stress

In the Figure 1 Total bolt force (F_b) required to generate the proper gasket seating stress

$$F_b = N_b \times S_b \times A_b$$

F_b = Total bolt force
N_b = No of bolts
S_b = Bolt stress
A_b = Stress area
A_g = Area of gasket

The total bolt force calculated must be sufficient to seat the gasket in to the flange. In most cases flange geometry is fixed by design and gasket ID, OD are known. The actual compressive stress available to seat the gasket is then

$$S_g = F_b / A_g$$

Material must be selected, that will seat satisfactorily under this stress. Check for material that have minimum seating stress equal to or less than the available stress computed in the above equation. Minimum seating stress is available in different Codes and can also be available from manufacturer.

As a thumb rule the product of operating temperature and operating pressure give fair idea regarding gasket material selection. P × T value for Some of the material are given below.

Gasket material	P(psi)×T (°F)	Max temp
Rubber	15000	300 °F
CAF	250000	850 °F
Metallic	⊃250000	Depend on metal

Once seated, a gasket must be capable of overcoming minor alignment issues, flange sealing face imperfections and operating variations such as/Thermal Cycling, Pressure Cycling, Misaligned flanges, surface imperfections, Startup and shutdown processing variations etc.

Some of the Commonly Used Gasket Material

1) **Graphite Foil**- It's physical and chemical properties make it suitable as a sealing material for more arduous operating condition. In oxidizing environment it can be used in the temperature range of −200 °C to +500 °C and in reducing atmosphere −200 °C to +2500 °C. As it has no binder it has excellent chemical resistance being unaffected by most of the chemicals. It has very good stress relaxation properties.
2) **Spiralwound Gasket**- It is constructed by spirally winding a preformed metal strip and an asbestos filler on the periphery of a metal winding mandrel. All spiral wound gaskets are furnished with a centering ring. In addition to controlling compression, these rings serve to locate the gasket centrally within the bolt circle. Inner rings are used where material have shown inward buckling of gasket (gasket with PTFE filler material). It also

prevent build up of solids between inside dia of gasket and bore of pipe. Under vacuum condition it protects when broken components could be drawn in to the system. They can operate at −250 °C to +1000 °C and pressure from vacuum to 350 bar. Spiral wound gaskets up tp 1″ size and up to 600# require uniform bolt stress of 25 000 psi to compress the gasket. All other sizes and classes require 30 000 psi compress the gasket.

3) **Ring Joint Gasket**- Oval or octagonal-shaped gaskets are commonly used in grooved flanges in high pressure piping systems and vessels. Their applicable pressure range is from 1000 to 15 000 psi. These gaskets are designed to give very high gasket pressure with moderate bolt load. These joints are not general pressure actuated and the hardness is lesser than that of flange material so that proper flow of material occurs without damaging surface of the flange.
   a) **Oval type**- Oval type gaskets contacts flange faces at the curved surface and provide high reliability seal. But due to its shape it is harder to achieve accuracy of dimensions and surface finish in oval type than octagonal type. Reuse of this type of gasket is not possible.
   b) **Octagonal type**- It is more economical to make and more accurate in dimension and surface finish than oval type because it consist of straight faces only. But more torque load is required to flow the gasket material into imperfections on the flange facings. Reuse is possible.
4) **Corrugated Metal Gasket**- These are formed in a range of metals including brass, copper, cupro-nickel, steel, monel, aluminium etc. and produced to almost any shape and size required The thickness of the metal is normally 0.25 or 0.3 mm with corrugations 1.6, 3.2 and 6.4 mm pitch. The sealing mechanism is based on point contact between the peaks of the corrugations and the mating flanges.

Gasket Reliability

Temperature and time can affect gasket materials in a variety of ways. Gasket material degradation due to thermal or chemical exposure can result in a change of material properties, making them less resilient and reducing their ability to spring back under fluctuating loads. Temperature and compressive load can cause creep relaxation in the gasket material itself. Gasket materials that have good creep relaxation and are less susceptible to temperature or chemical exposure may be chosen. Safety factors are generally recommended to insure the residual gasket load is sufficiently higher than the load required to maintain the seal. Safety factors are generally applied to allow for any relaxation of the gasket compressive forces and to accommodate the application process involved.

Gasket Performance Depends Upon the Following Factors

a) **Surface Finish**- Surface finish governs the thickness and compressibility necessary in the gasket material to complete a physical barrier in the clearance gap between the flanges. Too fine a surface finish can be undesirable since the surface may lack grip, especially on a harder gasket material allowing extrusion to take place. Larger flange surface finishes require higher bolt loads. So fine machining marks tangential to the direction of applied fluid pressure can be helpful. Flange faces with non-slip grooves of approximately

0.125 mm depth for gaskets more than 0.5 mm thick and for thinner gaskets 0.065 mm deep grooves are recommended. Under no circumstances should the flange sealing surface be machined so that tool marks extend radially across the gasket sealing surface. In that case leakage of the sealed fluid may occur.

Recommended flange surface finish	
PTFE-	250–500 μ″
SPIRAL WOUND-	125–250 μ″
RING JOINT (R & RX)-	63 μ″
BX-	32 μ″

b) **Thickness of Gasket-** Thinner materials are capable of taking higher compressive stresses than thicker materials of same type. A better quality surface finish is required for thinner materials. Gasket shall be at least four times as thick as the maximum surface roughness of the flange faces. It will be thick enough to take up the shape of the flange faces and still compress under bolt load. But where vibration is unavoidable, a gasket thicker than the minimum thickness should be employed.

Thinner gaskets offer many advantages:

- Greater blow-out resistance due to the smaller cross sectional area exposed to the internal medium pressure.
- Lower leakage rates again due to the smaller cross sectional area.
- Better compressive strength and therefore higher gasket surface loads (pressure) can be safely applied to a thinner gasket.
- Better fastener torque retention due to the lower creep relaxation characteristics of thinner gaskets.
- Lower cost of the gasket itself.

Thicker gaskets create a wider path for permeation to occur, and therefore give higher leakage rates, but note that the reverse can also occur. If a gasket is too thin to conform to flange irregularities, the media can leak over instead of through the gasket. This can lead to even higher leakage rates than with the thicker gasket. Therefore flanges which are flat and pristine enough to handle thin gaskets seal much tighter with a thinner gasket.

c) **Bolt Loading-** Bolt loads also affects the choice of materials thickness. Basically the material must be thick enough to deform sufficiently to accommodate any irregularities or in-equalities in the flange faces under the available bolt load. The lower this load, the greater the thickness which may be required and vice versa although this will also depend on the compressibility of material. It is better that multiple bolts are loaded equally and that the gasket stress is uniform over the gasket area. It is advisable to put larger number of small dia bolts rather than fewer bolts of larger diameter.

d) **Width of Gasket-** In order to reduce the bolt load required to produce a particular gasket pressure, it is advisable not to have the gasket wider than is necessary. For a given gasket stress raised face flange with a narrow gasket will require less preload and hence less flange strength than a full face gasket. So the tendency is for high pressure gaskets to be narrow.

e) **Stress Relaxation**- It is a measure of the resiliency of the material over a period of time expressed in percentage. All gasket material will loose some amount of resiliency over time, both due to flow or thinning of the material due to applied pressure and loss in its springiness.

After some initial relaxation the residual stress should remain constant for the gasket.

f) **External Dia of Gasket**- It is advisable to have the external dia of the gasket as large as possible. For the same gasket width gasket having higher OD shall withstand high leakage pressure. O.D. of gasket = bolt circle dia – bolt dia-1/4 inch.

g) **Temperature**- An increase in temperature will degrade the physical strength of the material and deform it so that bolt load and thus residual stress shall be modified. A poor gasket material which suffers high deformation with increasing temperature will show high relaxation and will collapse or extrude at high temperatures under moderate internal pressure.

h) **Media**- The gasket material must be resistant to corrosive attack from the media. It should chemically resist the system fluid to prevent serious impairment of its physical properties.

Failure Related to Flange

There are three types of flange facings commonly found on a plant. The surface finish of the facings is specified in the Flange Standards.

Flat Face (FF)- Here sealing happen by compression of a flat non-metallic gasket (very rarely a flat metallic gasket), between the phonographic/concentric grooved surfaces of the mating FF flanges. The gasket fits over the entire face of the flange. FF flanges are normally used on the least arduous of duties such as low pressure water drains and in particular when using cast iron, or bronze alloy, where the large gasket contact area spreads the flange loading and reduces flange bending against.

Raised Face (RF)- Sealing on a RF flange is by a flat non-metallic gasket (or a flat metallic gasket for special applications), which fits within the bolts of the flange. The facing on a RF flange has a concentric or phonographic groove with a controlled surface finish. If the grooves are too deep (or a rough surface finish), then high compression is required to flow the relatively soft gasket material into the grooves. Too shallow (exceptionally smooth surface finish) and again high compression is required as a leak path then becomes more possible. It is important to always check the flange surface finish for imperfections which would make sealing difficult. A radial groove for example is virtually impossible to seal.

Ring Type Joint (RTJ)- Typically found on the most severe duties, for example high pressure gas pipework. Ring type metal gaskets must be used on this type of flange facing. RTJs are manufactured according to API 6A Type B, BS 1560 and ANSI B16.5: The seal is made by metal-to-metal contact between the gasket and the flange groove. The faces of the two opposing flanges do not come into contact and a gap is maintained by the presence of the gasket.

Surface Finish

One of the major factor to take into consideration must be the surface finish of the gasket seating surface. As a general rule, it is necessary to have a relatively rough gasket seating surface for elastomeric and PTFE gaskets on the order of magnitude of 500 micro inches. Solid metal gaskets normally require a surface finish not rougher than 63 micro inches. Semi-metallic gaskets such as spiral wound fall between these two general types. The reason for the difference is that with non-metallic gaskets such as rubber, there must be sufficient roughness on the gasket seating surfaces to bite into the gasket thereby preventing excessive extrusion and increasing resistance to gasket blowout. In the case of solid metal gaskets, extremely high unit loads are required to flow the gasket into imperfections on the gasket seating surfaces. This requires that the gasket seating surfaces be as smooth as possible to ensure an effective seal. Spiral-wound gaskets, which have become extremely popular in the last 15–20 years, do require some surface roughness to prevent excessive radial slippage of the gasket under compression. The characteristics of the type of gasket being used dictate the proper flange surface finish that must be taken in to consideration by the flange designer and there is no such thing as a single optimum gasket surface finish for all types of gaskets. The problem of the proper finish for gasket seating surface is further complicated by the type of the flange design. For example a totally enclosed facing such as tongue and groove will permit the use of a much smoother gasket seating surface than can be tolerated with a raised face. Recommendations for normal finishes for the various types of gaskets are given below.

Spiral wound-	125–250 μ″ AARH
Solid metal-	63 μ″ AARH
Metal jacketed-	63–80 μ″ AARH
Flat – Non-Metallic-	250–500 μ″ AARH

Gasket, Flange, Fastener as a System

In order for a bolted flanged joint to perform satisfactorily, individual components must be suitable for the design operating conditions (chemical compatibility, temperature, pressure, etc.). These same components must be well suited to maintain a seal as an assembly. More specifically, the fasteners (size, quantity, material, etc.) must be capable of applying adequate load to the gasket to achieve the desired tightness under designed operating conditions. The gasket must be capable of operation within the design operating conditions and achieve the seal, given the available bolt load. The flange must be rated for the design operating conditions and of adequate stiffness to ensure loading on the gasket is maintained to expected levels.

Proper alignment of all joint members is the essential element of flange joint assembly. It results in maximum sealing surface contact, maximum opportunity for uniform and design-level gasket loading, and reduced friction between the nut and the flange. Flange misalignment creates additional loads that the bolts have to overcome before the bolt loads can be applied to the gasket. Minimizing alignment issues will help to minimize bending issues, but external forces such as weight of fluid in the piping, or thermal expansion, can create external bending forces. The external bending forces and moments tend to unload one side of the flanged joint and increase load on the other side creating load changes that can result in leakage. Misalignment of piping and flanges should be minimized. Proper piping supports and piping design can help minimize these external bending forces.

Most applications undergo pressure and thermal cycles which work to decrease the compressive load on the gasket. Pressure cycles change the hydrostatic end load. Thermal cycles can change the stretch in the bolts through various mechanisms, including bolt material modulus changes due to temperature, differential thermal expansion of the flange component materials and thermal transients, in heat-up and cool-down. These pressure and thermal cycles cause fluctuations on the compressive load of the gasket and may increase the potential for a leak.

Gasket Removal and Installation

The flanged joint should be carefully opened using a special flange spreading tool or soft wedges so as not to damage the flange seating surfaces. The old gasket can then be removed. After removal of the old gasket, the flange facings should be inspected for imperfections that can adversely affect the new gasket's ability to seal. If surface damage exists, it may be necessary to re-machine or replace the flange.

Removal of spent gaskets can be a tedious, laborious, and sometimes dangerous task.

Many a time flanged joints can be hard to separate, making it difficult to see where the gasket is located and remove it. Under such conditions it is not uncommon to have gaskets installed over old gaskets, or piece of the old gasket that were not removed. Even a small fragment of gasket left on a flange can make it impossible for the new gasket to conform properly, resulting in an immediate leak or a blowout.

After the old gasket has been removed and the flange faces cleaned and conditioned, the replacement gasket can be installed. Once the replacement gasket has been selected, it should be inspected for correct I.D., O.D. and appropriate thickness.

If the joint is boxed up for the first time, verify compliance with bolt and nut specifications materials, diameter, length of bolts, thread pitch, and nut thickness equal to the nominal bolt diameter. Check bolts for adequate length. Section VIII, Division1 of the ASME Boiler and Pressure Vessel Code requires that nuts engage the threads for the full depth of the nut (see para. UG-13). The ASME B31.3, Process Piping Code, has a similar provision but considers the nut to be acceptably engaged if the lack of complete engagement is not more than one thread.

For ease of installation, all fasteners should be lubricated with an oil-and-graphite mixture or other suitable thread lubricant, being careful not to allow it to come into contact with the gasket. Flanges with vertical seating surfaces should have at least two fasteners inserted into the bottom holes to support the gasket.

To avoid damaging the gasket, do not use an instrument to push it into place. The flange spreader can then be carefully used. The pattern in which the bolts are tightened is extremely important. If done improperly, the flange can move out of parallel. In most cases the preload is required to make a fastening. By controlling torque, turn, or stretch, one can control the build up of tension. Using calibrated torque and box-end type wrenches, tighten each fastener to no more than one-third of the desired torque value to uniformly compress the gasket. Repeating the same pattern, increase the torque wrench setting to two-thirds of the desired value. The final step of gasket installation that is commonly overlooked is the re-torque, which should be performed ideally 12–24 h after initial installation. However, if it is not possible to wait 12–24 h before the system is started/pressurized, retorquing an hour or two later can still be beneficial.

Common Causes of Flange Joint Failure

Bolt Related

Galvanic corrosion: Galvanic corrosion is set up when two dissimilar metals are in the presence of an electrolyte, such as moisture. A galvanic cell is created and the most active (anode) of the two materials is eroded and deposited on the least active (cathode). Because the anode is eroded in a galvanic cell, it should be the larger mass in the cell. Therefore it is not advisable to use carbon steel fasteners in a stainless steel or copper assembly. But stainless steel fasteners can be used in carbon steel assemblies, since carbon steel mass is the anode.

Stress corrosion: Stress corrosion occurs when a tensile–stressed part placed in a corrosive environment. A ductile material will fail at a stress much lower than its yield stress because of surface imperfections created by a corrosive environment. In general, the lower the ductility the more susceptible it is to stress corrosion cracking. There are materials which are less sensitive to stress corrosion. For example A286 is not susceptible to stress corrosion cracking.

Hydrogen embrittlement: Hydrogen embrittlement occurs whenever there is free hydrogen in close association with the material. Since most plating processes are electrolytic bath type, free hydrogen is present. Hydrogen reacts with the carbon in steel to form methane gas which can lead to crack development and strength reduction. Hydrogen can also react with alloying elements such as titanium to form hydrides. Because the hydrides are not as strong as the parent alloy, they reduce the overall strength of the metal.

Nut dilation: The tensile force present in the fastener during tightening acts on the threads to produce a wedging action which result in a diametral displacement. This diametral displacement s generally known as nut dilation. Theoretical and practical studies of this phenomena indicate that the top face of the nut contracts in a diametral direction while its bearing surface expands. The net effect of this dilation is to reduce the shear area of both internal and external thread. The stripping strength of an assembly can be improved by increasing the width across flats of the nut. This increases the stiffness locally around the internal thread and reduces diametral expansion.

Thread Bending: The tensile force in the fastener during the tightening process results in a degree of thread bending between internal and external threads. Thread bending reduces the shear area of both internal and external threads. The important factor controlling the degree of thread bending is the ratio between the strength of internal and external threads. The strength ratio is the ratio between the force necessary to cause the nut thread to strip, divided by the force required to cause the bolt thread to strip.

Vibration Loosening: Though bolted joints have the advantage of dismantling, this feature becomes a problem when it occurs unintentionally as a result of operational conditions. Such unintentional loosening is called vibration loosening. Vibration loosening occurs because of side sliding of the nut or bolt head relative to the joint, resulting in relative motion occurring in the threads. If this does not occur, the bolt will not loosen, even if the joint is subjected to severe vibration. The main causes of relative motion occurring in threads are:

- Bending of parts which results in forces being induced at the friction surface. If slip occurs, the head and threads will slip which can lead to loosening.
- Differential thermal effects caused as a result of either differences in temperature or differences in clamped materials.

- Applied forces on the joint can lead to shifting of the joint surface leading to bolt loosening.
- It is observed that transversely applied alternating forces generate the most severe conditions for self loosening. The key to prevent self loosening of fasteners is to ensure that:
  - i) There is significant clamp force present on the joint interface to prevent relative motion between the bolt head or nut and the joint.
  - ii) The joint is designed to allow for the effects of embedding and stress relaxation.
  - iii) Proven thread locking devices are used (e.g., Loctite).

Relaxation: In a joint at an elevated temperature, a fastener with a fixed distance between bearing surface of head and nut will produce less and less clamping force over time. This characteristics called relaxation. It is different from creep because the stress changes without change in elongation. If the relaxation is not compensated for, it will lead to fatigue failure or a loose joint even though it was properly tightened initially. As the temperature environment and the materials of the structure are normally fixed the design objective is to select a bolt material that will give the desired clamping force at all critical points in the operating range.

Torque: Joints often fail, for low bolting torques or for over-tight bolt loads. Over torquing of the above joint would lead to an uneven flange load and weaken the effectiveness of the flange/bolt load. This weakening leads to an increase in blow-out pressure and hydrostatic force leading to an eventual joint failure even though the gasket material is suited to the application.

Gasket Related

After knowing the temperature, pressure and media variables of the sealing application the correct gasket material are chosen. Thickness is another important factor in gasket design. For gaskets, thinner is always better. Far too often we have seen gaskets fail because it was mistakenly assumed that more gasket will be better. You definitely need enough thickness to allow for compression, filling of imperfections in flange faces and the creation of a good seal but you shouldn't over do it. Extra material will require additional loading for compression leading to undue stress on fasteners, flanges and increased chance of failure.

Environmental Factors

Even though correct gasket's material and style have been selected and installed perfectly, sometimes leaks can develop over time. These can be caused by a wide variety of environmental factors including, but not limited to, corrosion, vibration, reuse, thermal cycling and flange damage.

Depending on the application area, the development of corrosion can cause serious problems for an effective seal. Of course, the gasket itself becoming corroded (either by the media or surrounding environment) is the most obvious candidate for seal compromising but damage to the flanges or fasteners are also real concerns. Should either occur issues such as flange misalignment and relaxing of bolt torque can eventuate.

Temperature: Temperature causes the metals in the joint to expand or contract, causing the gasket to be over compressed or too relaxed.

Medium: The possible corrosive effect that the gases or liquids the gasket is sealing against may have on its material make up.

Flange Related

Joint Must Compensate for Wide Temperature Variations: Temperature causes the metals in the joint to expand or contract leading to relaxation. This will lead to looseness of the joint.

Solution: Consider use of sleeve around bolts to increase effective bolt length:

Or consider use of conical spring washers in place of sleeve to eliminate torque losses over wide temperature ranges.

Flanges Badly Cocked or Separated Too Far:
Solution: Do not try to correct problem with flange bolts – can overstress.

Do use spacers to correct problem with gasket on each side.

Flanges Out of Parallel:
Total allowable out of parallel: ~1 + ~2 = 0.015″. Note – Deviation on right is less critical than deviation on left since bolt tightening will tend to bring flanges parallel due to flange bending.

Wavy Surface Finish
Note:

1) If using jacketed or spiral wound gaskets – deviation should not exceed 0.015″.
2) If using solid metal gaskets – deviation should not exceed 0.005″.
3) If using rubber, more leeway is possible – perhaps total of 0.030″.

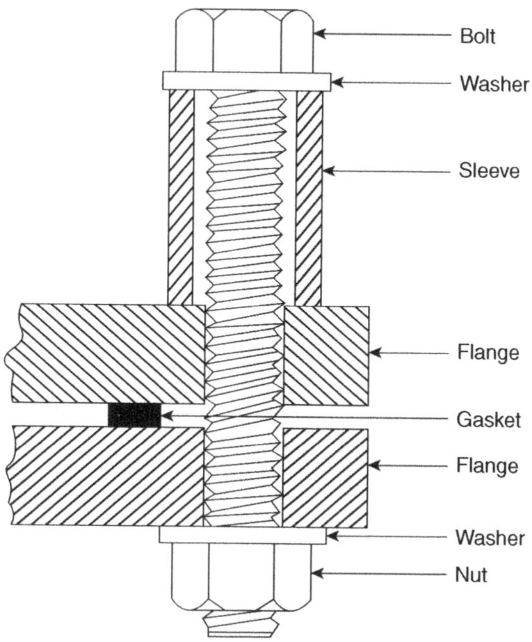

Figure 7.3 Use of sleeve around bolt.

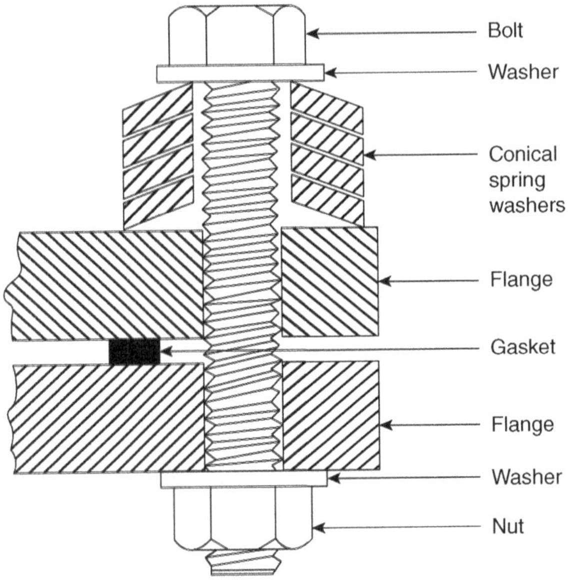

Figure 7.4 Use of conical washer around bolt.

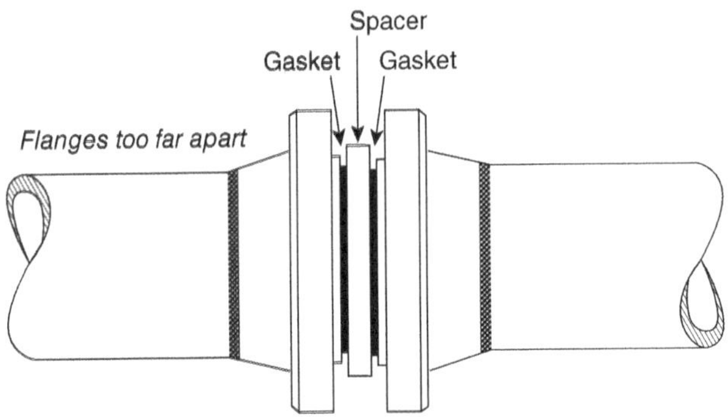

Figure 7.5 Gap in flange.

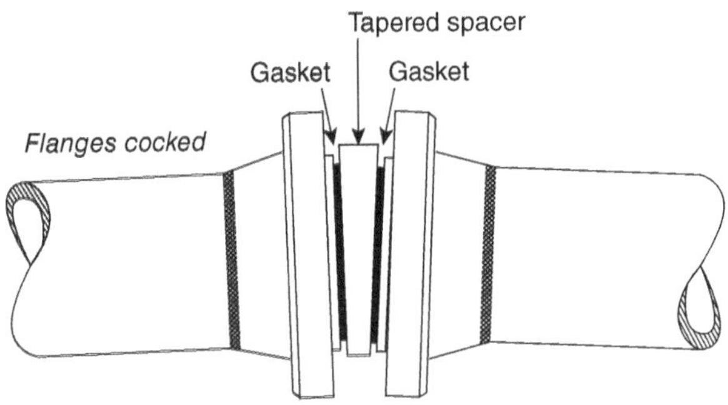

Figure 7.6 Flange cocked.

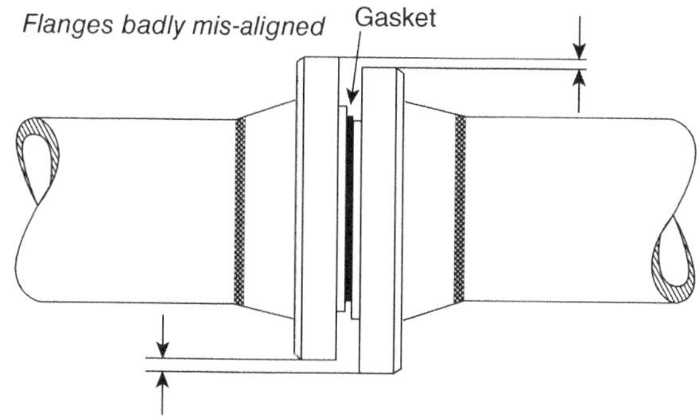

Figure 7.7 Mis-aligned flange.

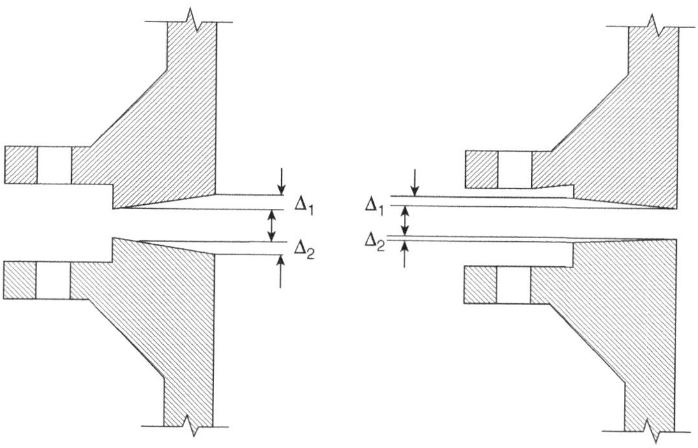

Figure 7.8 Out of parallel flange.

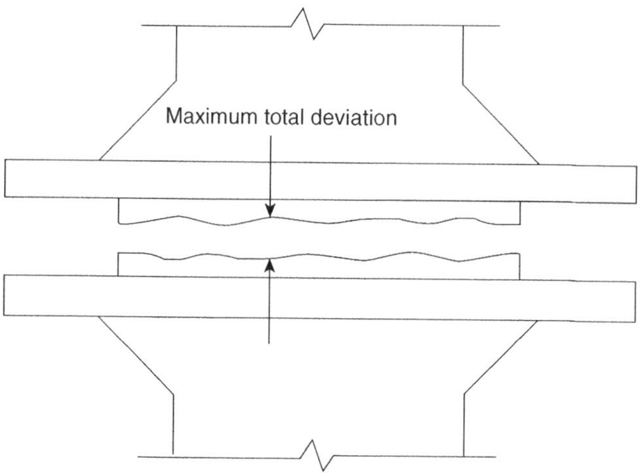

Figure 7.9 Wrong surface finish of flange.

Problems with Installation

In my experience, the majority of gasket failure incidents can be traced back to issues with installation. Gaskets over 70% failed due to issues with the installation process with the vast majority being due to insufficient sealing load being applied. Correct installation of a gasket varies according to the gasket type, flange type and various other factors. There are some best practices that can be adopted. The first of these is ensuring that bolts are tightened as per specifications (the use of torque wrenches or other controlled tightening equipment is suggested whenever practicable) and tightened in the correct order. You'd be shocked to hear how many gasket failures we've seen over the years that had their origins in bolts being tightened sequentially around a flange rather than in the advised alternating pattern. The diagrams below show the suggested tightening sequence for a few common styles of gasket.

It is should also be noted that, in recent years, we've seen a significant increase in the amount of low quality bolts available on the market. Many of these can be prone to over-stretching instead of clamping as they are being tightened which will, of course, lead to issues with creating a good seal. Failure to follow the sequence or correctly tighten the bolts can lead to an uneven loading of the flange faces creating weak areas where the above mentioned blow-out and hydrostatic pressures will have greatest effect on the gasket.

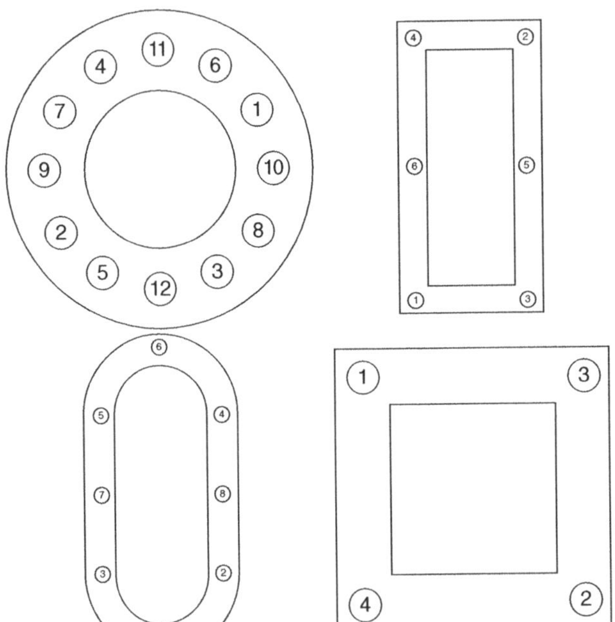

Figure 7.10 Bolting sequence of flange. *Source*: Round-12-Hole-Gasket-Bolting-Sequence-Chart, Rectangular-6-Hole-Gasket-Bolting-Sequence-Chart, Obround-8-Hole-Gasket-Bolting-Sequence-Chart & Square-4-Hole-Gasket-Bolting-Sequence-Chart. © 2016, Associated Gaskets.

Conclusion

Unless the bolted joint is properly designed for all the operating conditions in a process, including system upsets, sealing reliability cannot be ensured. The proper selection and condition of each component, including bolts, gasket and flange surfaces, their correct assembly and installation are fundamentally important to preventing leakage and joint failure. The proper training and qualification of personnel safely employing the appropriate tools, also is essential to joint integrity. To avoid leakage, the joint must be assembled to a bolt stress level that provides adequate stress to seat the gasket, adequate stress to provide sufficient residual stress on the gasket during operation and also be sufficiently low such that no damage occurs to any of the joint components during assembly or operation.

8

Failure of Coupling

There are many types of couplings in use but all of them can be put into two classes, rigid or flexible. Rigid couplings should be used only when the equipment has virtually no misalignment and/or when the shafts of the equipment or the rigid coupling (quill shaft rigid coupling) are long and slender enough that they can flex and accept the forces and moments produced by the mechanical deflection on these parts due to the misalignment imposed by the connected equipment.

The four basic types of rigid couplings are

1) The flanged rigid coupling;
2) The ribbed rigid coupling;
3) The sleeve rigid coupling; and
4) The quill shaft rigid coupling.

Flanged Rigid Couplings

The flanged rigid coupling is probably the most common type of rigid connection. It is available from almost every flexible coupling manufacturer in several standard series. One of the most common is interchangeable with standard flanged gear couplings. They can handle large amounts of torque for their sizes; The flanged rigid coupling can be used wherever no misalignment is present. Some applications include pumps (vertical and horizontal) and crane drives (see Figures 8.1 and 8.2).

Ribbed Rigid Couplings

The ribbed rigid coupling is used where ease of assembly and disassembly is required. The coupling clamps onto the shaft. The shaft and the coupling hubs are usually keyed. The two halves are held together by radial bolts at the split. The number of bolts can vary depending on the size of the coupling; typically, four to eight are used. Torque is transferred from one half to the other by the frictional force produced by the bolts rather than by direct loading of the bolts. Rbbed rigid couplings are usually used for low-speed, low-torque applications. They are typically used on vertical pumps, agitators, and winch drives (see Figure 8.3) and in many other types of applications.

Root Cause Failure Analysis: A Guide to Improve Plant Reliability, First Edition. Trinath Sahoo.

Figure 8.1 Flanged rigid coupling. *Source*: Sandesh Patoliya.

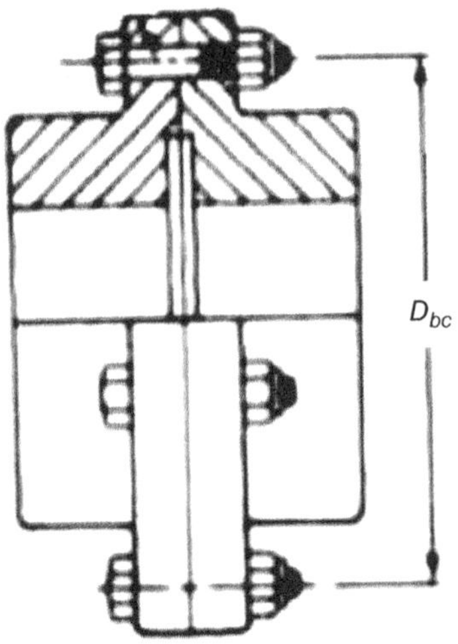

Figure 8.2 Cutaway view (DBC- Bolt circle dia.)

Sleeve Rigid Couplings

The sleeve rigid coupling is one of the simplest forms of coupling available. On small (usually fractional horsepower) drives where the equipment shafts are of the same diameter, a sleeve rigid coupling can be slid onto the shaft of one piece of equipment, the equipment put in place, and the sleeve rigid coupling slid onto the shaft of the other piece of equipment. The sleeve is usually locked to the shafts with two Setscrews, one for each piece of equipment.

Sleeve rigid couplings with setscrews are used on motor-driven pumps. The hydraulic sleeve rigid coupling is used on high-torque applications such as marine propulsion shafting (see Figure 8.4).

Quill Shaft Rigid Couplings

The quill shaft rigid coupling gets its name from the fact that it looks like the quill shafts that go through some large gears in gearboxes; they are usually long and slender. These rigid couplings do accommodate for some misalignment; they do this through the flexing of their long slender shafts. Quill shaft rigid couplings are usually much smaller than flexible couplings, generally 25–50% smaller. For example, a 40 in. quill shaft rigid coupling is approximately equivalent to a 60 in. Flexible coupling.

Quill shaft rigid couplings are usually made of high-grade alloy steels. This is because the stresses imposed on them when they are misaligned(flexed) are usually quite high and they must be designed for the cyclic loads imposed on them when they are flexed due to misalignment. These couplings are

Figure 8.3 Ribbed rigid coupling. *Source*: ABB Inc.

usually used on large, high horsepower steam and gas turbines (see Figure 8.5).

Flexible Coupling

Figure 8.4 Sleeve rigid coupling.

Historically, rotating equipment was first connected by means of rigid flanges. Experience indicates that this method did not accommodate the motions and excursions (that is, misalignment) experienced by the equipment. Shaft and flange fatigue failures were frequent. Then flanges were made thinner, which allowed them to flex. From this start, the design of flexible couplings has evolved to the many types and styles of today, all used to transmit the maximum amount of power while accepting the required amount of misalignment. (Note that nowadays, rigid flange couplings are still used to connect equipment that experiences very small shaft excursions.)

Functions of Flexible Couplings

Flexible couplings join two pieces of rotating equipment while permitting some degree of misalignment or end movement or both. The three basic functions of a flexible coupling are to

- Transmit power;
- Accommodate misalignment; and
- Compensate for end movement.

Transmit Power

Couplings are primarily used to transmit mechanical power from one machine to another. The power is in the form of mechanical torque at some speed, or work per unit of time. In general, the power lost by a flexible coupling is small, although some couplings are more efficient than others.

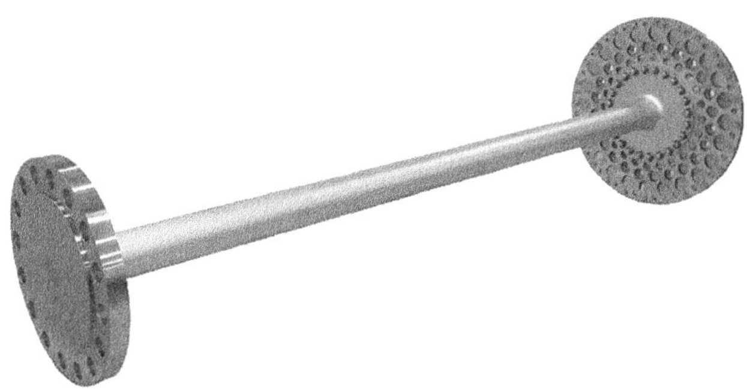

Figure 8.5 Quill shaft rigid coupling. *Source*: Coupling Corporation of America.

Flexible couplings must accommodate three types of misalignment (Figure 8.2).

- **Parallel offset** – the axes of connected shafts are parallel but not in the same straight-line
- **Angular** -the axes of shafts intersect at the center point of the coupling, but not in the same straight-line.
- **Combined angular and offset** – the axes of shafts do not interest at the center point of the coupling and are not parallel.

It is important to recognize that while the equipment may see these three types of misalignment, the coupling sees only angular and axial displacement. The flexible elements see only angular misalignment and axial movement; therefore, most equipment needs couplings with more than one flexible element to accommodate offset. The exception being some elastomeric element couplings that can accommodate all three types of misalignments.

End Movement

Most flexible couplings are designed to accommodate axial movement of equipment or shaft ends. In pumps that are driven by sleeve bearings motors, the couplings are usually required to limit the axial float of the equipment to prevent internal rubbing.

Types of Flexible Couplings (General)

There are three basic types of couplings:

- Mechanical element;
- Elastomeric element; and
- Metallic element.

The mechanical element types generally obtain their flexibility from loose-fitting parts or rolling or sliding of mating parts or from both. The most common types are the gear coupling and the grid coupling. They usually require lubrication unless one moving part is made of a material that supplies its own lubrication need. The elastomeric element types obtain their flexibility from stretching or compressing a resilient material (rubber, plastic, etc.). There are two basic types: the shear type and the compression type. The metallic element types obtain their flexibility from the flexing of thin metallic, disc, or diaphragms.

The flexible couplings can also be divided into general-purpose coupling and special-purpose coupling.

Types of General-Purpose Couplings

General-purpose couplings are used on pumps and other equipment that if shut down will not shutdown the plant or the process. They are mainly low speed, generally motor speed designs. Like any other coupling, these will transmit torque from one shaft to another while allowing misalignment and axial motion between the ends of the coupled shafts.

General-purpose types are more standardized and less sophisticated in design and are substantially cheaper and are used in quantities substantially greater than special-purpose types.

General-purpose equipment uses couplings where the flexible element can be easily inspected and replaced, sometimes considered "throw away parts." These types of couplings are usually very flexible and require simple alignment techniques. It is usually sufficient to align equipment with these couplings to within 0.001 in/in of shaft separation. Therefore, a coupling with 10-inch shaft separation should be aligned to be within 0.010 inches. A failure for this type of coupling occurs at the flex element, and little or no damage usually occurs to other components.

A few examples of general-purpose couplings are the gear, grid, elastomeric, and the disc.

General purpose couplings can generally be classified according "size." Under 100 hp is classified as small. Between 100 and 1000 hp is characterized as medium. Usually hp over 1000 is considered critical.

Special-Purpose Coupling

If a small pump shuts down, it usually does not affect the plant operation. Hence this equipment uses a general-purpose coupling type where the flexible element can be easily inspected and replaced.

A coupling moves from the general purpose to the special-purpose category once it is applied to very critical equipment within the production or process system. Thus, a pump coupling on a spared, redundant system, in any process plant, is likely not a special-purpose unit. On the other hand, a high horsepower un-spared boiler feed pump in a base station or the 10 000 rpm compressor train is certainly special-purpose and critical to trouble-free operation. A turbine generator set of 40 200 hp at 3600 rpm (606 615 in-lb) is also critical.

Most gas turbine-driven generators on peaking or cogen systems are special-purpose machines that require special-purpose couplings. In general, special-purpose machines can be identified as expensive, high powered, and high speed. They are driven by various types of motors, especially synchronous ones, and also gas or steam turbines at 3600 rpm or more. The horsepower is usually in excess of 1000. Usually, for the reason of expense, they are not spared. Another point is that although these machines are high powered, they are also sensitive to almost everything in their environment. That is forces or moments that would seem insignificant to high powered mill machinery become life-threatening to sensitive machines. As a result of that sensitivity and the speed and the power, coupling criteria for the machines take on an entirely different perspective.

When the critical application is found in a refinery or refinery-related setting, the coupling comes under the API 671 specification. That specification has definite requirements for coupling construction as well as coupling selection. For example, the specification calls out certain service factors and certain torque selection variables. A disc or diaphragm coupling selected for the continuous operating torque might have a service factor as high as 1.5. If selected by motor size rather than driven equipment output, it could be as low as 1.2. Transitory torque may also be used for coupling selection.

How General-Purpose Couplings Work

The various types of general-purpose couplings, include gear, grid, compression donut, block jaw, urethane tire, corded tire, and disc.

Gear Couplings

Gear couplings is used for transmission of power above 100 HP. As shown in the above figure, gear-type couplings consist of two hubs with external teeth that engage internal teeth

Figure 8.6 Gear coupling. *Source:* Focus Technology Co., Ltd.

on a one or two-piece sleeve. The teeth may be straight or curved (crowned). Torque is transmitted from one hub through the sleeve to the opposite hub through the meshing of the teeth. These couplings obtain their flexibility due to the looseness (backlash) between the mating teeth. These couplings accommodate axial (in–out) shaft movement better than other competing designs because their hub teeth easily slide along their sleeve teeth with no effect on coupling operation.

As shown in the above figure, with straight hub teeth, there is a high concentration of load under misaligned conditions. As misalignment increases, more of the load is carried by the ends of the teeth, resulting in premature breakdown and coupling failure.

In a crown tooth, the contour of the tooth is a segment of an arc. The crowning can include tip crowns, flank crowns, and chamfers on the sharp edges. Under all operating conditions, contact areas equal or similar to contact areas shown in the figure exist between the hub teeth and the sleeve teeth. The larger contact area results in reduced stress in the crown tooth form. In addition, it prevents the sharp squared edges of the tooth from digging in and locking the coupling. Note that crowning applies to hub teeth only; sleeve teeth are straight except for a chamfer on the minor diameter edge. For application requiring over 0.25° angular misalignment, crowned/curved teeth may be more appropriate.

There are several methods of lubricating gear couplings. These are grease pack, oil fill, and continuous oil flow. The vast majority of drives operate at 3600 rpm or less and use grease as the lubricant. Both grease and oil are used at speeds of 3600–6000 rpm. Oil is normally used as the lubricant in couplings operation over 6000 rpm. Most high-speed couplings use a continuous oil flow to carry away the heat generated within the coupling.

The major failure mode with gear couplings is lubrication. Grease tends to separate under centrifugal force (to prevent oil and soap separation, use grease recommended by the coupling manufacturer only) and, overtime, O-ring type of seals begins to leak. If proper lubrication is maintained, these couplings will operate successfully for many years.

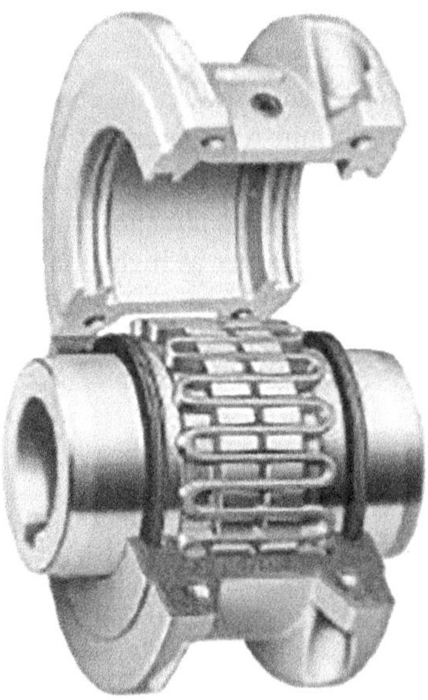

Figure 8.7 Grid coupling.

Grid Couplings

Grid type couplings are very similar to gear couplings. Usually composed of all metal they have some degree of resilience. They have two hubs with serrations (grooves) rather than teeth. The grooves are connected by a steel grid. Flexibility is achieved by sliding movement of the grid in the slots. Flexure of the grid in the curved slots provides some torsional resilience. A cover keeps the lubrication contained inside the housing. The covers are either vertically split or horizontally split.

Grid couplings are used where both high torque levels and dampening requirements exist. Unlike gear and disc couplings (alternative metallic type couplings capable of transmitting high torque levels), grid couplings have a unique ability to reduce vibration by as much as 30%, and cushions shock loads to safeguard driving and driven equipment. However, grid couplings have a limited ability to accommodate parallel shaft misalignment. These couplings do not transmit as much power (per the same outside diameter) as gear couplings but are usually less costly. Grid couplings are used for medium and small equipment applications.

Like gear type coupling, proper lubrication is essential for grid type coupling also. In view of many problems that could arise with improper lubrication, it is necessary that the coupling manufacturer's instructions be adhered to closely.

Disc Couplings

Disc couplings transmit torque by a simple tensile force between alternating driving and driven bolts on a common bolt circle. Misalignment is accomplished from the flexibility that comes from the length of material between the bolts. Disc couplings have been around for years, but with the use of finite element analysis, this type can and has been optimized for optimum characteristics. These couplings are composed of all metal, and do not require lubrication. The discs are usually continuous, but can be individual links. Most disc couplings use multiple thin discs rather than one thick disc/link. These couplings are used in medium-size equipment applications. If the misalignment is beyond 1/2–3/4° during operation, the flexible element will probably fail in fatigue.

Elastomeric Couplings

Elastomeric flexible shaft couplings are limited to general-purpose applications. These couplings are typically used in applications below 100 HP. In these couplings, the elastomer is placed in shear or compression. Their ability to compensate for shaft misalignment is obtained by flexure and/or displacement of the elastomeric element. It may be

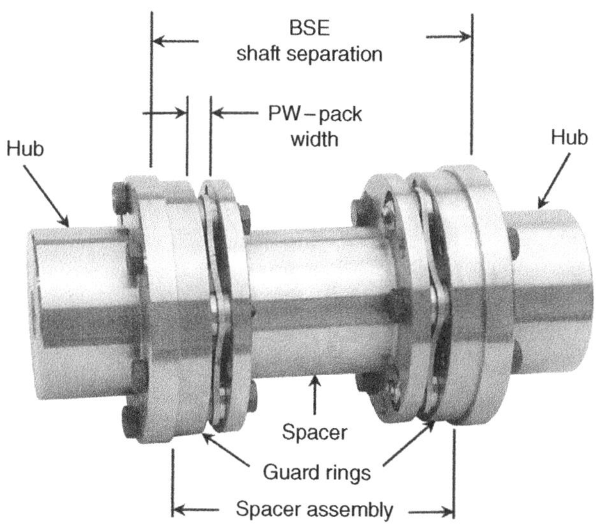

Figure 8.8 Disc coupling. *Source*: Elliot Wilson. Disc Couplings – Common Terminology (Part 2), Monday, 28 September 2015, CouplingAnswers.com.

noted that in an elastomeric coupling, elastomeric element in a single plane distorts enough for parallel/radial as well as angular misalignment. These couplings require no lubrication.

These couplings basically consist of two hubs separated and connected by elastomeric element. In shear-type couplings, the elastomer may be bonded, clamped, or fitted to matching sections of the hubs. The compression type couplings usually utilize projecting pins, bolts, or lugs to connect the components. The elastomeric flexing elements may be natural rubber, neoprene, nitrile, urethane.

Elastomeric type couplings are torsionally soft. This means that they will deform under torque and misalignment and also provide damping to some extent.

Synthetic gear, pin and bush, jaw/spider, and tire (urethane or corded) type couplings fall under this category of couplings.

Pin and Bush Type Coupling

A pin and bush type coupling comprises two hub sections which are mounted on the machine shafts. The hubs are connected positively by steel pins and elastomer bushes. As shown in the above figure, many times BARREL-shaped bushes are used instead of plain (cylindrical) bushes. The barrel-shaped bushes permit extra torsional flexibility as compared to plain bushes. Pin and bush couplings are very popular couplings for wide-ranging general engineering applications.

This type of coupling is generally suitable for use at ambient temperature between −40 and +80°C. However, by using alternative elastomer bushes, the permissible ambient temperature range can be extended to between −50 and +100°C.

These couplings are economical, but the rubber bushes are subjected to aging and consequent implications.

Jaw/Spider Type Coupling

Jaw/spider type couplings have their elastomers in compression. This type of coupling is most often referred to as a spider coupling. The flex element can be one-piece or split to facilitate replacement. In case of split element, it can be wrap around the outside of the coupling, and then be held in place with a retaining ring. Small and medium-size equipment employ these types of couplings. These couplings are fail-safe type, that is, they will perform

Figure 8.9 Pin bush type coupling. *Source:* Shah, K. P. The Hand Book on Mechanical Maintenance.

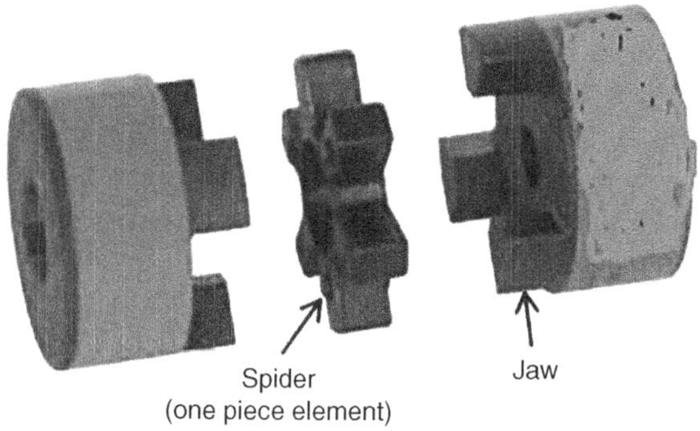

Figure 8.10 Jaw coupling. *Source*: Shah, K. P. The Hand Book on Mechanical Maintenance.

with failed elastomer. Flex elements are made of many types of elastomeric materials, such as Nitrile Butadiene Rubber (Buna N), urethane.

The most standard jaw coupling spider material is Nitrile Butadiene (Buna N) rubber. This elastomer is oil resistant and has good chemical resistance. It has high dampening capacity. It operates effectively in a temperature range of −40 to +100°C. It is well suited for cyclic loads (stop/start or reversing).

Urethane has a greater torque capability relative to Buna N (approximately 1.5 times), but provides less dampening effect. Its operational temperature range is also reduced on both ends of the spectrum, −34 to +71°C. (The in-shear type spider has a slightly different type of urethane and is rated for −34 to +93°C). It does have good resistance to oil and chemicals, but is not recommended for cyclic or start/stop applications. Spiders are having shore hardness 40D or 55D. Their shelf life is 5 years

Corded Tire Couplings

In these types of couplings, the tire bends/flexes to accommodate the misalignment. Such couplings have their elastomeric elements in shear. Many times, they use a reinforced element (similar to auto tires). These couplings are highly flexible. Because of their low torsional stiffness and damping capacity, these couplings are especially suitable for coupling machines with a highly nonuniform torque pattern. These couplings are also suitable for connecting machines with high shaft misalignment. These couplings can typically accommodate up to 4° of misalignment.

The coupling can be fitted with tires made of natural rubber for ambient temperatures of −50 to +50°C. Chloroprene rubber (Neoprene) should be used in applications where

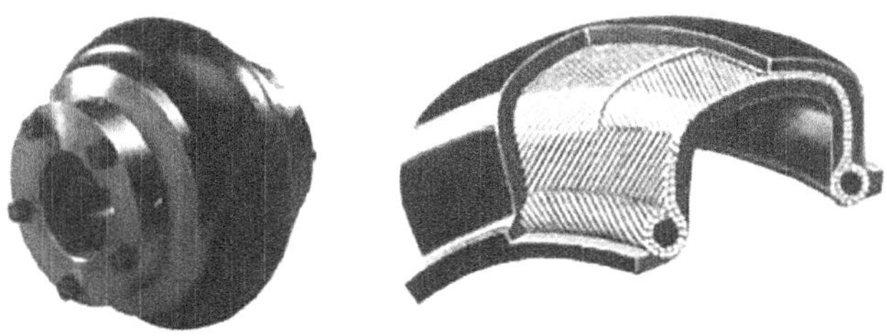

Figure 8.11 Corded tire couplings. *Source*: Shah, K. P. The Hand Book on Mechanical Maintenance.

exposure to greases and oils are likely. Chloroprene rubber can accommodate temperatures ranging from −15 to +70°C. Urethane used for making tire is slightly different type than used to make spiders for jaw couplings and is rated for −34 to 93°C.

The tire is slit at the circumference and can be changed without having to move the coupled machines. If the tire gets damaged or worn, the metal parts can rotate freely against one another because they are not in contact with one another. Most small-size and some medium-size equipment applications use these couplings.

Special-Purpose Couplings

Special-purpose couplings are thoroughly covered in API 671. There are two categories "lubricated" and "nonlubricated." There is some older equipment that still use "lubricated" couplings. Most new equipment or upgrade equipment will have "nonlubricated" couplings. Generally, gear coupling (mechanical element) requires oil or grease lubrication and metallic element types that require no lubrication. There are two primary types of nonlubricated special purpose, metallic element couplings, the diaphragm type and the disc type. For some large equipment, elastomeric element type is used.

High-Performance Gear Couplings (Mechanical Element Type)

The gear coupling consists of two sets of meshing gear teeth (Figure 8.6). Each set becomes a flexible point or pivot point. The gear coupling was chosen for its high-power density. It provides more horsepower capability per pound of weight and cubic inch of space than any other coupling. The gear coupling is also very rugged, which means it can take the type of beating that might come from torque spikes or starts and stops. If kept well lubricated these couplings can be very reliable for many years.

The gear coupling transmits torque and accepts misalignment by the meshing and movement of gear teeth. This movement, though, leads to wear on the mating surfaces. A gear coupling has its most significant effect not only on itself but on the system components from the forces and moments generated when it slides and/or misaligns. When a gear coupling accommodates the shaft float from thermal growth, axial forces react back onto the thrust bearings and other equipment. When misaligned, a gear coupling will produce a bending moment that will load equipment shafts, bearings, and other system components. Both the axial forces and bending moment are significantly affected by the lubrication and the coefficient of friction between the mating members.

Lubrication or lack of it is the biggest cause of failure. The coupling itself tends to act as a centrifuge and separates dirt particles out of oil. This produces "sludge", which can cause the coupling to "lock-up," which causes increased equipment vibrations. The most common mode of failure for a gear coupling is wear due to lubrication problems. If proper lubrication is maintained, these couplings will, and have, operated successfully for years.

Metallic Element Couplings

Metallic element coupling use for special-purpose applications started in the late 1960s and early 1970s. There are two basic types of disc type and diaphragm type.

Disc

The disc coupling is one style of coupling used to replace gear couplings on special-purpose machinery. The principle of operation is that torque is transmitted through a flexible element by tensile loading between alternate bolts that are on a common bolt circle. One of the alternate bolts is the load transmitter, and the other the load receiver. They are fastened to opposite sides of the torque path. The misalignment is accommodated by the flexing of the elements between adjacent bolts. The element must be thin to be flexible. Stacks of elements provide parallel load paths, and the diameter of the bolt circle is an indicator of the amount of torque to be carried.

Figure 8.12 Disc couplings. *Source*: Shah, K. P. The Hand Book on Mechanical Maintenance.

Disc Type Coupling

The amount of misalignment is related to the chord length between bolts and the thickness of the discs and disc packs. Since the discs are almost always put together into packs, one of the benefits of the design is multiplicity. If one or more discs fail, the rest can still carry the load until the equipment is shut down.

The thin element has both tensile stresses and cyclic bending stresses imposed on the element so that a fatigue life analysis is required. Once the fatigue life is determined, an infinite life coupling can be designed to keep loading below that fatigue limit. It took some time to develop high fatigue strength materials. Also, advances in material load analysis, such as finite element analysis, were necessary to speed the development of reliable couplings.

The modern high-performance disc coupling is nonlubricated and designed for infinite life. The amount of misalignment available is a function of the bending that can be accommodated while under tension. The unit will allow axial movement such as seen with thermal growth. However, that is not unlimited. Also, axial capability and angular capability are interrelated.

An important feature of disc couplings, and couplings that work in a similar manner, is the low reactionary load that is transmitted to the machinery that it couples, compared to the relatively high load of gear couplings. Disc couplings can be more expensive in the first cost compared to gear couplings.

As is the case for other types of couplings, additional considerations are needed for the special-purpose application. In order to save space and reduce cost, the coupling designer will select disc packs to closely match the application. In some cases, a disc will be designed specifically for the equipment. That design will involve the bolt circle diameter, number of bolts, size of bolts, and the number of discs needed. Once the disc pack unit is designed and built at the factory, the pieces should not be disassembled. Piloted disc packs or factory assembled disc packs help to ensure against fatigue failure.

Figure 8.13 Diaphragm type couplings. *Source*: Shah, K. P. The Hand Book on Mechanical Maintenance.

Diaphragm Type

This coupling uses two flexing elements called diaphragms, separated by an intermediate member. The diaphragm is normally attached at the outside and inside diameter by bolts to connect the hubs to the intermediate member. The torque goes through the diaphragm assembly from the outside to inside diameter, or vice versa. Each flexing element is made up of one (tapered profile) or more (contoured or cut-out profiles) diaphragm elements depending on the design. The flexibility of the diaphragm design accommodates angular and parallel misalignment as well as a limited amount of axial misalignment (end float). Misalignment of the connected shafts should be restricted to within the manufacturer's recommendation for long coupling life.

This deflection of the outer diameter relative to the inner diameter is what occurs when the diaphragm is subject to angular and axial misalignment.

Angular misalignment twists the outer diameter, relative to the inner diameter, and produces a complex shape on the diaphragm where it must stretch one way at one point and then stretch the other way at 180 degrees. In between these points, the diaphragm is subject to a combination of stretching and twisting. Axial displacement attempts to stretch the diaphragm, which results in a combination of elongation and bending of the diaphragm profile.

The thickness of a diaphragm can be changed to permit a trade-off between torque capacity and flexibility. A thicker diaphragm has greater torque capacity, but is not as flexible and vice versa.

Elastomeric Block Couplings (Elastomeric Element Type)

Large critical equipment (such as synchronous motor-driven equipment) uses elastomeric block couplings in special-purpose applications to reduce vibratory torques or to torsionally "tune" a system (using its torsional "softness"). Block couplings use rubber in compression. The rubber blocks are installed in cavities formed by internal sleeve blades, external hub blade, and two end plates. This type is unique among couplings due to its "fail-safe" feature. If the elastomer fails, the coupling may run for sometime on the metal blades. These couplings can provide up to 1/2 degree of misalignment and parallel offset capabilities of 1/64 to 1/4 in. Sometimes medium and small equipment will use these couplings, because if properly aligned they require no maintenance (except for replacement of blocks every three to 5 years)

Coupling Selection for Reliability

The best coupling to use is the type that originally came with the equipment. But if the plant has been using one coupling for years and has had problems, failures, or has found that the cost to maintain them is high, lots of downtime due to lubrication and alignment problems, it is time to consider changing the coupling type.

While selecting a coupling– Ask Questions:

- Does the application require high torsional stiffness?
- What are the accuracy requirements?
- Does the application require dampening or shock absorption?
- How much misalignment is present in the design? Is it angular? Parallel? Axial? Complex?
- Does the coupling need to be the break-first point in the system? Does it need to be fail-safe?

- Is electrical isolation a requirement?
- What is the maximum torque applied to the coupling?
- At what speed or speeds will the coupling be operating?
- In what temperature will the coupling operate?
- Are there other environmental factors for the application (e.g.: chemicals, wash down, vacuum)?

In addition to above, important points like safety factor, service factor, balancing requirement, and coupling fit need to be considered during selection process.

Safety factor and service factor are essential parameters to ensure a proper coupling selection for a particular application. Safety factors are known and determined by the coupling manufacturers, since they are directly related to the coupling component design and materials. Service factors are known and determined by the makers and users of the equipment, since they are directly related to how the equipment is operated. There is no industry standard that deals with coupling ratings. Not only will one find differences in what and how ratings are established for different types of couplings, but also one may find the same type may even have different ratings from different suppliers. Therefore, it is important to obtain the coupling manufacturers' catalogues and use their recommended procedure and service factors, if one wants to assure success.

Safety Factors

Coupling designers use safety factors because there are uncertainties in the design. The designer's method of analysis uses approximations to model the loading and, therefore, the calculated stresses may not be exact. Likewise, the material properties such as modulus, ultimate strength, and fatigue strength have associated tolerances that must be considered.

Today, with the use of such computational tools as FEA, stress analysis is generally capable of more accurate results than in the past. In addition, the properties of the materials used in high-performance products are more controlled and better known. Therefore, couplings designed today vs those designed 20 years ago can indeed operate safely with lower calculated safety factors. Also, the design factor for flexible-element couplings can be lower than gear couplings simply because the "safeness" is more accurately predicted.

Service Factors

Service factors, on the other hand, are used to account for the higher operating torque conditions of the equipment to which the coupling is connected. In API 671, a service (or experience) factor is applied to the normal operating torque of, for instance, a turbine or compressor. This factor accounts for torque loads that are not normal, but which may be encountered continuously, such as low-temperature driver output, compressor fouling, or possible vibratory torques. Also, service factors are sometimes used to account for the real operating conditions, which may be five to 20 percent above the equipment rating.

Different service factors are used or recommended depending on the severity of the application. Is it a smooth-running gas turbine-driven compressor application or will the coupling be installed on a reciprocating pump application?

Balancing Requirements

One important reason for balancing a coupling is that the force created by unbalance could be detrimental to the equipment, bearings, and support structures. The amount of force generated by unbalance is proportional to the square of speed. Therefore, the key element to the balance level required for a coupling is how fast it rotates. Things that Contribute Coupling Unbalance are

- **Component Eccentricity.** Any eccentricity of mating surfaces that permit radial displacement of the mass axial of mating parts
- **Component Clearances.** This is the clearance that permits relative radial displacement of the mass axis of coupling components
- **Hardware unbalance**: This is clearance of hardware, weight difference in them, etc.
- **Balancing Errors.** Runouts of fixtures used, clearances, and eccentricity of assembly on fixtures.

There are two basic industry standards for balancing couplings. The two standards are as follows: AGMA 9000. This usually applied to general-purpose couplings. In fact, API 610 calls out an AGMA level 8 class requirement. Whereas, API 671 for special-purpose couplings calls out 4W/N, or 0.0008W (50 micro inches) or 0.01 oz-in, whichever is greatest for residual.

Coupling Fit

The type of fit between a shaft and a hub can be one of the three types listed below:

- Clearance fit with key
- Interference fit with key
- Keyless interference fit

Required fit between the hub and the shaft for various types of couplings is as under.

- Gear, disc, and diaphragm: interference
- Grid and chain: clearance
- Elastomeric couplings: clearance

Clearance fits are extremely popular on smaller couplings while interference fit becoming more dominant in the 4-inch (100 mm) to 6-inch (150 mm) diameter range.

Clearance Fit (with Keys and Setscrews)

Keyed clearance fit couplings are used on low power/low torque/low-speed applications with shafts under about 2.5 inches. The design is based on torque being transmitted through the key to minimize the play and resist the moments and forces reacted from misalignment and unbalance. Setscrews are usually provided over the key. These are normally used up to speeds up to 3600 rpm.

Historically, this type has been used for process pumps with straight fits. Many users successfully use clearance fits on all straight bored pump couplings, but they also generally employ high standards of alignment. They do this because they expect the user's maintenance forces to either intentionally open the fit for convenience or to allow it to open with time and pump manufacturers design their shaft ends for repeated assembly/disassembly.

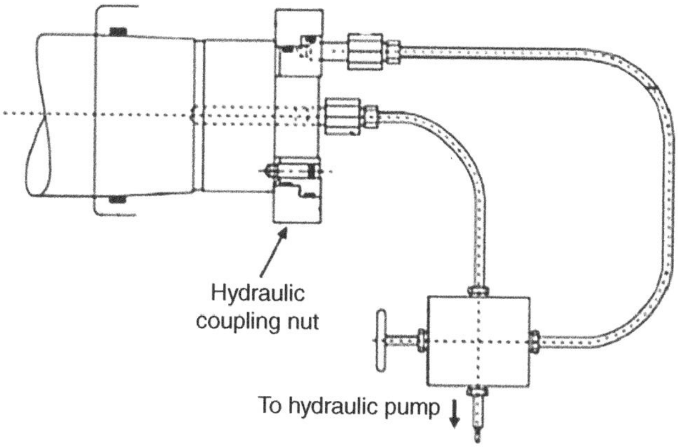

Figure 8.14 Hydraulic hub installation tool.

Shrink Fit

Couplings with **shrink fit** to shafts provide a tight fit, resist forces and moments applied across the coupling, and prevent their rocking on the shaft. This benefit subsequently prevents fretting, which can result in failure of the shaft and/or the coupling. Care should be taken not to have too tight a fit with keyed hubs, as this induces stresses and can cause cracking of the hub.

Shrink Fit on Tapered Shafts

Tapered shafts are common on many equipment applications. The taper provides for easy installation and removal while allowing for a proper and tight fit from the hub to the shaft. Tapered shafts are common on what are typically referred to as high-energy machines, such as boiler feedwater pumps, large fans, and crushers.

Keyless interference fit

A keyless interference fit coupling is usually referred to as a hydraulically dilated coupling. The hydraulically dilated coupling hub is used for ease of installation or in locations where heat or open flame is not acceptable

Above figures show a keyless hydraulically dilated coupling shaft and the hydraulic coupling hub installation tool. For installing the coupling, a hydraulic pump is used to pressurize inside of the coupling via drilled passages in the shaft. When the hub is pressurized, the valve is diverted to push the hydraulic coupling nut and advance the hub to the prescribed amount. The pressure is then released between the coupling and the shaft, and the hub shrinks to the shaft. Once this is accomplished, the pressure is relieved on the hydraulic coupling nut, and the pump is removed.

Cause of Coupling Failure

The keys to avoiding coupling failure are correct coupling selection utilizing all application design criteria, proper installation, and periodic system maintenance. Consider all of the application requirements early in design as this will reduce the risk of selecting the wrong type of coupling. Install the coupling properly, verifying that design considerations were correct. Regularly maintain the system to ensure that design parameters have been consistently maintained and that no wear, contamination, or other detrimental factors have been introduced to any system components. Investigating the causes of coupling failure and to eliminate them is important. The most critical factors affecting coupling performance and reliability are as below.

Improper Selection

A coupling must match its' intended purpose and application parameters, including many different performances, environmental, use, and service factors. Many a time coupling failures happen because of incorrect selection. Coupling selection involves a number of design criteria including: application, torque, misalignment, stiffness, inertia, RPM, shaft mounting, environmental factors, space limitations, service factors, cost, and others. All criteria must be considered and addressed in the selection process to ensure that the coupling will work properly without premature failure.

An oversized coupling is specified with the hope that it will last longer. However, this may not be true, as it will increase the radial loads on the shafts if misalignment exists. If the load transmitted is too low, the oversized coupling may actually wear out much faster. When oversized coupling is selected due to shaft sizes, care must be taken to align the coupling more accurately than usual.

Speed limits should be strictly adhered to for Elastomeric couplings. Speed generates centrifugal forces that can place high stress on the flexible components. Tire type elastomeric coupling can "Balloon" and eventually rupture at high speed. Compression type elastomeric couplings may whirl or vibrate laterally. Many of these couplings may need some type of lateral support for use at high speeds.

To avoid premature coupling failure, it is critically important in design selection to match the correct coupling to the misalignment condition or combination of conditions present. Shaft misalignment may be angular, parallel, or axial, with further complications when any

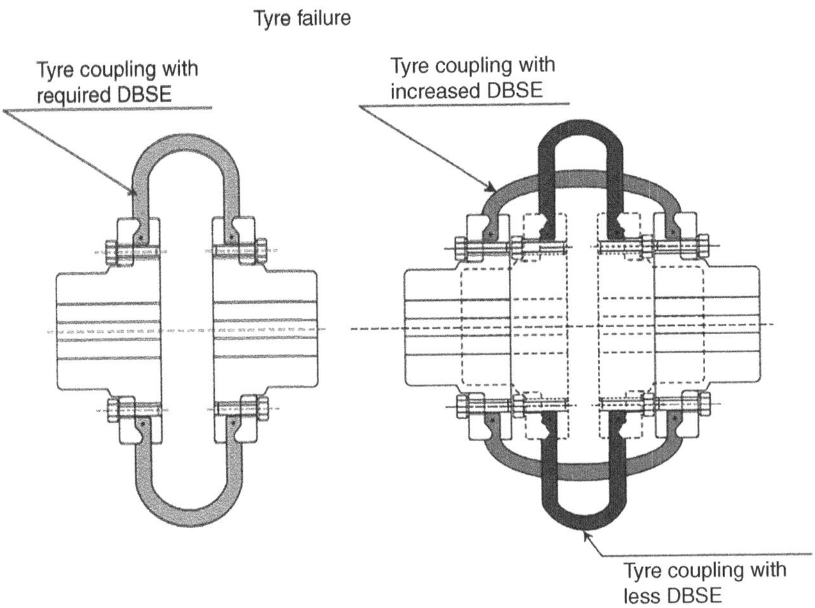

Figure 8.15 Couplings tire failure.

combination of these occurs (complex misalignment). Flexible couplings are typically designed to compensate for specific application misalignment conditions.

Application speed is another very important factor in selection. When a coupling's safe operating speed is not addressed in the design criteria, it can quickly result in failure, sometimes with tragic consequences. In high-speed applications, the use of a balanced coupling is essential. It is also important that consideration be given to coupling stiffness since speed also causes deflection. Pay particular attention to the manufacturer ratings for speed, never adversely alter the dynamic balance of a coupling before or after installation, and remember that any shaft misalignment can significantly affect a coupling's safe operating speed.

Effects of Thermal Variations and Misalignment

If the equipments are aligned when they are cold and no compensation for thermal growth is considered, the equipment will probably be running severely misaligned at running temperature. The couplings that accommodate misalignment through sliding, the useful life is determined by the type of wear. For example: Lubricated metallic flexible couplings.

The couplings that accommodate misalignment through flexing of metallic element, the useful life is determined by fatigue, in some cases by corrosion and fretting. For example, dry metallic flexible couplings.

The couplings that accommodate misalignment through flexing of elastomeric element, the useful life is determined by the deterioration in properties which is caused either by aging of the material or by damping. For example, elastomeric couplings in compression and Shear.

Corrosion Induced Failures:

Corrosion affects every type of couplings: salt, chlorides, hydrogen among others corrode ferrous components. Aluminum is corroded by alkali and salt. Rubber elements are affected by some hydraulic oils. There are applications which suffer from corrosion for example:

1) Flexible couplings create air movement through and around them. Air/gases/steam/hydrogen gas enters into a coupling above the shaft key depending on the environment in which the coupling is running, corrosion takes place. To avoid this before installing hubs on the shaft, a bead of sealant compound should be placed on the top of the key. The sealant will close the gap that must exist over the key, and prevent any corrosive gases to penetrate into the coupling.
2) Water can enter into a coupling over the key is often found in grease lubricated coupling that operate outdoors. Rain water can be sucked inside a coupling; it is retained by centrifugal effect.
3) Oil-borne corrosive elements are normally found in oil-lubricated gear couplings, corrosive agents carried into couplings by oil. Acids and water are found mixed in the lubricants.

Failures Caused by Wear

Abrasive wear can occur in all couplings that accommodate misalignment through sliding, whenever lubrication conditions become poor or marginal. Marginal lubrication causes tooth surfaces to wear. These wear particles mix with the remaining lubricant, creating an abrasive lubricant. If couplings are not periodically cleaned of the contaminated lubricant, the wear rate becomes higher and higher, and coupling teeth can become so thin that they fail in bending.

Steel grid couplings can wear at the hub teeth, at the grid spring, and also at the covers. The spring and hub teeth wear in a similar way as the gear coupling.

Fatigue Failure

All types of couplings suffer from fatigue failures. Fatigue is a type of failure that occurs gradually.

Lubricated coupling failure is caused by fatigue when operating the coupling at an excessive misalignment than specified. Fatigue failure of metallic disc coupling is very different than the gear coupling, and most often the failures of metallic discs occur without any kind of warning. Once the stresses in a flexible disc exceed the endurance limits, failure occurs within a short period without alarm of failure.

Fretting

Discs fail due to fretting. Fretting occurs because adjacent blades slide back-and-forth over each other under the influence of misalignments. Discs of metallic couplings can also fail if the bolts are loose and do not provide the required clamping force.

Brittle Fracture

All elastomers have a certain amount of damping property, which causes some of the energy that is used to flex them, which gets converted into heat. The amount of energy absorbed depends on the characteristics of the material, on the amplitude and frequency of the flexing. If the heat generated by damping cannot be dissipated at the same rate, the temperature of the elastomer will increase and the ability of the element to flex and transmit torque is decreased. Rubber becomes brittle with time, particularly in the presence of ozone. Brittle flexible element can accommodate less and less misalignment and eventually fail.

Hardware Failure

Coupling bolts' main function is to clamp together various components. Failures of coupling bolts occur exclusively in fatigue, which in turn occurs whenever bolts are not properly tightened. Bolts can fail if the holes in the coupling flanges are not equally spaced. Bolts can also fail if they have the wrong geometry, or too weak material.

Time -related Failure

Some failure modes can be experienced with the accumulation of operating time. Specifically, shaft misalignment can develop after many cycles of operation as a result of:

Settling Foundations – Once a coupled system's bed plate is grouted, it may experience settling due to foundation conditions. During welding operations of the bedplate, residual stresses may warp the base, causing difficulty during initial alignment.

Thermal Growth – Due to differences between component material thermal expansion coefficients, at elevated operating temperatures, shaft centerlines may "grow" to be farther apart than at room temperature. In situations where a system will normally operate at elevated temperatures (i.e. steam turbine driven equipment, etc.), the zero-misalignment condition should be set at these elevated temperatures. In addition, total operational scenarios must be considered. If the system has a substantial cooldown or warm-up period of operation, then consideration must be given as to whether or not operations can be sustained during these periods of misalignment.

Connecting Piping Reactions – If, during extended operation, piping braces loosen or fail, the coupled components may have to support excessive reaction loads from connecting pipes. This can put a severe strain on bearings and coupling alignment.

Vibration – Excessive vibration can act to bring about material fatigue, fastener loosening, or stress corrosion cracking. After extended operation, component wear can open clearances and augment vibration amplitudes. Increased vibration can act to worsen shaft misalignment. This may indirectly affect coupling life.

Bearing Wear – Lack of lubrication, contamination of the bearings, and wear can deteriorate a bearings performance over a period of time. Bearing failure can increase vibration, which can, in turn, add to shaft operational misalignment leading to coupling failure.

Incorrect Installation

If the coupling is installed improperly or if the actual application parameters are outside of original design criteria, it may fail early. There are many reasons why incorrect installation results in coupling failure.

Most important being somebody didn't follow the installation instructions supplied. The other most frequent reasons being improper tightening of bolts, use of low-quality bolts, keys made of soft steels, incorrect hub spacing, incorrect hub installation, guards that are too close to elastomer elements, improper coupling sealing etc.

The coupling is to be installed within the "Initial Alignment Limit" specified by the manufacturer. It has been observed that lack of installation and alignment procedure results in improper installation. This will result in vibration and additional loads which, depending on their severity, can produce premature wear, or even catastrophic failure of bearings, seals, the coupling itself, and other machine components.

Don't introduce additional stress on the coupling by compressing or stretching it upon installation. Couplings must always be installed in their free-state. Assure maintenance people are trained properly in installation and maintenance of the type of couplings being used. Follow

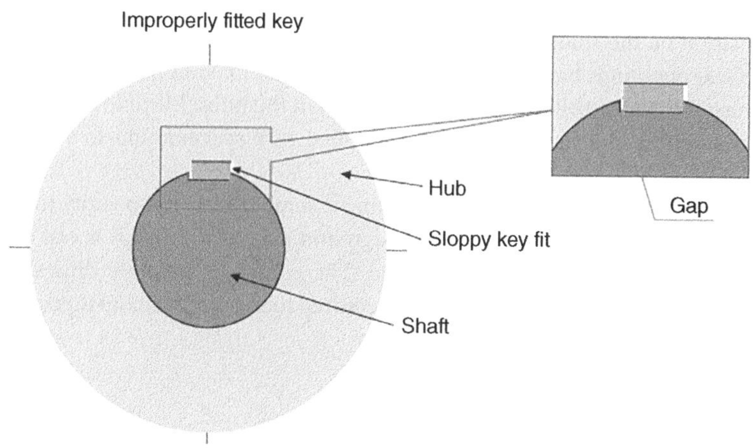

Figure 8.16 Improperly fitted key.

all instructions for fasteners, specifically the tightening sequence and torque requirements. Use the bolts supplied. If one gets lost, buy a spare set from your coupling supplier. Use a torque wrench to torque bolts. Lubricate, as required with the right stuff and correct interval.

These basic guidelines are intended to reinforce the importance of proper coupling installation, thus reducing the possibility of premature failure. Always refer to the specific manufacturers' instructions when performing a coupling installation.

Other important points to be considered during installation

Bolting

Bolts are used in couplings to connect flanges and to connect disc or diaphragm packs to hubs and spacers. Some coupling designs use close clearances on the bolt diameters to maintain concentricity between parts. Coupling bolts are subjected to bending, shear, and torsion. When bolts are tightened incorrectly, they are more likely to fail in fatigue. Loose bolts can induce fretting corrosion, as well as hammering and pounding which will eventually destroy the bolts and coupling discs.

Proper pretensioning of the bolts to maintain the clamping force is extremely important for coupling reliability. Couplings are supplied with high-grade fasteners, commonly with yield strengths greater than 100 000 psi. Bolts for high performance, high-speed couplings are usually supplied in weight-matched sets, with the weight of each bolt and nut assembly held to within 0.1 g tolerance. The diametrical clearance around coupling bolts is usually tighter than standard clearances. These tighter tolerances help maintain balance quality after repeated disassembly and reassembly.

Disc and diaphragm packs are usually supplied preassembled with the bolts pretensioned. Unless the coupling design requires that the disc pack bolts be removed for spacer installation, disc packs should not be disassembled in the field. The bolts in disc packs are subject to shear from torque and bending from misalignment (flexing of the disc packs). The maximum combined stress occurs at the flange surface, where most disc coupling bolt failures are from bending

Distance Between Shaft Ends

Coupling assemblies are designed based upon a given distance between shaft end (DBSE). It is rare that machines are set to the exact DBSE as specified on the coupling assembly drawing. Likewise, hub position on the shafts is almost never exact due to mounting and manufacturing tolerances. Gear couplings have a certain amount of DBSE tolerance built into the coupling design because the gear shrouds have longer teeth than the hubs. Flexible element couplings, like disc or diaphragm designs, usually are supplied with spacer shims to adjust the spacer length.

DBSE should be measured and compared to the coupling assembly drawing to verify that the dimension is within tolerance for the coupling design and size. If the DBSE is out of tolerance, one of the machines may need to be moved. The coupling assembly drawing should also be checked to see whether any axial offset is required to account for axial thermal expansion in the cold condition.

Spacer Installation

Most coupling assemblies are match marked, and the match marks need to be lined up during spacer installation. Spacers for disc and diaphragm couplings are often piloted in counterbores located at the face of the hub or flex element pack. Because of this, the spacer is somewhat longer than the distance between hubs or flex-element packs. Most coupling manufacturers use jacking bolts to compress the flex elements axially to allow spacer installation. Care should be taken so that the jacking bolts do not bear on the flex elements, as this will damage the flex elements. The amount that the flex elements are compressed should be measured so that the allowable axial compression (listed on the coupling drawing) is not exceeded.

Proper System Maintenance

Regular "system" maintenance is important, for the entire system in which the coupling is an integral component. System maintenance requirements and schedules are generally a function of the specific application, duty cycles, operating parameters, environment, and other factors. Any maintenance or service plan for the system as a whole is intended to avoid component failure anywhere within the system, including shafts, couplings, motors, bearings, etc. The coupling may be adversely affected if other component operating characteristics force operation outside of design specifications.

Basic system maintenance requirements might include:

- Performing visual inspection, checking for abnormal operating characteristics such as unusual noise, excessive component temperature, vibrations, and signs of wear or fatigue
- Checking and changing lubricant if the coupling is lubricated. This maintenance is required at regular interval say twice in a year for most couplings and more frequently for couplings operating in adverse environments or in demanding operating conditions
- Check for any signs of wear or looseness in fasteners; re-torque where necessary.
  - When using a jaw type coupling, consideration should be given to the duty-cycle of the center disc or spider. Wear on this component may result in backlash, thus introducing system performance issues. Replace center discs and spiders with the vendor specified part and material when the duty cycle has been exceeded or when excessive wear is noted. The discs are low-cost items, easily replaced, and will restore the coupling's original capabilities.

In most cases, these maintenance steps should be sufficient to keep couplings working smoothly and to enable them to reach their full-service life. In addition to these, it is also recommended that after the equipment has operated long enough to become temperature stabilized, it is best to shut it down and immediately recheck alignment. Due to thermal growth, equipments that are aligned in the "COLD" preoperating condition are almost always out of alignment when operating temperatures are attained.

Special-Purpose Coupling Failure Mode

Failure Modes of Gear Coupling

While failures in gear coupling caused by a combination of factors, the most common failures involve improper lubrication (approximately 75% of known failures). One of the main lubrication problems is an inadequate supply. This could lead to heavy pitting and spalling and/or excessive wear. Excessive misalignment will also lead to heavy spalling, called worm-tracking. Under extremely high misalignment, tremendous forces are transmitted to the connected shafts and bearings through the couplings. This is especially true for a gear coupling, which has up to 10 times the bending moment under misaligned conditions compared to a metallic flexible element coupling. Serious damage can result if the situation is not rectified.

Another important mode of failure in gear coupling is oil contamination and degradation. Foreign materials such as dust and metal particles can mix with the intended lubricant, or worse, separate from the lubricant and centrifuge out to the tooth area and form sludge. In extreme case, this sludge can lockup the coupling and prevent the movement necessary to accommodate misalignment. Due to oil degradation over time, the quantity of sludge increases and it can impair axial float, corrode the teeth thereby accelerating their wear rate, or reduce the circulation of the lubricant. In extreme cases, sludge can even lock-up the coupling and prevent the movement necessary to accommodate misalignment.

Also, excessive misalignment causes failures due to an increased bending stress toward the end of the gear teeth, or end loading. Under extremely high misalignment, tremendous forces are transmitted through the couplings to the connected shafts and bearings. As the bearings wear and the misalignment increases, bending stress at the end of the gear teeth is increased and wear accelerated.

Generally, it is recommended to carry out a periodic inspection as under:

- After every 3000 h – Carry out lubrication and check that sleeves are freely moving axially.
- After every 8000 h or two years – Check alignment. Inspect gear teeth and O-rings. Carry out lubrication and check that sleeves are freely moving axially.

Disc Coupling Failure Modes

Flexing metallic element couplings generally fail in either of two basic causes: over misalignment or overtorque. Over misalignment generally means excessive angular or parallel offset misalignment, with or without excessive axial misalignment. There are, of course, combination failures, misalignment, and torque, but there is usually only one that is primary.

An angular misalignment applies an alternating stress on the metallic flexible element or elements. The element(s) bends back and forth each revolution to accommodate the machinery

Figure 8.17 Gear teeth worn from excessive misalignment. *Source*: Paper from turbomachinery laboratory – conference paper, Texas A&M University Press.

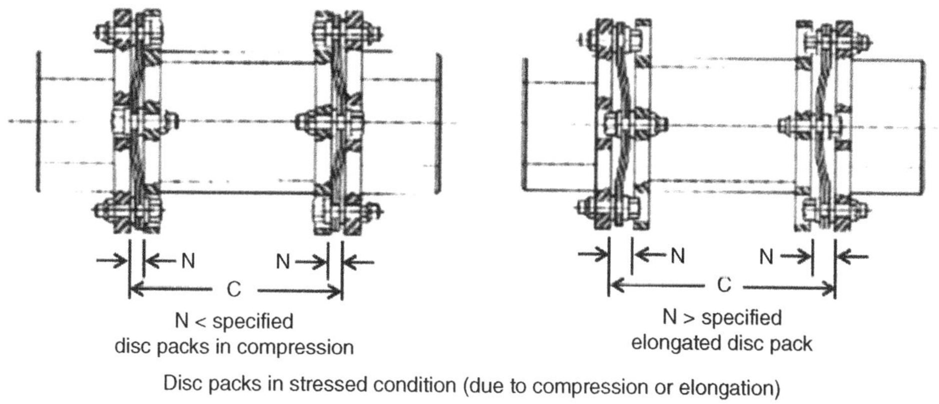

Figure 8.18 Disc pack coupling.

angular or parallel offset misalignments. So, the failure mode from these excessive misalignments is bending fatigue.

Many a time failure in disc coupling result from fatigue due to excessive flexure from greater than designed axial misalignment. As shown in the above figure, if a coupling is installed such that the distance between two shafts is too much or less as compared to designed distance, it leads to axial misalignment.

The other major cause of disc coupling failure is due to a torque overload. In the event of a torque overload – caused by a compressor ingesting a liquid slug, or a generator fault, etc., – the disc pack will yield. Distortion of the bushings, bent bolts and contact between the bushing and the disc pack flange may also be evident upon further inspection.

If the load is large enough, fractures in the disc pack can occur at the links or at the bolt holes. However, though yielded, the disc pack coupling will stay together, and typically cause high vibration from unbalance, unless the load is so high and monstrous that the whole coupling fails at various places catastrophically. More commonly, again, parts will yield but not break from potentially high over torques (Figure 8.19).

Torsional oscillations can also cause significant damage to couplings. Failures due to severe torsional oscillations can also occur with applications with reciprocating engines or pumps that have torque pulses from an engine firing cylinders or reciprocating pump action. Moderate oscillations may cause fatigue failures (either high or low cycle) at either the bolt holes accompanied by considerable fretting, or in adjacent tension and compression links. High magnitude torque fluctuations can result in bidirectional elongation of the disc pack at

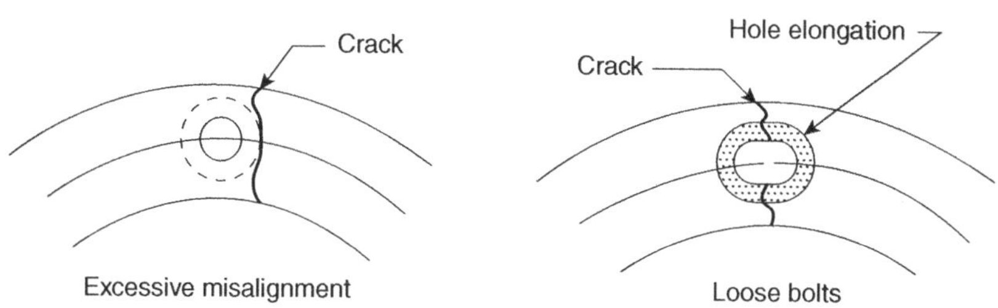

Figure 8.19 Disc pack failure.

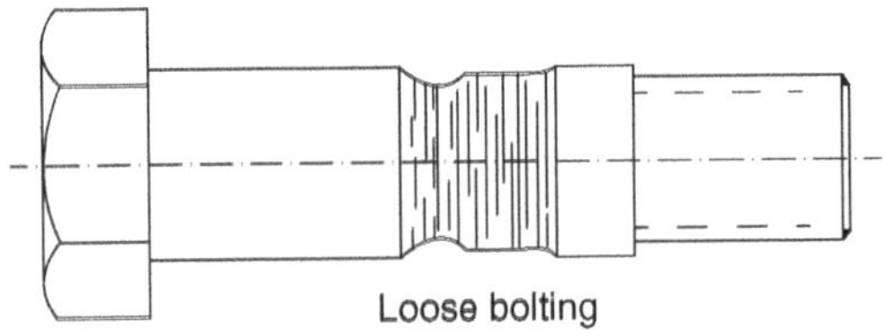

Figure 8.20 Bolt damage.

the bolt hole as well as serious fatigue failure to other coupling components if they are not designed for the torsional load.

Above figure shows disc pack failure due to excessive misalignment and loose bolts. Due to excessive misalignment, cracks usually start on the outer discs and progress inward. If hole elongation and cracking is found, it is most likely due to loose bolting.

Above figure shows how a coupling bolt would look after loose bolting. Note that the disc has tried to embed in the bolt, causing a reduction in diameter or cutting of the bolt. When this failure occurs, the bolts and disc pack should be replaced. Torque should be applied to the nut and not the bolt.

Operating in an environment where corrosives are present will also cause cracking of the disc pack leading to premature failure of the coupling. Please note that a disc coupling gives no warning of impending failure like a gear coupling which generates noise and vibration prior to failure.

If an outer disc breaks, the load is redistributed to the inner discs, which then might have a higher torque load, but a lesser misalignment load. After enough discs' break, there can be enough unbalance to cause higher machine vibrations, so that a decision can be made to shut the connected machines down and investigate the problem.

Diaphragm Coupling Failure Modes

Just like disc couplings, the most common failure modes for diaphragm couplings are due to either excess misalignment or a torque overload. Failure due to misalignment generally means excessive angular or parallel offset misalignment, with or without axial misalignment. Angular misalignment causes an alternating stress in the web of the diaphragm as it bends back and forth with each revolution. Failures occur due to bending fatigue and start with a crack in the flexible element. Axial misalignment stretches the diaphragm, resulting in an additional continuous stress.

Elastomeric Coupling Failure Mode

Misalignment is often a major cause for failure of elastomer type couplings. Large amounts of misalignment can cause the elastomers to be subjected to high loads, which can tear them apart.

Due to the damping effect of elastomeric couplings, heat is generated with the flexing or compressing of the material. If the temperature limits are in excess for the material used, the coupling will fail.

Couplings are designed to transmit torque from one shaft to another through the coupling hubs and spider. The weakest link in a jaw coupling is the insert, or spider, and the coupling's

Figure 8.21 Excessive angular misalignment and axial movement diaphragm failure. *Source*: Paper from turbomachinery laboratory – conference paper, Texas A&M University Press.

torque capacity is determined by the rated capacity of the elastomer to transmit torque. When the elastomers' capacity to transmit torque is exceeded, an over-torque, or overload, situation can exist. This happens when the coupling has not been correctly sized for an application and typically results in the failure of the coupling. A major factor influencing the selection of a coupling is the use of service factors to adjust the application torque for specific applications. When an incorrect service factor is selected or this calculation is not done properly, spider failure and often hub failure will occur. The elastomer, or spider, acts as a cushion between the metal jaws of the two coupling hubs. The driving hub pushes the driven hub through the spider resulting in an expected compression of the elastomer. Over a period of time, this compression result in wear also.

Applying excessive torque or excessive interference fit to a coupling can cause the hub to break starting at the top corner of the keyway where the thinnest cross-section occurs. Ultimately, the hub will split in half.

When a spider is exposed to a chemical, it is not compatible with, the elastomer material will break down quickly, often causing a rapid failure of the spider. Care should be taken at the time of coupling/spider selection to avoid issues related to chemical exposure.

Exposure to extreme temperature has always been a factor in selecting an elastomeric insert or spider for a jaw coupling. Many applications require adjustments in application designs to move away from elastomeric couplings to all metal coupling designs to counter issues raised due to temperature.

When making coupling selections, the user will need to take into consideration both chemical exposure and the temperature of the environment where the coupling will be operating. This will result in better coupling selection and longer life.

Conclusion

The coupling is often one of the last items selected in the system. Selecting a coupling for worst-case operating conditions can save money, and downtime. Nominal torques, peak torques, misalignment, heat transfer, torsional vibration, transmittable torques, key stress, shaft tolerances, assembly, and maintenance should all be factored into coupling selection. Selecting the right coupling will help improve an equipment efficiency and reduce overall downtime. Wrong applications for each type are characterized by the conditions that shorten operating life. In metallic couplings, premature failure of the torque-transmitting element most often results from metal fatigue, usually due to flexing caused by excessive shaft misalignment or erratic/pulsating/high-inertia loads. In elastomeric couplings, breakdown of the torque-transmitting element most often results from excessive heat – either from ambient temperatures or from hysteresis (internal buildup in the elastomer) – or from deterioration due to contact with certain oils or chemicals. By properly understanding the mode of failure and implementing the design criteria, longer coupling life will be achieved.

9

Bearing Failure

The primary function of a bearing in a turbomachinery is to carry load between a rotor and the case with as little wear as possible. They are used to positions the components and prevent friction between parts during relative motion. Two fundamentally different approaches to bearing design have evolved over years. Among tribologists, it is common to refer to them as antifriction bearings and fluid film bearings. Antifriction bearings operate using rolling components inside the bearing (i.e. balls, rollers), journal bearings operate by using a self-generated hydrodynamic oil film pressure to support the shaft while preventing the shaft from contacting the bearing surface. Antifriction bearings have predictable life cycles before they will fail that are known based on statistical testing of many similar bearings. In contrast, a journal bearing that uses hydrodynamic lubrication has an infinite expected life unless there is a loss of the oil film due to lack of lubrication or excessive forces. In small machinery, e.g., low horsepower electric motors, both rolling element and fluid-film bearings are frequently found. In large turbomachineries such as steam and industrial gas turbines, centrifugal compressors, etc., the fluid-film bearing is the predominant choice of designers.

Anti-Friction Bearings

Antifriction bearing or rolling element (or R. E.) bearings essentially rely on surfaces-of-revolution and metal-to-metal contact to provide the necessary positioning and load transmission capability. The surfaces of revolution may be balls, rollers, needles, etc. Bearings of this type are found in a wide variety of machinery automobile wheel bearings, electric motors, centrifugal pumps, fans, etc. They are generally applied in machinery with low power ratings or in high power applications where their advantages are judged to outweigh their drawbacks, such as aircraft gas turbines.

Antifriction bearings have well defined operating limits based on shaft speed and diameter. The factor used for general design selection is the DN number, which is calculated as follows:

$DN = \text{Diameter}\,(\text{mm}) \times \text{RPM}$

Limits for DN numbers are published based on the type of lubrication provided:

Root Cause Failure Analysis: A Guide to Improve Plant Reliability, First Edition. Trinath Sahoo.

Table 9.1 Max DN value of bearing.

	MaxDN value	
Bearing Type	**Grease**	**Oil**
Single row ball	200 000	300 000
Double row ball	160 000	220 000
Cylindrical roller	150 000	200 000
Spherical roller	120 000	170 000

The fundamental advantages of R. E. bearings are as below:

- They are highly developed and standardized.
- They may be designed to provide simultaneous axial and radial load capacities.
- Compactness -low length/diameter ratios are possible.
- The load capacity is relatively independent of speed.
- Low power loss and required lubricant flow rates.
- The shaft position is relatively fixed and independent of speed.

The fundamental disadvantages of R.E. Bearings are:

- They are inherently fatigue limited even under steady load.
- Antifriction bearings inherently have little or no damping. Therefore, very high vibration amplification at the critical speeds can occur during starting or coast down. The amplification factor can be as high as 50 % which may result in machine failure during a single start/stop cycle

AntiFriction Bearing – Type, Selection, and Failure Mode

Rolling element bearings will use either balls or rollers as the rolling element component of the bearing. There are other types of bearings used to overcome friction such as journal and fluid bearings, but rolling element bearings have become the most widely accepted, cost-effective, and easily adapted antifriction bearing solution.

Bearing Basics

Types of Rolling Element Bearings

There are many different types of rolling element bearings. The key in selecting a bearing type is the style of the rolling element itself. Although ball bearings are available in different configurations and designs, they are still identified as ball bearings because of the rolling element. Bearings using rollers as the rolling element can be in various shapes. As illustrated below, roller bearings are available as cylindrical, needle, tapered (trapezoid), and spherical (barrel-shaped). Roller bearings will have a line contact area between-rollers and rings and will provide heavier load capacities compared-to a point contact in ball bearing. However, the point contact ball will generate less rolling friction than a roller element bearing, making it better suited for higher operating speeds.

In addition to the rolling element, the majority of bearings consist of an inner ring, outer ring, and retainer or cage to guide and separate the rolling elements. In tapered roller bearings, the inner ring is normally called the cone, while the outer ring is referred to as the cup. The rolling paths of the elements on the rings are called raceways. With ball bearings, since the ball path is actually a groove, it is referred to as a raceway groove. In thrust bearings, the inner and outer rings are called shaft and housing washers, respectively.

The inside diameter of the inner ring is normally referred to as the bore and is mounted on a shaft or spindle. The outside diameter of the outer ring is usually in contact with a housing or hub in most applications. Radial bearings are designed primarily for carrying radial load. Most radial bearings can carry some thrust loading with the thrust carrying ability dependent on the contact angle and type of rolling element (steeper angle-tapered rollers have relatively high thrust capacities). However, with the exception of spherical thrust bearings, which can carry a small amount of radial load, thrust bearings can only carry thrust loads.

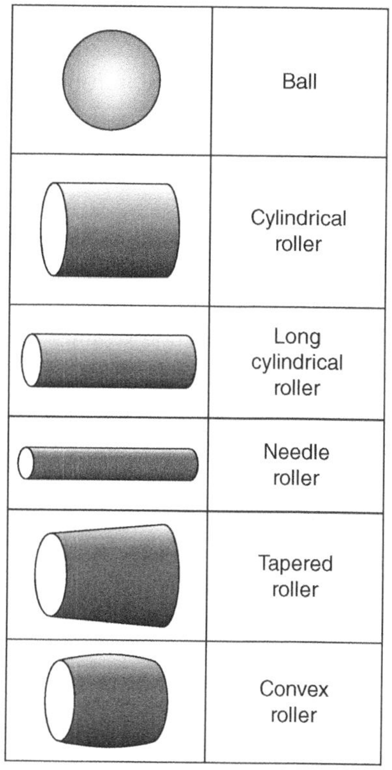

Figure 9.1 Type of antifriction bearings.

Radial bearing Designs, Features, and Applications

Features of Deep Groove Radial Bearings

The most popular rolling bearing type used in a wide variety of industries and applications

- The inner and outer rings have deep uninterrupted grooves
- They can support radial loading and certain degree of axial loading in either direction
- High-speed capability

Angular Contact Ball Bearings

The contact angle between the bearing balls and rings is normally 15, 30, or 40°

- The larger contact angle bearings have greater thrust load capacity.
- Single-row bearings can accommodate radial load and axial load in one direction only
- Double row matched bearings can accommodate radial load and axial load in either
- Direction

Self-aligning Ball Bearings

The self-aligning ball bearing has two rows of balls

- The balls roll on a spherical outer ring raceway
- The inner ring, balls, and retainer can align themselves and accommodate for shaft and housing misalignment and deflections

Radial Roller Designs, Features, and Applications

The design allows linear contact of the cylindrical rollers with the raceways resulting in excellent radial load and impact load capacity.

- The rollers have a slight crown to relieve the ends and reduce stress concentration
- The geometry allows for accurate machining to precision grade quality and use at high speeds
- The inner or outer rings can be separated, simplifying mounting and removal of the bearing

Tapered Roller Bearings

The inner ring assembly of a tapered roller is referred to as the cone, while the outer ring is called the cup.

- Tapered rollers are guided by the cone back-face rib.
- This bearing design is suitable for heavy radial loading and a sizeable amount of axial loading.

Spherical Roller Bearings

The bearing is self-aligning and forgiving to errors of up to 1.5° of misalignment between the shaft and housing due to shaft bending.

- Spherical bearings can carry radial and axial load in both directions.
- Suitable for applications where there are heavy radial and impact loading applied

Needle Roller Bearings

For those applications, where minimal cross-section height is required

- The rollers of needle bearings are longer and smaller in diameter than cylindrical rollers
- Needle roller bearings cannot carry thrust loading but have relatively high radial capacity

Bearing Selection Process

In order to select the most appropriate bearing for an application, it is very important to understand the expected operating conditions of the bearing. The main bearing selection criteria will be covered in this section.

Mounting Space

When an application is designed, a primary consideration is for the shaft to have the proper strength and rigidity. Consequently, the minimum required shaft diameter is determined followed by a determination of the allowable housing size, weight, and material needs based on the application environment and loads. Once these shafts and housing size restrictions have been determined, the allowable bearing envelope dimensions can be specified.

Loading

Load type, magnitude, and direction of loads are all key in determining the proper bearing for a particular application. Knowing the type of loading, load combinations, and if shock loading is possible are important factors in bearing selection. The load types are normally described as being either pure radial (a load perpendicular to the shaft centerline), pure thrust (a load parallel to the shaft centerline), and moment or overturning load (a load offset from a bearing arrangement which causes an overturning motion).

If a bearing's load requirements are only to be able to carry a radial load, there are numerous ball, cylindrical roller, spherical roller, and tapered roller possibilities. However, if the application load requirements are for load-carrying ability of all three types of loading, tapered roller bearings or possibly multiple rows of angular contact ball bearings may be the only options. Of course, when only pure thrust loading is involved, there are various types of thrust bearings that can be used.

Speed Requirements

The limiting speed for bearings refers to the rotating speed at which the bearing can be rotated continuously without an excessive increase in operating temperature. In general, point contact ball bearings such as the deep groove and angular contact, which generate less heat than line contact roller bearings, have the highest speed capabilities.

Rigidity Requirements

The higher the bearing rigidity, the better the resistance to load-induced deformation. Bearing rigidity can be enhanced by setting up the bearings with preload or negative clearance. Using preload to increase rigidity is suitable for use with angular contact and tapered roller bearings. Preload is provided by applying an axial load to remove any clearance in the bearings. The additional preload force on the bearings results in more rolling elements sharing the loading on the bearing thereby increasing the bearings resistance to elastic deformation.

Misalignment Considerations

Bearing misalignment can be caused by a number of different conditions. Frequently misalignment is introduced by loading that causes a shaft to bend resulting in angular deflections in the bearings. Misalignment can also be induced by various machining inaccuracies of bearing seats and backing shoulders. In general, whenever misalignment angles greater than 3-4 minutes are anticipated, consideration should be given to using a self-aligning bearing such as a spherical roller bearing or self-aligning ball bearing.

Mounting and DisMounting Considerations

If application requirements call for periodic inspections that require mounting and dismounting of the bearings, the ease and methods required for these bearing procedures should be a bearing selection consideration. Bearing mounting and removal is simplified by the use of bearings that have separable races. Bearings such as cylindrical roller bearings, needle roller bearings, and tapered roller bearings have separable races and should be considered for applications requiring frequent inspections and removal of the bearings.

Bearing Service Life

When bearings are rotated under load and subjected to repeated contact stresses, wear will eventually result in the raceway material spalling and flaking off. The total number of revolutions until flaking occurs is described as the bearing service or fatigue life. Bearing service (fatigue) life varies depending on design, size, materials, manufacturing methods, and operating conditions.

If a group of identical bearings is rotated under the same conditions, the total numbers of revolutions until 10% of the bearings exhibit flaking or spalling damage is defined as the basic rating life or "L10" life. This rated life can be expressed in terms of time when the rotating speed is constant.

The calculation of the bearing service (fatigue) life is normally used in the selection of a bearing for a particular application and is a criterion for acceptability of a bearing. However, other factors and environmental effects can result in premature bearing damage and a reduction in fatigue life. Factors such as improper bearing mounting procedures, lubrication methods and type, fits of the inner and outer rings, and the wrong bearing clearances for the expected operating conditions can all result in fewer hours of bearing life.

Generally speaking, bearings in an application have a calculated life. Whether or not bearings reach or exceed that calculated life depends on a number of factors.

Antifriction bearing failure mechanism

Most bearing damage can be classified into two damage categories: preoperational and operational. Preoperational damage occurs before or during bearing installation, while operational damage occurs while the bearing is in operation.

Causes of preoperational damage:

- Bearing quality and application
- Storage and handling
- incorrect shaft and housing fits
- defective bearing seats on shafts and in housings
- static misalignment
- faulty mounting practices
- passage of electric current through the bearing (excessive voltage)
- transportation, handling, and storage

Causes of operational damage:

- material fatigue
- ineffective lubrication
- ineffective sealing
- vibration (false brinelling)
- operational misalignment
- passage of electric current through the bearing (current leakage)

Preoperational Causes of Damage

Bearing quality- Only bearings manufactured to the highest quality standards can provide a long service life.

Application

The application utilizes the appropriate bearings.

Storage of Bearings

The conditions under which bearings, seals, and lubricants are stored can have an adverse effect on their performance. Inventory control can also play an important role in performance, particularly if seals and lubricants are involved. To maximize the service life of bearings, the following basic housekeeping practices:

- Store bearings flat, in a vibration-free, dry area with a cool, steady temperature. There should not be a draft in the stocking area.
- Bearings should not be stored near a window where direct sun and moisture could be a problem.
- Bearings should be stored at a room temperature of 20°C/70° F and relative humidity levels of less than 65%.
- Keep bearings in their original unopened packages until just before mounting, to prevent dust and moisture contamination as well as corrosion of the bearing components.

Bearing Handling Precautions

Bearings like other precision machine components can be damaged by improper handling procedures. The following are general precautions for proper bearing handling:

- Keep the bearing and assembly area clean, even very small particles of dirt can cause bearing contamination and result in bearing damage and a reduction in bearing life.
- Avoid abusing the bearings; they are precision ground and heat treated. If a bearing is dropped or subjected to impacts or excessive force due to rough handling, they can be cracked or Brinelled before they are mounted resulting in a point of origin for premature damage and failure.
- Do not expose the bearings to high temperatures since a standard bearing can be tempered if they are heated to temperatures over 120 °C (250° F), which could result in a reduction in hardness and bearing life.
- The proper tools need to be used for handling. Using whatever tool is at hand could result in bearing damage.
- The bearing and assembly area should be well protected from dampness and moisture to avoid the possibility of bearing corrosion.
- The bearings should only be handled by experienced or well-trained operators.

Bearing Tolerances, Fits, and Clearances

The purpose of fitting an inner or outer ring onto a shaft or into housing is to prevent circumferential sliding or spinning of the fitted ring. Bearing ring sliding or "creep" will have a detrimental effect on bearing operation. Bearing creep can cause excessive heat generation, wear, and contamination of the bearing from wear particles, vibration, and misalignment problems. Therefore, the proper fit is critical to the service life of the bearing and the performance of the application. If a bearing ring rotates and the load is unidirectional and constant, interference fit is required. The degree of interference or tightness is governed by the magnitude of the load and the bearing type and size. Typically, the heavier the applied load, the tighter the required fit. If a bearing ring is stationary and the load unidirectional and constant, it is typically fitted with a loose fit (clearance fit). The presence of shock loads or continuous vibration requires a heavier interference fit on the ring that rotates relative to the load.

Fits that are too loose can result in a speed difference between contact surfaces. Improper shaft or housing fits, or fits that are unnecessarily loose, can enable the inner or outer ring to rotate on its seat. This relative movement is called ring creep. The relative movement generates friction and can result in wear or smearing. An interference fit between an inner ring and shaft will induce hoop (tensile) stresses in the ring. If the interference fit is excessive, the resultant hoop stresses can exceed the strength of the ring, causing it to fracture.

Bearing Internal Clearance

It is necessary to distinguish between the internal clearance of a bearing before mounting and the internal clearance in a mounted bearing that has reached its operating temperature (operational clearance). The initial internal clearance (before mounting) is greater than the operational clearance because different degrees of interference in the fits and differences in thermal expansion of the bearing rings and the associated components cause the rings to be expanded or compressed. The radial internal clearance of a bearing is of considerable importance to achieve satisfactory operation. As a general rule:

- Ball bearings should always have an operational clearance that is virtually zero, or there may be a slight preload.
- Cylindrical, spherical, and CARB toroidal roller bearings should always have some residual clearance during operation.
- Tapered roller bearings should always have some residual clearance.

Static Misalignment

Deep groove and angular contact ball bearings as well as cylindrical and tapered roller bearings can accommodate only very small misalignments. Misalignment in these bearings typically causes edge loading, which can result in premature fatigue. The load was carried only over a small area at the edge. The resulting very high stresses in this area led to material fatigue and premature subsurface initiated spalling.

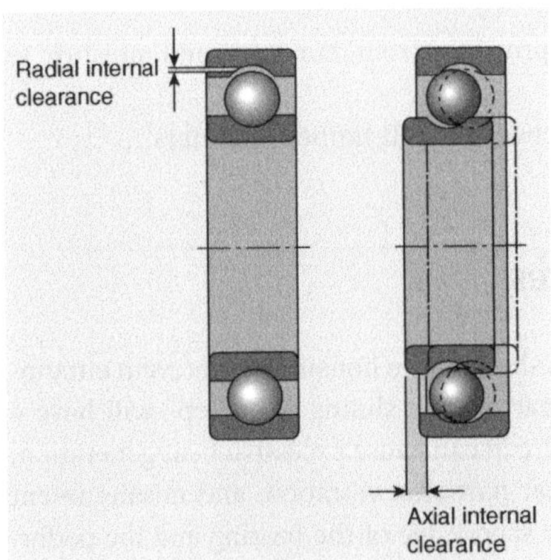

Figure 9.2 Bearing internal clearance.

Faulty mounting practices

Abuse and neglect during mounting often lead to damage and premature fatigue or failure. One of the primary causes of early fatigue failures is impact damage during handling, mounting, storage, and/or operation. In these cases, the impact is higher than the strength of the material (overload), which plastically deforms. Damage starts at the point of deformation and ultimately results in premature bearing failure.

Figure 9.3 shows that the mounting force, applied to the wrong ring, passed through the rolling elements. This can also occur if the bearing is subjected to abnormal loading while not

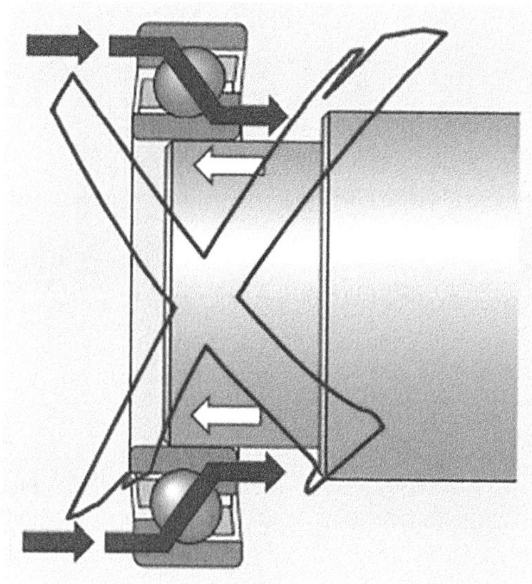

Figure 9.3 Mounting force applied to the wrong ring.

running. As the impact load is an axial load, dents can be found in the rings that are axially displaced from the center. The distance between the dents is the same as the rolling element spacing.

Passage of excessive electric voltage through the bearing

Under certain conditions, electric current will pass through a bearing. For example, when repairing a shaft, excessive voltage potentials can result from improperly grounding the welding equipment. As electricity arcs from one bearing ring to the rolling elements and from there to the other ring, severe damage occurs.

Material fatigue (subsurface-initiated)

In operation, load is transmitted from one ring to the other through the rolling elements. Every time a rolling element comes into the load zone, the load transmitted in the contact area goes from zero to a maximum and back to zero. This leads to a build-up of residual stresses in the material. Depending on the load, temperature, and the number of stress cycles, these stresses will lead to structural changes in the material and result in the formation of cracks underneath the surface. These cracks will finally propagate to the surface and spalling will occur. A bearing is damaged as soon as the first spall occurs. This does not mean that the bearing cannot continue to operate. Spalls will gradually increase in size and number and give rise to increased noise and vibration levels in the machinery. The machine should be stopped and repaired before the bearing fails catastrophically. To avoid premature subsurface initiated fatigue, three major conditions must exist:

- clean bearing steel – best quality bearing
- good lubrication conditions (no contamination)
- good load distribution over the rolling elements.

Ineffective lubrication

Bearings give satisfactory life if the right lubricant in the right quantity will reach the bearing at the right time. All bearings require adequate lubrication for reliable operation. Without effective lubrication, metal-to-metal contact occurs between the rolling elements and the

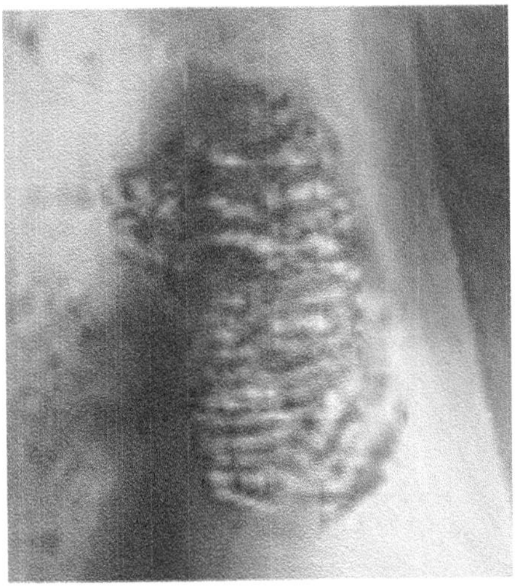

Figure 9.4 Advanced spalling due to subsurface initiated fatigue of the material.

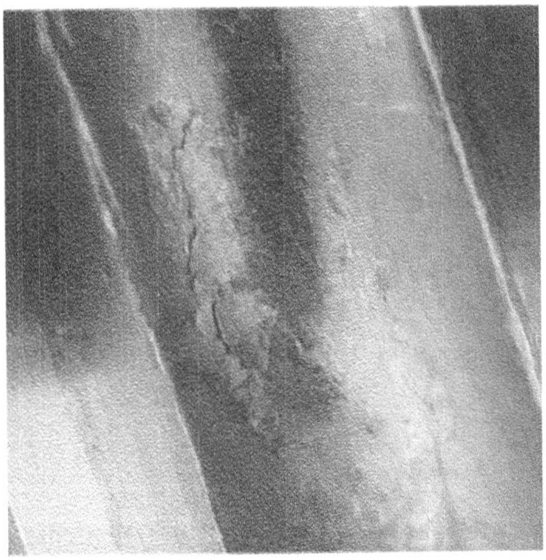

Figure 9.5 Spalling (surface distress) caused by ineffective lubrication.

raceways and other contact surfaces, causing damage to these surfaces. Many damage cases are the result of insufficient lubricant viscosity, over-lubrication, inadequate lubricant quantity, contaminated lubricant, or the wrong lubricant being used in the application. When lubrication is ineffective, damage in the form of surface fatigue will result. The first visible indication due to inefficient lubrication is usually a fine roughening or waviness on the surface. Later, fine cracks develop, followed by spalling.

Ineffective Sealing

The purpose of a seal is to keep lubricants in and contaminants out of the bearing. Premature bearing failure could result if the application is not sealed adequately. For example, take an application with inadequate sealing. When contaminants in the form of particles get into the bearing through the seal, they can be over-rolled by the rolling elements. The over-rolling creates dents in the raceways (Figure 9.1). Hard particles may cause dents with sharp corners. When the area around the dents is stressed, surface fatigue is initiated and metal will start to break away from the raceway. This is called spalling. Once spalling has occurred, damage will progress until the bearing becomes unserviceable.

Sealing Devices

Sealing devices not only prevent foreign material from entering a bearing cavity area they prevent lubricant from leaking from the bearing cavity. Consequently, when a sealing device is selected, it is important to consider the application operating conditions and type of lubrication to be used.

The performance of a sealing arrangement is vital to the cleanliness of the lubricant and the service life of the bearings. Where seals for rolling bearings are concerned, a distinction is made between seals that are integral to the bearing and those that are positioned outside the bearing.

Integral Bearing Sealing Solutions

There are two categories of integral bearing sealing solutions:

- shields
- seals

Shields – Shields, produced from sheet steel, are noncontacting and are used in applications where contamination is limited. They are also used in applications where bearings are lubricated for life and should not be relubricated.

Seals – Seals integrated in many bearings are generally made of elastomer materials and reinforced by sheet steel. Bearings with contact seals are preferred for arrangements where resistance to contamination is needed, where the presence of moisture or water spray cannot be ruled out, or where a long service life with minimal maintenance is required.

External seals

There are two broad categories of external seals:

- contact seals
- noncontact seals

Seals in contact with stationary surfaces are known as static seals and their effectiveness depends on the radial or axial deformation of their cross-section when installed. Typical examples include gaskets and O-rings. Seals in contact with sliding surfaces are called dynamic seals and are used to seal passages between a stationary component, e.g. a housing, and a rotating component, normally the shaft. Their function is to keep lubricant in and contaminants out of the bearing arrangement. The most common contact seal is the radial shaft seal.

Cleanliness- Contamination can adversely affect bearing and seal service life. It also can have a negative influence on the service life of the lubricant. Therefore, it is important that rolling bearings are lubricated with clean grease or oil and that the lubricant is fully protected from contaminants by an effective sealing system. Cleanliness should be observed during all maintenance activities from mounting and relubrication to inspection and dismounting. Keep bearings in their original package, where they are well-protected, until immediately before mounting.

- Mount bearings in an area that is free from dirt, dust, and moisture.
- Use professional tools for all maintenance activities.
- Clean up grease and oil spills immediately.

Figure 9.6 Moisture acids in a spherical roller bearing **– Moisture corrosion**.

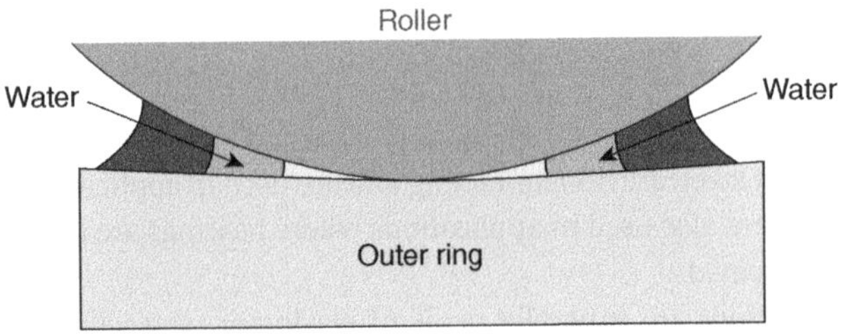

Figure 9.7 Water in bearing.

Corrosion- Corrosion is another problem that occurs as a result of an ineffective sealing arrangement, mostly at standstill. Water, acid, and many cleaning agents deteriorate lubricants, resulting in corrosion. When water, acid, or cleaning agents enter an application, it negatively affects the ability of the lubricant to protect steel surfaces from oxidation. As a result, when a machine is at a standstill, deep-seated rust forms easily. Over time, the excessive moisture will form an acid in the lubricant and etch the surface black, as shown in Figure 9.6. In the presence of water and due to capillary action, the area next to the rolling element contact zone might become corroded.

Figure 9.8 False brinelling. *Source*: SKF.

Vibration (false brinelling) – Bearing damage often occurs in the standby auxiliary equipment due to vibration caused by nearby machinery that is in operation. Depending on the proximity of the idle unit to the operating one(s), vibration created from the running equipment causes the rolling elements in the bearing of the static machine to vibrate. Depending on the intensity and frequency of the vibration, the condition of the lubricant and the load, a combination of corrosion and wear occurs, forming shallow depressions in the raceway. The magnitude and duration of the vibration and the bearing internal clearance can influence the damage. Roller bearings seem more susceptible to this type of damage than ball bearings.

Figure 9.9 Fatigue fracture of the outer ring flange in a double row full complement cylindrical roller: Fatigue fracture.

Operational misalignment: Causes of operational misalignment include shaft deflections caused by heavy loads or load amplitude changes during operation (imbalanced load). When operational misalignment is present, load zones are not parallel to the raceway grooves. The result is induced axial loads, which can be dangerous, because they can lead to fatigue fracture. Figure 9.9 shows the outer ring of a double row full complement cylindrical roller bearing. The outer ring flange is almost completely broken off, due to fatigue from induced axial loads resulting from shaft deflection.

Proper alignment of drives and driven machinery depends largely on the quality of the machine installation. An optimal installation contributes to a quick and easy alignment process with precision results. To achieve optimal installation, several aspects deserve consideration:

- foundation quality
- alignment targets
- soft foot
- shimming
- bolt tightening

Journal Bearing

Hydrodynamic journal bearings have been widely used to support high-speed rotating machinery such as turbines and compressors because of their superior durability and load-carrying capacity. Though antifriction bearings are best for, smaller motors, pumps, and blowers but as size of a pump (or fan or motor, etc.) gets large enough and fast enough, temperatures rise, rotor dynamics often become a concern and critical speeds are encountered. This is when damping is required and fluid film bearings become increasingly necessary. Turbomachinery designers (and users) normally consider using fluid film bearings if machine is running above 3,000 RPM or the machine exceeds 500 HP except very special cases.

The main deciding factors for using a journal bearing are as follows:

- If the machine will operate above a critical speed.
- If the physical size and speed produce a peripheral speed that exceeds reasonable design limits for an antifriction bearing.

In a journal bearing, the shaft or "journal" rotates in the bearing with a layer of lubricant separating the two parts through fluid dynamic effects. Since the bearing surface and the journal are moving with relative speed, lubricant will be drawn into the gap between them

Figure 9.10 Typical plain journal bearing.

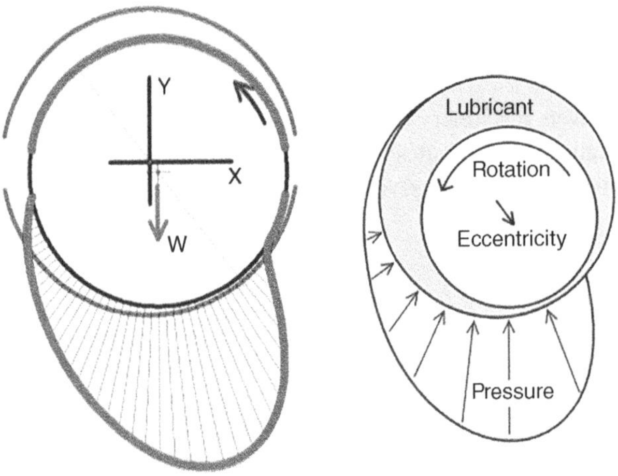

Figure 9.11 Pressure profile in a journal bearing.

and forced to squirt out the sides of the bearing while the gap is converging. Figure 9.10 shows a plain bearing in which a steel base material is overlaid with a babbitt material and bored to a circular diameter equal to the shaft diameter plus the desired clearance. At zero speed, the shaft rests on the bearing at bottom dead center. As soon as shaft rotation begins the shaft "lifts off" on a layer of oil. In fluid film bearings, lubrication is required between a pair of surfaces with relative motion between them. There is always a convergent wedge developed that is formed due to the relative surface speeds and the lubricant viscosity to carry the applied load. An oil pressure film develops with equal and opposite force vectors to the applied load. One surface drags the lubricant, usually an oil, into a converging gap. As the space available in this gap decreases, the fluid develops a pressure gradient, or pressure hill. As the fluid leaves the gap, the high pressure helps expel it out the other side. A simple diagram of this is shown in Figure 9.11.

The oil film pressure is shown in the figure that is generated around the bearing. This self-generated pressure is what supports the load imposed by the shaft. While the oil film is converging (getting thinner), the pressure will be positive which will support the load of the shaft. If the oil film is diverging (getting thicker), the pressure will attempt to decrease.

The pressure cannot fall below the vapor pressure of the oil, so the film will tend to cavitate in the diverging section. However, special designs can accommodate operation at higher temperatures. When the bearing operates in the hydrodynamic regime, a fluid pressure builds up in the bearing by a converging geometry effect in the lower portion of the bearing, shown in Figure 9.11. Because of the pressure imbalance between the inlet side and the outlet side of the bearing, the shaft shifts toward the outlet side which allows the pressure profile to balance with a resulting total force in the vertical direction.

To prevent contact between a journal and its bearing, the minimum oil film thickness must at all times be greater than the combined mean surface roughness of the journal and bearing. This film thickness depends on the following four factors:

1) The lubricant viscosity.
2) Speed of journal rotation.
3) Load on the journal.
4) Operating temperature

Slight changes in load, lubricant flow to the bearing or temperature will alter the oil film thickness, the most common of these being load changes due to some vibration or harmonic in the system. A vibrating load alternately will decrease the film thickness and increase the bearing offset; thus, the journal center will follow an elliptical path within the bearing if the vibration is a constant or may follow a complex path, as in the case of an engine main bearing.

There are several sources of bearing load which may be present in a machine. These are generally the dead weight of the rotor, unbalance or inertia forces, those caused by misalignment, unbalanced pressure forces, and those caused by axial or radial changes in the momentum of the fluid passing through the machine. In some situations, load may be transmitted to a turbomachinery driver from its driven equipment by coupling friction; it is possible to transmit both radial and thrust loads even through a gear-type flexible coupling. The fundamental requirement is that of maintaining the desired relative positions, despite the forces or loads that must be transmitted. It is therefore always important to minimize the wear by reducing the friction factor which may be created by these surfaces under relative motion.

Lubricated contacts can be divided into three distinct zones according to the Stribeck curve, Figure 9.12. A lubricated contact at rest is most often in the boundary or mixed lubrication regime in which the surface contours (asperities) of the two surfaces are well-interconnected resulting in high friction. At the onset of relative motion between the two surfaces, pressure begins to build in the lubricant. Building lubricant pressure causes the load to shift from the asperities to the lubricant. The lubricant pressure continues to build with increased speed, decreased load, or increased lubricant viscosity which results in greater separation between the surfaces. As these factors are further increased, the lubricant pressure reaches a point where the surfaces' asperities no longer make contact and the load shifts completely to the lubricant. This condition is termed as hydrodynamic lubrication.

So long as the hydrodynamic conditions exist to create a lubricating film between the two rotating surfaces, the life of a journal bearing is virtually limitless. If the film breaks down for any reason (e.g., high loads, low speeds, low viscosity), the bearing surfaces will come into contact with each other, and a phenomenon called boundary lubrication takes place. Film thickness decreases as rotational speed decreases or as bearing loading increases.

The Stribeck curve analyses the transition from boundary to mixed and finally to hydrodynamic lubrication regimes with changing speed, viscosity, or load on the contact. Richard

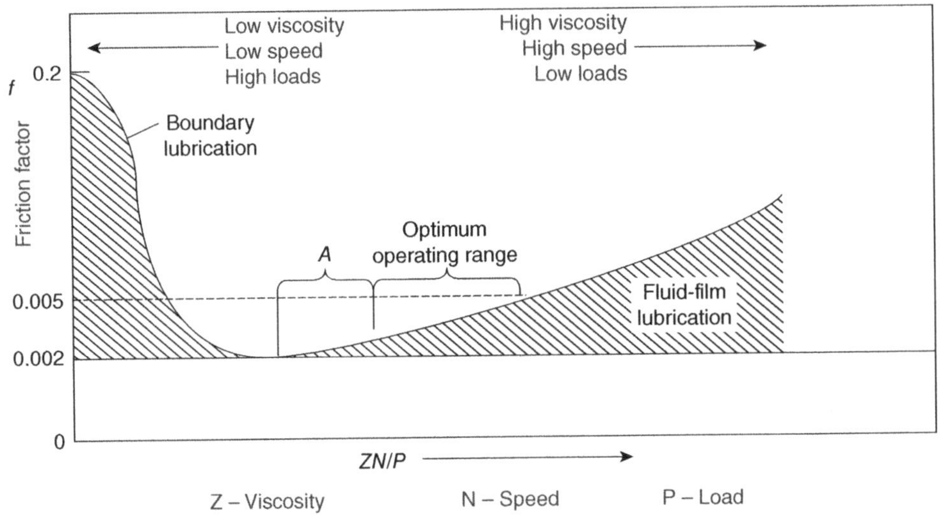

Figure 9.12 Stribeck curve relating friction factor to viscosity, speed, and load.

Stribeck, a German engineer, who did extensive friction testing in 1902 to find the friction factor as a function of viscosity, load, and speed. There was a great deal of research at that time trying to find the best combinations of materials and lubricants that would give the lowest coefficient of friction. Figure 9.12 is the Stribeckcurve. The friction factor is plotted as a function of ZN/P where Z is the viscosity, N is the speed and P is the load. This is a nondimensional equation. According to the friction factor value, the lubrication regime can be predicted as boundary or hydrodynamic type.

In hydrodynamic lubrication as observed from Stribeck curve, heat generated is a function of lubricant viscosity, applied load, and relative speed between cone and journal. As bearing temperature increases, lubricant viscosity and film thickness decrease creating a potential for journal bearing malfunction. Heat generated varies with the square of the speed. A twofold increase in bearing rotational speed produces a fourfold increase in the generation of heat in the bearings. In larger bits, heat generation can be significant, and large bearing design must include provisions for removal of heat from the bearing area.

The stiffness and damping provided by a journal bearing are crucial. The other design factors that affect bearings performance is eccentricity. For example, if the eccentricity is too high there is a risk of metal-to-metal contact and higher dynamic loads being imparted to the babbitt causing premature fatigue. If the eccentricity is too low (journal is nearly centered) then the machine could more easily become unstable. Eccentricity is a function of both speed and load. With a constant load, as speed increases, the eccentricity decreases. While it is possible to operate journal bearings above 200degF, typically a bearing designer will seek to keep the maximum oil film temperature below that due to loss of babbitt fatigue strength. If possible, a good bearing design will have the maximum oil film temperature less than 175 degF to allow for some margin for transient events.

Journal bearings are classified on the type of the sliding surfaces:

If the journal bearing has fixed sliding surfaces, then it is defined as a "sleeve bearing" (Figure 9.13); if the bearing has several pivoted pads, which can tilt freely and it is defined as a "tilting pad journal bearing" (TPJB).If the bearing cross-section is composed of two to four circular arcs, forming the so-called "lobes", it is called multilobe bearing.

Figure 9.13 Sleeve journal bearing. *Source*: Pennacchi P. (2017). In: Advances in Steam Turbines for Modern Power Plants. Elsevier.

Sleeve Journal Bearings

The basic shape of sleeve bearings is the cylindrical bearing, in which the cross-section of the bearing surface is a circle. Sleeve bearings offer several advantages, including lower cost, less maintenance, greatly reduce noise at low speeds, and easier installation. Sleeve bearing performance is also marked by a higher load capacity, especially shock load resistance due to their greater contact zone. Sleeve bearings with a conformal liner compensate for misalignment issues and vibration damping, these one-piece design low-friction bearings reduce housing dimensions and weight by having a thin wall section.

Sleeve bearings can be made from bronze, steel or stainless steel. Brass bearings and plastic sleeve bearings are also manufactured for certain applications Sleeve bearings are prone to causing instability, in particular oil-whip [5], and is has been replaced for a long time by other designs, i.e., by multilobe bearings or by TPJBs

Multilobe Bearings

Due to the operation at high speed, the problems of instability are encountered in sleeve bearing. Instability of the bearings will ruin the bearings and machine itself. Multi lobe journal bearings maintain the stability of the bearings at high speeds and different loading conditions. Multilobe bearings have a cross-section composed of two (Figure 9.14) to four circular arcs, forming the so-called "lobes." In the case of two lobes, the bearing is sometimes dubbed as "lemon-shaped." Two-lobe bearings may have a pocket machined in the upper half, called the pressure-dam, whose aim is to impose an additional downward load on the shaft, which contributes to stabilizing the rotor (and to increasing the bearing dynamic stiffness).

They accommodate clearance, preload, offset, and provision for lubricants to take away the heat during running conditions as a result produces a stabilizing effect on the shaft. The multilobe bearings are more stable in major considerations for dynamic characteristics. Non-circular bearing geometry enhances shaft stability under proper working conditions; they reduce power losses and increase the oil flow rate, due to which bearing temperature is reduced. Noncircular lobe bearings are used in high-speed machinery.

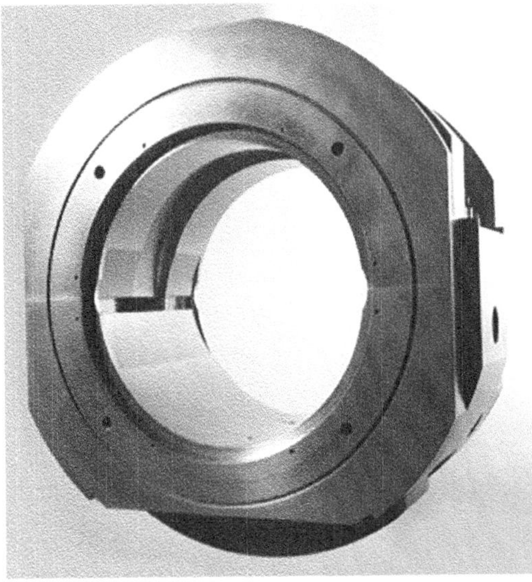

Figure 9.14 Two-lobe lemon-shaped sleeve bearing with pressure-dam. *Source*: Pennacchi P. (2017). In: Advances in Steam Turbines for Modern Power Plants. Elsevier.

Tilting Pad Journal Bearings (TPJB)

TPJBs have several pads, around both the halves of the bearing shell (Figure 9.15) or only in the lower one. Moreover, the pads may be equal or different from each other. Equal pads are the most common cause, while, for instance, asymmetric three-pad TPJBs have been used in large steam turbines employed in nuclear power plants.

The free tilting of the pad is about a pivot, which can be ideal, i.e., the pad rocks about a straight line on the backside (which has a geometrically ruled surface) of the pad that is in contact with the bearing shell. In this case, the TPJB is said to be of the rocker type. If the pivot is machined on the pad back or it is realized by hardened metal inserts in the pad back and in the shell, the TPJB is said to be of the pivoted type.

TPJBs have also some drawbacks, like hot oil carryover, risk of flutter of unloaded pads (i.e., those in the upper half), higher costs, and more difficult determination of clearances

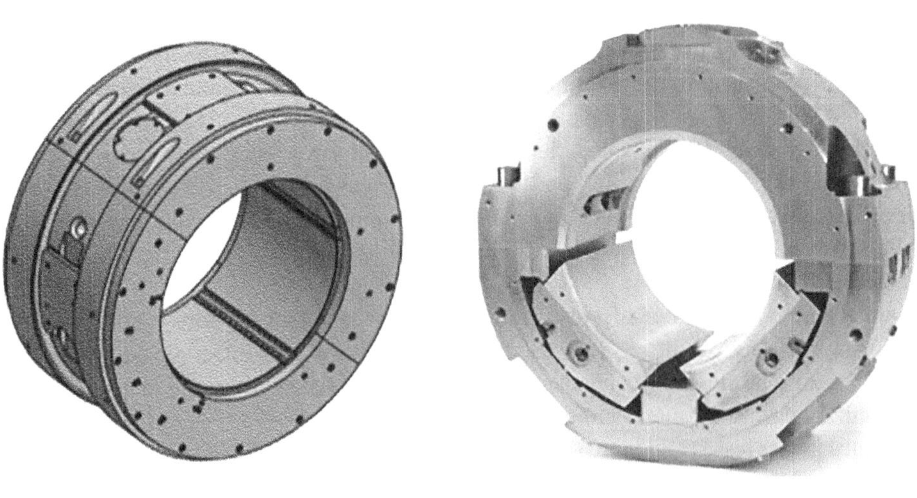

Figure 9.15 Tilting pad journal bearings. *Source*: Pennacchi P. (2017). In: Advances in Steam Turbines for Modern Power Plants. Elsevier.

than sleeve bearings. Pad fluttering is a somewhat difficult phenomenon to overcome and is defined as the unstable vibration of the pad floating back and forth between the pivot point and the journal continuously during shaft rotation. It is related to another phenomenon known as "spragging" of bearing pads. However, the advantages of TPJBs outweigh their disadvantages and their use is growing in steam turbines.

Journal Bearing Failure Mechanisms

Journal bearings operate in the boundary regime (metal-to-metal contact) only during the star-tup and shutdown of the equipment when the rotational speed of the shaft (journal) is insufficient to create an oil film. It is during star-tup and shutdown when almost all of the damage to the bearing occurs. Hydrostatic lift, created by an external pressurized oil feed, may be employed to float large, heavy journals before start-up (shaft rotation) to prevent this type of damage. During normal operation, the shaft rotates at sufficient speed to force oil between the conforming curved surfaces of the shaft and shell, thus creating an oil wedge and a hydrodynamic oil film. This full hydrodynamic fluid film allows these bearings to support extremely heavy loads and operate at high rotational speeds. Surface speeds of 175–250 m/s (30 000–50 000 ft/min) are common. Temperatures are often limited by the lubricant used, as the lead and tin babbitt is capable of temperatures reaching 150°C (300°F).

There are many reasons a journal bearing might fail. A journal bearing will normally not fail gently, they either work or don't.

Lack of Lubricant

An unfortunately common cause of journal bearing failure occurs as it runs dry of oil, resulting in a diminished flow of lubricant to the bearings. Since the shaft rides on a thin film of oil that could be as thin as 0.0005" at the minimum film thickness, loss of lubricant (or Inadequate flow) will result in the shaft and bearing contacting. If this occurs, a tremendous amount of heat is generated due to the sliding friction between the shaft and liner which results in a major failure in a very short time. For fluid film bearings, viscosity is also the most important factor.

Typically, minimum oil film thicknesses in the load zone during operation ranges from 1.0 to 300 μ, but values of 5–75 μ are more common in midsized industrial equipment. The film thickness will be greater in equipment which has a larger diameter shaft.

Lubricant Contamination

Most contaminants carried by a lubricant will be harder than the babbitt surface, and in many cases, harder than the journal (thrust collar). Scratches from these contaminants will follow the flow path of the lubricant. Circumferential scratches would be expected from normal operation of a journal or thrust bearing, although a radial component will also be observed on a babbitt scratch in a thrust bearing. Scratches associated with lubricant exiting a hydrostatic lift pocket would generate scratches distributed outward from the pocket, although possibly with a circumferential component if the machine rotor is turning when the scratching occurred.

Wrong viscosity grade

Viscosity is the lubricant's resistance to shear. The viscosity grade required is dependent upon bearing RPM, oil temperature and load. Higher viscosity lubricants and bearings

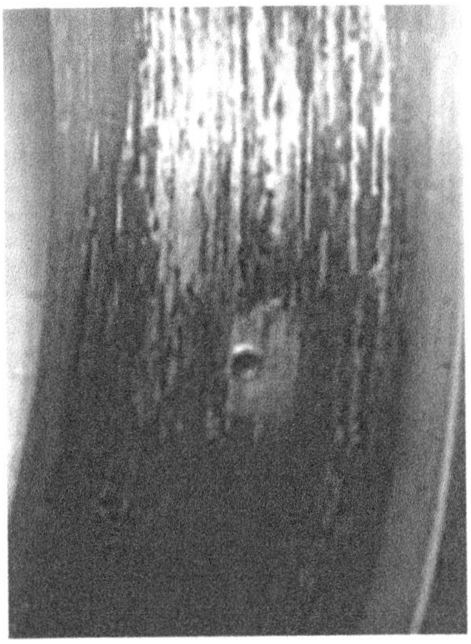

Figure 9.16 Bearing melting. *Source*: Larry Carley (2017). Engine Bearing Technology – The Spin on Spun Bearings, 25 March 2017. Babcox Media Inc.

operating under higher loads stabilize at higher operating temperatures. Viscosity varies significantly with temperature and the variation is highly non-linear. The higher the operating temperature of the unit, the higher the oil viscosity that is required. If vibration or minor shock loading is possible, a higher grade of oil than the one normally used in similar machines should be considered. If the oil selected is too low in viscosity, heat will generate due to an insufficient film thickness and some metal-to-metal contact will occur. If the oil is too high in viscosity, heat will again be generated, but due to the internal fluid friction created within the oil. Selecting an oil which is too high in viscosity can also increase the likelihood of cavitation.

Bearing melting

Heat is generated in bearings by friction in the lubricant as it forms an oil wedge. The bearing material chosen for any operation must have a melting point higher than the operating temperature. Figure 9.16 shows a bearing where the operating temperature was above the overlay material melting point. The extra friction in a section of the load zone from a high spot in the bearing will cause a localized failure. Since bearings are generally precision components, high spots are caused by a solid object lodged behind the bearing shell.

Abnormal load

Bearing wear due to a vibrating load is shown in Figure 9.17. Note the undulating wear pattern formed on the surface of the unloaded bearing half. The loaded half of this bearing was excessively worn.

Fatigue

Fatigue damage may represent itself as intergranular or hairline cracks in the babbitt. The cracks may appear to open in the direction of rotation. Pieces of babbitt may spall out or appear to be pulled away in the direction of rotation.

Fatigue can occur when conditions produce concentrated cyclic loads, such as misalignment, journal eccentricity, imbalance, bent shaft, thermal cycling, and vibration. High-bearing temperature may also be considered as a contributing factor to fatigue damage.

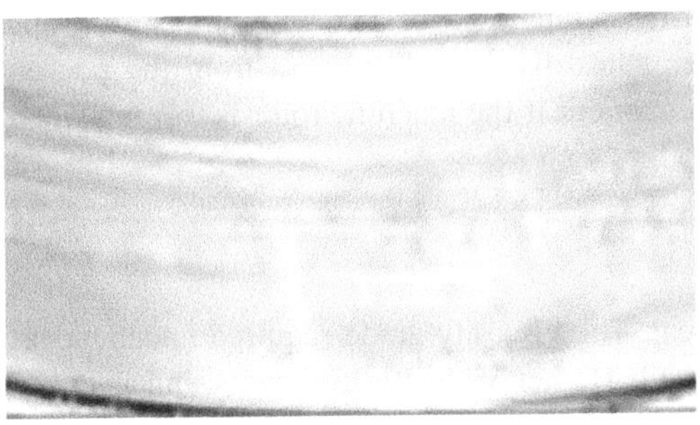

Figure 9.17 Bearing damage due to abnormal load. *Source*: Larry Carley (2017). Engine Bearing Technology – The Spin on Spun Bearings, 25 March 2017. Babcox Media Inc.

Cavitation

Cavitation damage is caused by the formation and implosion of vapor bubbles in areas of rapid pressure change. Damage often occurs due to the existence of higher velocities oil lubricating oil. Also, changes in the pressure wedge from grooves cut across the bearing caused entrained air to be released at a microlevel, causing the cavitation. Cavitation initially erodes the bearing overlay material but over a period of time will progress into the bearing material, causing eventual failure if the bearing material has a poor resistance to fatigue.

Corrosion

Corrosion damage is characterized by the widespread removal of the bearing lining by chemical attack. Corrosive materials may appear in the lubricating oil through:

- Decomposition of oil additives
- Acidic oxidation products formed in service
- Water or coolant in lube oil
- Direct corrosive contamination

Bearing housing seals, oil additive packages, and oil reservoir operating temperatures should be evaluated as an initial step in eliminating corrosion. The integrity of cooling coils should also be examined. Corrosion can be eliminated by replacing the lubricating oil. In addition, the entire bearing assembly, oil reservoir and piping should be flushed and cleaned.

Surface Corrosion – After time, aging stagnant oil and some acids may lead to corrosion of the surface of a tin-based babbitt. The resulting attach will leave small angular deficits in the babbitt, as the corrosion selectively attacks the tin, copper, or antimony. Closer inspection is required to discriminate this appearance from pitting, which would be rounded.

Erosion – Areas of very high lubricant velocity on the babbitt surface may lead to erosion, such as at the corners of a machined step or at the end of a section of wiped babbitt. Some difficulty may arise in discriminating between erosion and corrosion, which is driven by flow velocity, and cavitation, where the flow velocity leads to a change from vapor to gas of the lubricant, or some fraction of the lubricant.

Surface Displacement – A surface of layer can be disturbed due to a combination of thermal and mechanical effects.

Creep

Creep would be the surface movement of the babbitt due to a combination of high local pressure and high local temperature which exceeds the local yield strength of the babbitt. Movement of babbitt occurs from the loaded region (high material compressive stress) towards an unloaded region (lower compressive stress) (see region circled in Figure 9.18). Creep is often associated with displacement due to higher-than-expected temperatures over time.

Wiping

Wiping involves rapid heating of a region of the babbitt surface due to contact between the journal and the babbitt surface. This contact represents a local breakdown of the hydrodynamic film and results in a sharp increase in friction from the contact. The friction energy may be dissipated by displacement of some of the babbitt from regions away from the contact. While often self-correcting, wiping may become progressively more severe, Where the

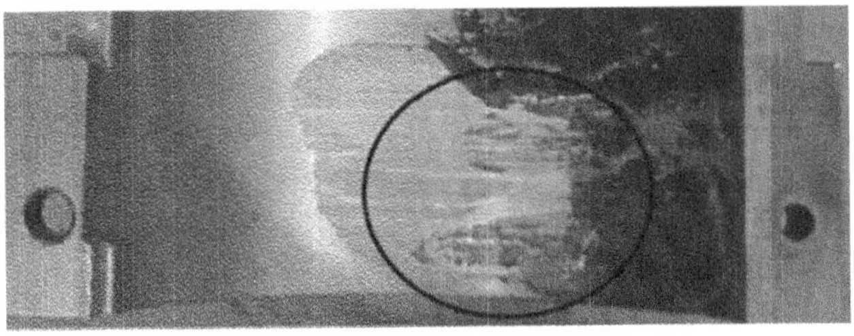

Figure 9.18 Regions of babbitt material creep in a hot-running region of a cylindrical bearing. *Source*: Branagan, Lyle A. (2015). Survey of damage investigation of babbitted industrial bearings, 1 April 2015. Lubricants, MDPI.

friction heating is due to a loss of the hydrodynamic film and operation under boundary lubrication conditions. If the wiping impedes the oil distribution slot or groove and reduces the oil flow to the hydrodynamic film seizure may occur.

Surface Displacement Can Be Termed Creep or Wiping

Occurring more often than creep, wiping represents surface displacement due to contact between the rotating journal (thrust collar) and the babbitt surface (Figure 9.19). Babbitt may be only slightly displaced or may be moved and deposited to a different region of the bearing (Figure 9.18). Movement of the babbitt occurs from the region of contact (minimum film thickness) towards a region of larger film thickness

Surface changes can arise on the babbitt which impact the local operation of the hydrodynamic film:

Tin Oxide – Tin oxide is a hard material that can form on a babbitt surface in the presence of chlorides, generally from saltwater contamination of the lubricant. This formation in a bearing generates hard regions on the babbitt surface which may build in thickness up to 100 μm (0.004″). On thrust bearings, this can disrupt the hydrodynamic film. Operation of either radial or thrust bearings with tin oxide build up can release hard particles into the lubricant which causes abrasive damage.

Varnish – Degraded lubricant, due to either excessive service time or heat, can form polar molecules which bond with the babbitt surface, typically in the hotter regions of the hydrodynamic film. In most cases, this build up, which intrudes into the hydrodynamic film, is insignificant in thickness. With heavy varnish formation, the varnish layer may continue to build in thickness or to expand in area. However, the combination of intrusion into the hydrodynamic film and poor heat transfer from the film through the varnish may disrupt the operation of heavily loaded bearings.

"Loss of Material", or Scratches

Scratches can be attributed to:

Handling Scratches – non-circumferential scratches, for example axial scratches on journal bearings, often arise during handling of the babbitted components, either during assembly or disassembly, or during shipping of the bearing. Scratches during handling may be quite deep as the damaging components may be quite large.

Cracking – cracks limited in length or number can arise from local overheating of the babbitt or initial cracking around a region of poor babbitt bond.

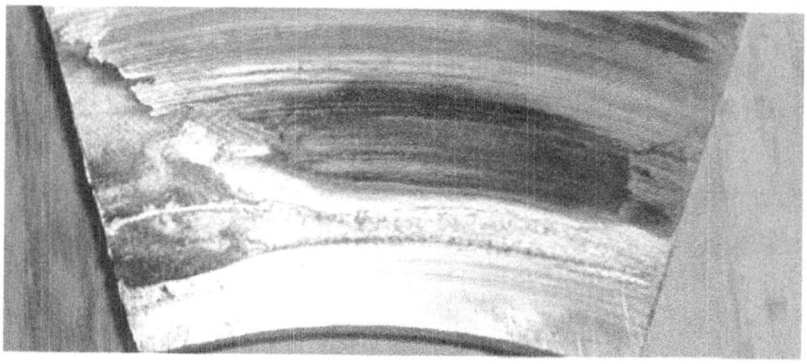

Figure 9.19 Severe wiping on a thrust shoe (circumferential scratching, narrow lines, also visible). *Source*: Branagan, Lyle A. (2015). Survey of damage investigation of babbitted industrial bearings, 1 April 2015. Lubricants, MDPI.

loss of material"- pot holes" in the babbitt can arise from several causes.

- Severe electrical damage, such as due to improper grounding of a welder, leads to a wide deep area of loss. The edges of the babbitt loss will generally be smooth rather than angular.
- A large section of babbitt may become separated due to an underlying broken bond, where inadequate bond is the underlying cause. The fracture surfaces on the remaining intact babbitt will initially be along grain boundaries. Over subsequent operation, some rounding of the fracture surfaces will occur.
- A region of fatigue cracking may eventually lead to a large loss of material as the network of cracking propagates and pieces of babbitt are released to be carried out of the bearing by the lubricant. Such fatigue damage is encountered in radial bearings where the dynamic loads exceed the static loads.

Problem with bearing installation

Try to align the journal to its bearing with uniform clearance. Errors in assembly or changes between cold setup and hot operation can lead to misalignment. This misalignment overloads one region of the hydrodynamic film with the potential for heavy polishing, wear, or cracking. Changes in machine vibration also result from misalignment.

Distortion of the structure supporting the bearing, whether thermal distortion or mechanical distortion (e.g., from condenser vacuum) can lead to misalignment of the journal (thrust collar) with respect to the babbitt surface of the bearing. This misalignment overloads one region of the hydrodynamic film with the potential for heavy polishing, wear or cracking. In radial bearings, this may appear as edge loading. In segmented thrust bearings, this may result in heavy wear on a few shoes, unless an equalizing arrangement is present.

Conclusion

If a damaged bearing goes undiagnosed, and is not replaced before it fails catastrophically, secondary damage to the machine and its components can result. The amount of time from the first (initial) damage until the bearing becomes unserviceable can vary considerably. At higher speeds, it can take a matter of seconds. In large, slow rotating machines, it can take months. The question, "When should I replace the bearing?" is best answered by monitoring the condition of the bearing.

Most damage resulting from these mechanisms can be detected and monitored easily when condition monitoring is part of a comprehensive maintenance program. Using vibration analysis, and measuring temperature the first signs of bearing damage can be detected, enabling maintenance personnel to take corrective actions in a timely manner. This can significantly reduce costly, unexpected downtime and can avoid catastrophic failures that damage adjacent components. It also enables maintenance staff to examine the damaged bearing at an early stage to determine the root cause, and take the necessary steps to prevent the problem from recurring.

10

Mechanical Seals Failure

The basics Mechanical seals are leakage control devices, which are found on rotating equipment such as pumps and mixers to prevent the leakage of liquids and gases from escaping into the environment. Figure 10.1 shows a typical centrifugal pump, which highlights its constituent parts, including the mechanical seal.

The main components of a mechanical seal are the seal rings on which a mechanical force is acting, generated by springs or bellows, and an hydraulic force, generated by the process fluid pressure. The seal ring which rotates with the shaft is called the "rotary ring;" the seal ring fixed on the casing of the machinery is called the "stationary ring." Secondary seals are required to perform static sealing between rotary rings and shafts and also between _stationary rings and the casing of the machinery. Elastomeric O-Rings are usually used as secondary seals but alternative systems can be used, as described in the following sections (Figure 10.2).

A mechanical seal consists of two principal components. One component is stationary and the other rotates against it to achieve a seal (Figure 10.3). There are many types of mechanical seal, ranging from simple single spring designs to considerably more complex cartridge seal types. The design, arrangement and materials of construction are essentially determined by the pressure, temperature, speed of rotation and product being sealed (the product media).

A simple mechanical seal design has seven components (Figure 10.4):

1) Stationary component; commonly referred to as the seat.
2) Stationary component sealing member.
3) Rotating component.
4) Rotating component sealing member.
5) Spring.
6) Gland plate.
7) Clamp ring.

A mechanical seal has four main sealing points (indicated as per Figure 10.4):

1) The seal between the rotating (3) and stationary faces (1). This is known as the primary seal.
2) The seal between the stationary member (1) and stuffing box face, i.e. Gasket (2).
3) The seal between the rotating member and shaft or shaft sleeve (4). This is known as the secondary seal and may be an *o* -ring as shown, a *v* -ring, a wedge or any similar sealing ring.
4) The seal between the gland plate and stuffing box, this is usually a gasket, or *o* -ring.

Root Cause Failure Analysis: A Guide to Improve Plant Reliability, First Edition. Trinath Sahoo.

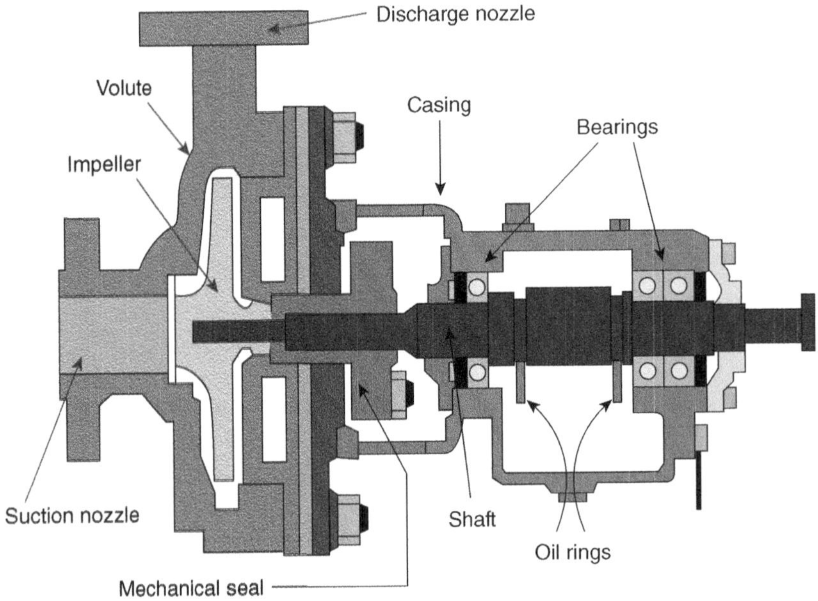

Figure 10.1 Cross section view of centrifugal pump.

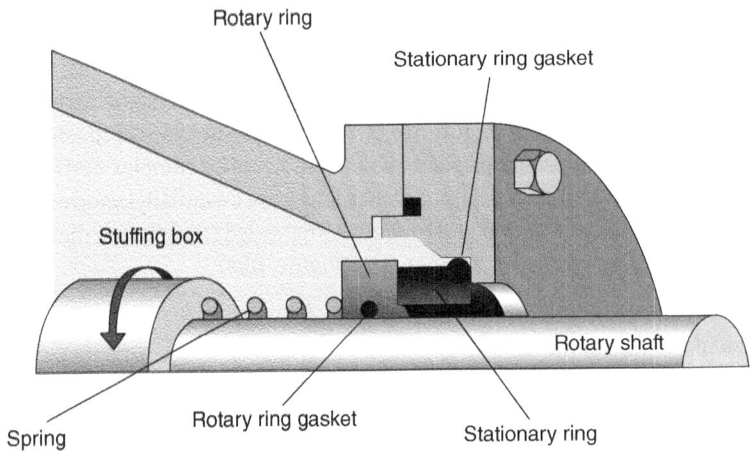

Figure 10.2 Cross section view of Mechanical seal.

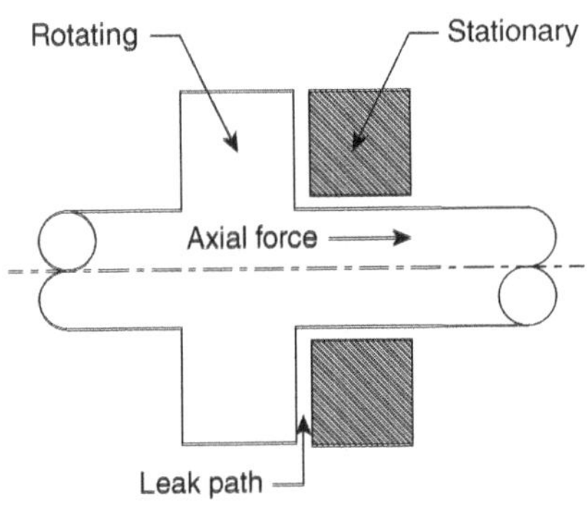

Figure 10.3 Leakage path in seal.

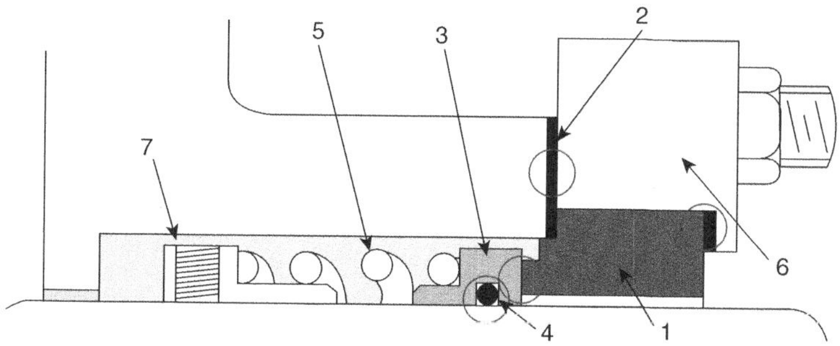

Figure 10.4 Components of Mechanical seal.

Sealing points 3 of the four main sealing points need little explanation, but consideration is required for the sealing point between the rotating and stationary components (faces). This primary seal is the basis of a mechanical seal design, and is what makes it work.

The rotating component (3) and stationary component (1) are pressed against each other, usually by means of spring force. The mating faces of both components are precision machined (lapped) to be extremely flat (usually to within two light bands, which is an optical method of measuring flatness). This flatness minimizes leakage to a degree where it is essentially negligible. In fact, there is leakage between these faces but it is minute and (for immediate consideration) appears as a vapour.

Spring compression (usually) provides initial face pressure. This pressure is maintained when the seal is at rest via the spring(s) thus preventing leakage between the faces.

If the mechanical seal faces rotated against each other without some form of lubrication they would wear out (and the seal would fail) due to face friction and the resultant heat generated. So, lubrication is required which for simplicity, is supplied by the product media. This is known as fluid film and maintaining its stability is of prime importance if the seal is to provide satisfactory and reliable service.

The liquid film in order to minimize the amount of friction between the seal rings an efficient lubrication is required. Seal faces can be lubricated by the process fluid or, with double mechanical seals, by a proper auxiliary fluid. An stable and complete layer of lubrication greatly affects the performance and the life of a mechanical seal. In order to insure good lubrication and sufficient cooling of the seal rings, the correct selection of a mechanical seal shall take into consideration the following parameters:

- Process fluid temperature;
- Vaporisation pressure at operating temperature; and
- Process fluid characteristics.

Concepts and principles above discussed are valid for all mechanical seal operating with a liquid. Dry-running seals and gas-seals operate on different principles and shall not be discussed here.

Type of Mechanical Seal

The variety of seal designs available results from the various mounting methods used, the flexibility provided, and the needs of the applications. The most common designs are pusher and nonpusherseals.

Figure 10.5 Pusher seal. *Source*: Flowserve Corporation.

A pusher-type seal (Figure 10.5) consists of a primary sealing ring assembled with an "O" ring and springs (can be one or multiple). The purpose of this is to force the sealing fluid across the face and keep it from leaking to the ID (atmospheric) side of the seal (Figure 10.6). The dynamic "O" ring is designed to move axially (be pushed) along the shaft or sleeve (in a cartridge seal). The axial movement of the dynamic secondary seal, such as an O-ring, compensate for face wear and keep the seal faces in contact. The surface underneath the dynamic "O" ring must therefore be very smooth (<32 RMS) to allow for this axial movement (hang up).Pusher type seals are used more commonly in low S.G.(<0.7) services.

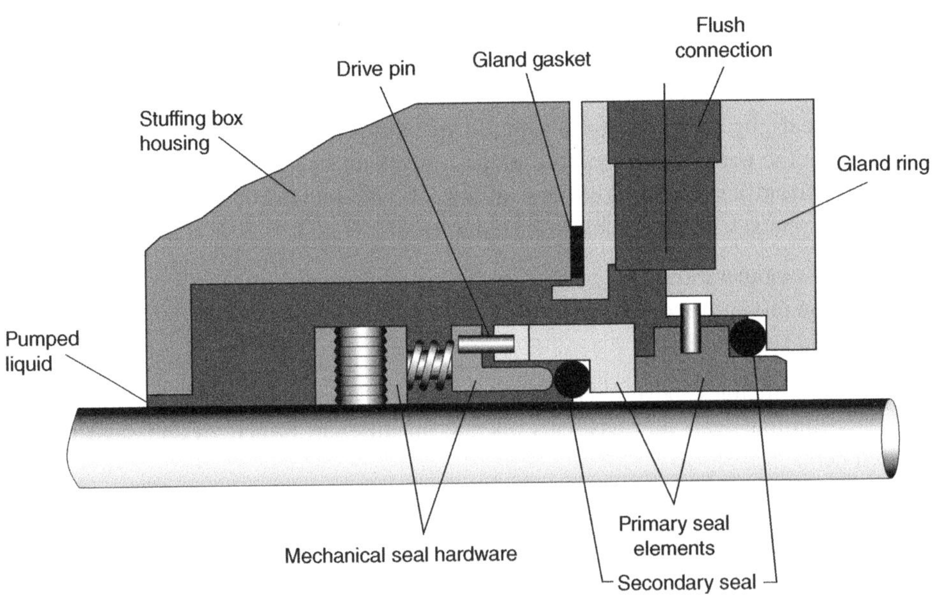

Figure 10.6 Cross section view of pusher seal.

Pusher Seal

Facts About Pusher Seal

- Closing force is supplied by the springs.
- Used in low temperature services.
- "O" ring is used as the secondary seals.
- Used in light end services (ethylene, propane, methane, butane etc.)

Non-Pusher Seal

A non-pusher type seal consists of a bellows assembly. The bellows is a component that acts as both the load element (like a spring in a pusher type) and a secondary sealing element like an "O" ring in a pusher type). Because the bellows prevents any leakage to the atmospheric side of the seal, and has a large clearance between itself and the shaft or sleeve it can move freely in the axial direction (no dynamic "O" ring), reducing the potential for hang up.

Facts About Non-Pusher Seal

- Closing force supplied by bellows (no dynamic "O" ring).
- Can be used in high temp services (metal bellows).
- Metal bellows use "grafoil" secondary seals to handle high temperature.

Bellows seal

Figure 10.7 depicts the schematic of a high-speed stationary welded metal bellows seal. This has no moving parts or sliding elastomer in contact with the shaft or sleeve and, therefore, eliminates wear on these parts. The absence of sliding elastomers eliminates seal drag or hang-up, which is the major cause of failure in conventional seals with dynamic contacting secondary seals. A bellows seal is inherently balanced and there is, therefore, no need for "stepping" the shaft or sleeve. The hydraulic balance range of the bellows seal is variable by changing the contact nose profile. The specified balance will depend upon the duty

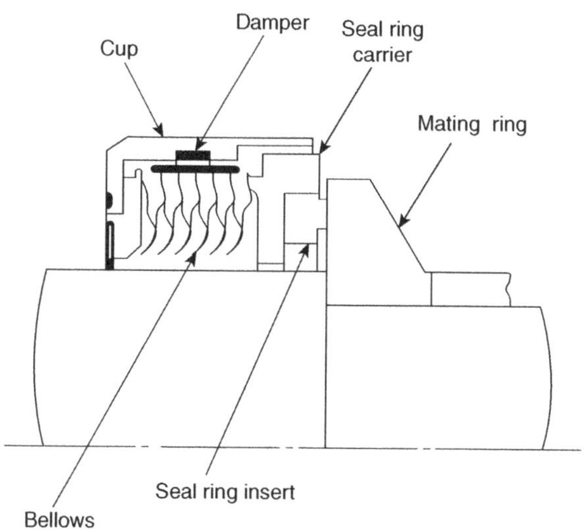

Figure 10.7 High-speed stationary welded metal bellow seal schematic.

(media, direction of pressure differential) but a typical standard product for the process industries will be valued at around 70%. Materials of construction can withstand extremes of temperature (cryogenic to hot hydrocarbon). In these designs, the static elastomer seals are replaced by graphitic packings.

In addition to improved performance, certain secondary advantages can now be claimed. The bellows seals are generally very compact in design (particularly with larger sizes). Since there is no sliding elastomer and the seals are inherently balanced, producing less heat, flushes are generally not required. In many cases, single metal bellows seals have replaced double conventional seals. It is important to remember that any system which changes the operating environment introduces an additional threat to performance reliability. Since bellows seals are of one piece construction and low profile, they are easily fitted into almost any equipment.

Application

In fact, the bellows seal is ideally suited to temperature extremes from liquid ethylene at −105 °C to bottom duties of 400 °C, and as extremes of corrosion and speed are comfortably handled by standardized "brochure" products. The bellows seal has gained reputation as the elimination of "hang-up" and for use at high temperature where the elastomeric secondary seal is eliminated altogether.

Temperature Control in seal

Controlling the temperature at the seal faces is desirable because wear is a direct function of temperature. Heat at the seal faces also causes thermal distortion, which will contribute to increased seal leakage. Many applications require some type of cooling.

The temperature of the sealing surfaces is a function of the heat generated by the seal, plus the heat gained or lost to the pumpage. The heat generated at the faces from sliding contact is the mechanical power consumption of the seal being transferred into heat. If the heat is removed at the same rate it is produced, the temperature will not increase.

If the amount of heat removed is less than that generated, the seal face temperature will increase to a point where seal face damage will occur.

Heat removal from a single seal is accomplished by a seal flush. The seal flush is usually a bypass from the discharge line on the pump or an injection from an external source.

Different methods are used to supply cool liquid to the seal chamber. When the liquid is clean, an internal flush connection can be used to cool the seal. When the liquid is dirty, an external flush can be used. This will allow the flush, a bypass from the discharge line, to pass through a filter or centrifugal separator. The seal faces will be flushed with clean, cool liquid. Increased pressure from the flush provides positive circulation and prevents flashing at the seal faces caused by the heat generation.

When a pump handles liquids near their boiling point, additional cooling of the seal chamber is required. This seal require a pumping ring and a heat exchanger. The pumping ring acts as a miniature pump, causing the liquid to flow through the outlet piping at the top of the seal chamber. The liquid passes through the heat exchanger and returns directly to the faces at the bottom inlet in the end plate. As the liquid is circulated, heat is removed from the seal and seal chamber. A closed loop system is commonly used on hot water pumps. This method is extremely efficient since the coolant is circulated only in the seal chamber and does not reduce the temperature of the liquid in the pump.

Cooling System and API Plans

A suitable cooling system should be implemented in mechanical seal system to limit the operating temperature of the seal. Many different lay-outs can be used, depending on the configuration and the required service.

API standard has supplied an exhaustive collection of flushing and pressurisation lay-outs, each intended for a specific service. The various connection lay-outs are identified by a specific number which gives the possibility to simply define all possible configurations.

Selection of Flushing Plans

Clean, Not Harmful, Neutral, Not Flammable Products

Example: Water, Vegetal oil, Glycol.

API Plan 11 or 01 is the recommended lay-out, in order to dissipate the heating produced by the seal rings and to carry out a proper venting of the stuffing box. In the case of a conical stuffing box also API Plan 02 can be used.

Fluids Crystallizing When in Contact with Atmosphere

Example: Sulphates, fosfates, saline solutions, alcaline solutions.

A single configuration is recommended, combined with API Plan 11 or 01 in order to dissipate the heating produced by the seal rings and to carry out a proper venting of the stuffing box. Implementing an additional API Plan 62 with water or steam at low pressure (max 0.3 barg), an efficient removal of crystallization deposits can be insured, preventing locking of the rotary ring.

Acid Products

A single internal seal is recommended, API Plan 11/61 or 01/61 with proper connection. In case of conical stuffing box use API Plan 02/61.

Hot Liquids

Example: Heavy hydrocarbons, diathermic oils.

Temperatures over 200 °C up to 400 °C are typical applications in refinery plants or pumps for diathermic oil. It is important to evaluate the effective operating temperature in the stuffing box. Many pumps come with a cooling system which reduces the temperature in the stuffing box, in order to avoid very expensive configurations of the mechanical seals.

The selection of the materials and the configuration will mainly depend on the operating temperature. The recommended configuration is a single internal seal, with **API Plan 02.**

A complete venting of the stuffing box is required and then the installation of a suitable system has to be verified. Implementing an additional API Plan 62 with water or steam at low pressure (max 0.3 barg), an efficient removal of crystallization deposits can be insured, preventing locking of the rotary ring.

Aqueous Solutions Prone to Solidify or Produce Sediments

Example: lime, paper pulp, slurry.

A single internal seal recommended, installed with the **API Plan 32** flushing system in order to supply a clean fluid, compatible with the process fluid for a good lubrication and cooling of the seal faces (auxiliary fluid should have a pressure higher than the process fluid).

A throat bushing, properly dimensioned, provides a barrier flushing equivalent to a pressurised system.

A valid alternative, if solid particles are in low percentage, is an **API Plan 02/62** in a conical stuffing box. A quench with water provides an efficient washing of the seal rings and cools them as well.

Toxic, Poisonous or Highly Viscous Fluids

Example: Solvent based varnishes, inks, creams, glues, lattice. The back-to-back configuration is recommended with a pressurised **API Plan 53**.

The lubrication of the seal faces is provided by the auxiliary fluid. Suitable instruments (i.e. level switch) installed on the pressurisation system can detect an eventual leakage.

Abrasive Fluids

Example: Water mixed with sand, slurries.

A double configuration is recommended with a pressurised **API Plan 54.**

The best lay-out is a stationary seal with the product outside the seal rings.

Less used but sometimes suitable is a single internal seal with **API Plan 31**, where the pumped liquid is passed through a cyclone separator and then injected into the stuffing box.

Flammable Fluids

Example: Hydrocarbons, solvents.

A tandem configuration is recommended with an unpressurised **API Plan 52**.

An auxiliary tank, complete with level and/or pressure switch can provide an efficient flushing of the seal and prevent emissions into the atmosphere.

Hot Water

Example: boiled feed water, condensate recovery.

These kinds of applications are more difficult than expected at an initial evaluation. Viscosity of water consistently decreases at high temperatures, supplying a poor lubrication of seal faces. At temperatures over 90 °C the consistency of the lubrication film is so reduced that a progressive and fast wearing of seal faces can be expected.

The recommended choice is a single internal seal with **API plan 23**. The hot water is cooled while it follows the path of a closed loop around a heat exchanger, all with the help of a suitable pumping device, typically a pumping ring. With correct dimensioning of the flushing system, operating temperatures lower than 90 °C can be obtained, insuring good performance of the mechanical seal.

Seal Failure Causes

To arrive at the true cause of seal failure it is necessary to look at the following points.

1) Review the pump's design.
2) Review the piping system and vessels.
3) Define the fluid characteristics.
4) Review of environmental control.
5) Understand the pump's operating envelope and off design performance.
6) Review the operator skill and training.
7) Review the mechanical seal installation procedure.

Pump Design Affect Seal Reliability

This is primarily necessary to determine to what approximate percentage of pump discharge pressure the seal cavity will be pressurized, if the seal is left without a flush (dead ended seal cavity). The pump needs to be reviewed for the extent of hydraulic balance designed into the impeller in single stage pumps, or if adequate facilities are provided to relieve the seal cavity from the effect of the pump discharge pressure in multistage pumps.

Typical mechanical design areas, which affect the choice of mechanical seal and/or the seal flush plan, are:

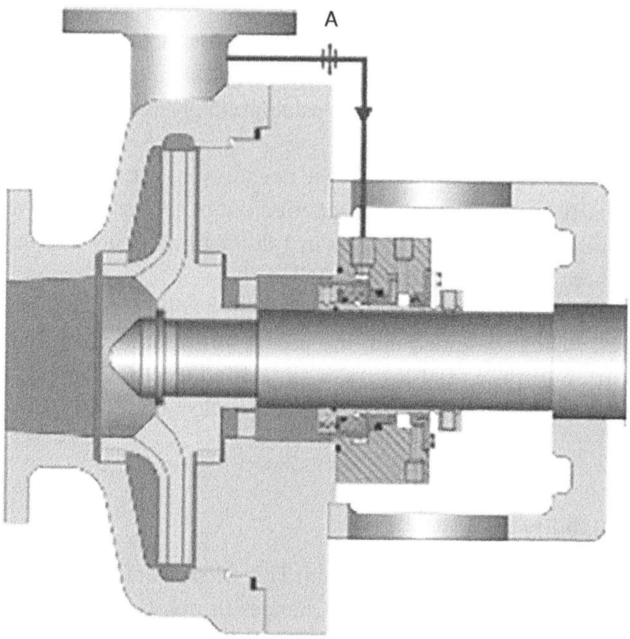

Figure 10.8 API seal Plan 11 (Recirculation from the pump discharge through a flow control orifice into the seal chamber).

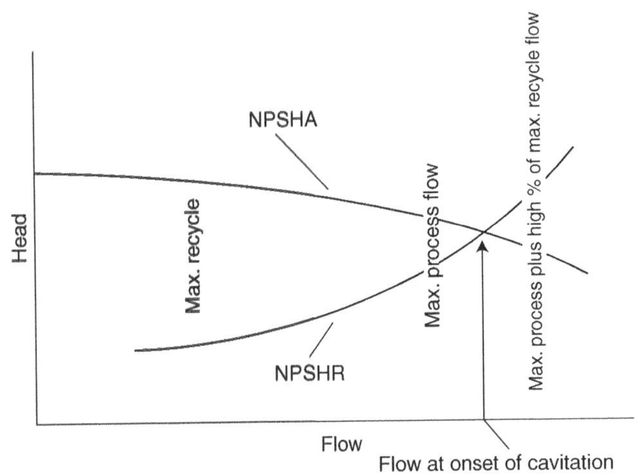

Figure 10.9 H–Q curve of a centrifugal pump.

Shaft flexibility

In single stage overhung pumps, where excessive shaft deflection under low flow conditions in volute pumps (other than concentric volute) would create a loss of perpendicularity between the seal faces.

Diffuser design

Pumps do not suffer appreciably from excessive radial thrust, regardless of flow rate.

A single volute pump will experience greater shaft deflection under low flow conditions than a double volute pump. For this reason pumps of single volute design, above 30 HP (22.4 kW), should be reviewed for the operating range and consideration given to recirculation flow control to maintain flow at or above a minimum.

The pump orientation, horizontal or vertical, can impact the need to vent the seal cavity prior to start-up. Vertical pumps, particularly after a lengthy period out of service or after a repair should always have the seal cavity vented before start-up, although this may also be required for a horizontal pump, where a throat bushing is installed.

Vertical inline pumps are very often coupled to the motor through a solid type of coupling, without any bearing support system designed into the pump. This requires that the motor bearings carry a major portion of the radial load and all of the thrust load experienced by the pump. As with horizontal pumps, the shaft will tend to deflect under low flow conditions.

Chamber Design Affects Seal Reliability

A critical part of the sealing system is the seal chamber design. The selection of the proper seal chamber can increase the life of the mechanical seal. Changes in pump design have resulted in three chambers. These are referred to as the standard bore, the enlarged bore, and the tapered bore.

Seal chambers can influence the seal environment in four basic ways:

1) Pressure
2) Solids handling

3) Vapor removal
4) Temperature

To understand how the seal chamber affects the sealing environment, it is helpful to understand each basic design:

Standard Bore. More commonly known as a stuffing box, this seal chamber was originally designed for soft packing and therefore has radial and axial dimensions based upon standard packing sizes. This seal chamber is dependent on the application of the proper piping plan to remove heat or abrasives.

Enlarged Bore. It has increased radial clearances that create a larger volume of fluid encasing the seal than does the standard bore. The increased cross-section increases the volume of the liquid for cooling. The bolt circle for the gland is also enlarged. This seal is also dependent on the proper piping plan to remove heat.

Tapered Bore. This seal chamber features the increased radial clearance of the enlarged bore; however, the bottom of the chamber is exposed to the impeller, and the walls of the seal chamber are tapered to promote self venting upon shutdown and self draining during disassembly. Internal flow within the chamber eliminates the need for external piping for cooling. For some applications, operating at a higher pressure will require an external flush.

Seal Chamber Selection Criteria

Pressure- Pressure is perhaps the most important seal and seal chamber selection criterion. To prevent flashing of the product, chamber pressure should be about 25 psi higher than the vapour pressure of the process fluid. If this pressure differential cannot be assured through pump design, a seal chamber flush should be used to provide it. This is especially important in low NPSHa applications

Solids Handling- Solids within the pumpage are a major obstacle in achieving maximum mean time between planned seal maintenance. The ideal seal chamber prevents solids from entering the seal environment and thus eliminates erosion problems from the equation. To accomplish this requires a seal chamber with a restricted throat, such as the standard bore or enlarged bore with a flush.

The flush provides a flow from the seal chamber into the pumpage and prevents solids from entering. In applications where the concentration of solids is above 10% by weight, this is the preferred method of seal environment control. The enlarged bore seal chamber is preferred over the standard bore because increased radial clearances allow any solids that may infiltrate the seal chamber to be centrifuged to the periphery of the chamber and kept away from the seal. The standard bore seal chamber cannot keep the solids away as effectively as the enlarged bore.

Vapor Removal- Entrained air or gases create problems for the mechanical seal as well. While solids are spun to the periphery of the seal chamber and cause erosion, gases accumulate around the shaft and seal preventing proper seal face lubrication. If these gases are not removed, or a flush is not provided directly on the seal faces, dry running and seal failure will occur. Venting the seal chamber prior to start-up is essential for enhanced seal life. Tapered bore seal chambers inherently provide this by design while the standard bore and enlarged bore seal chambers require manual venting through a high point.

Temperature- Research shows that enlarged bore seal chambers run substantially cooler than the standard bore seal chamber when no flush is applied. From what has been observed so far, it is clear that the restricted bore seal chambers require a seal face flush for all the conditions discussed, whether or not seal chamber temperature will allow it. In situations where the seal and seal chamber temperatures must be controlled closely, the enlarged bore seal chamber should be selected. The enlarged bore chamber has a greater fluid volume to dampen temperature fluctuations. The seal faces operate in a non-confined volume and tend not to experience drastic temperature excursions due to poor circulation as they would in a standard bore seal chamber.

Review of Suction Piping System

A suction vessel or reservoir that has the potential for a marked reduction in liquid level may cause intermittent or chronic cavitation. Where a mechanical seal shows evidence of dry running this potential cause should be considered. Fouling of process piping can also lead to cavitation, if the fouling is in the suction piping. This can include gradual surface fouling of the suction piping, as well as plugging of any installed suction strainer.

Sudden control valve opening can lead to excessive flow rate with the potential for cavitation. This is more probable where the system resistance curve is not steep and the pump curve is relatively flat. It also holds more potential for cavitation where the net positive suction head margin (net positive suction head available minus net positive suction head required) is low under normal flow rate.

Sudden suction vessel pressure reduction, typical of what can occur if a boiler feed water de-aerator receives an influx of colder water, which suddenly reduces the pressure in the de-aerator. This can result in the existing hotter water still in the long vertical section of pump suction piping reaching a pressure at pump suction at which it will vaporize, leading to temporary, sometimes severe, cavitation. Any suction vessel that contains liquid at a boiling point may be susceptible to such an occurrence.

Pump distortion caused by piping can result in misalignment of the pump/driver set, which will cause erratic seal face motion, usually caused by loss of perpendicularity of the faces and difficulty in face tracking due to increased vibration levels. Suction piping installed such that the pump inlet flow pattern is uneven can lead to cavitation, even when sufficient net positive suction head (NPSH) is available to meet the pump's net positive suction head required (NPSHR) curve. This is particularly susceptible in pumps with double suction impellers, where one side of the impeller will receive a disproportionate amount of the total flow rate.

Fluid Characteristics Affect Seal Reliability

Understanding the fluid properties that the seal will encounter is essential to the correct selection of the seal. The key fluid characteristics are: Specific gravity, Lubricity, Viscosity, Pressure, Pressure increase versus improved temperature margin, Particulates and Corrosiveness.

Low specific gravity fluids are often volatile liquids with vapour pressures above atmospheric. They may also have poor lubrication qualities, due to their very low viscosity. While very low viscosity fluids may be poor lubricants, higher viscosity fluids, typically above 32 cSt, may cause a degree of sticking of the seal faces at start-up. Such sticking can cause pull-out of part of the carbon face during the high start-up torque.

Fluid pressure refers to the pressure to be expected in the seal cavity during pump operation. While this pressure will impact the choice of seal flush plan, it will also allow the seal vendor to optimize the seal balance ratio and face geometry, particularly in very high pressure applications.

The NPSH margin is an important consideration when reviewing the need to accommodate the heat generation at the seal faces and/or any heat conducted to a seal flush in hot applications. The presence of particulates in the fluid will dictate the need for either an external flush of clean compatible fluid or the insertion of a throat device, which can evacuate the particulates to the suction side of the impeller.

Corrosiveness must be given serious consideration with respect to the choice of materials.

Since we are dealing here with liquid lubricated mechanical seals the seal faces must have a liquid film over at least part of the face contact area to achieve reasonable seal life. For single mechanical seals this is normally achieved through control of liquid pressure and temperature within the seal cavity. For dual seals the pressure and temperature of the barrier fluid, which fills the space between the inner and outer seal, is important. For the purposes of discussion it is assumed that the seal faces and components are chemically compatible with the fluid they come in contact with.

Maintaining a clean liquid at the ingress point (point of higher liquid pressure at the seal faces) requires that the liquid be maintained, in the seal cavity, at a temperature which can accommodate the heat generated at the seal faces and the heat conducted by the pump shaft and seal cavity (in hot applications). It is also imperative that the viscosity of the fluid in the seal cavity is maintained within a range that allows for adequate seal face lubrication, while being at a low enough viscosity to prevent seal face damage during start-up (face breakaway). Generally speaking, a viscosity greater than 14 cSt and less than 32 cSt would be ideal.

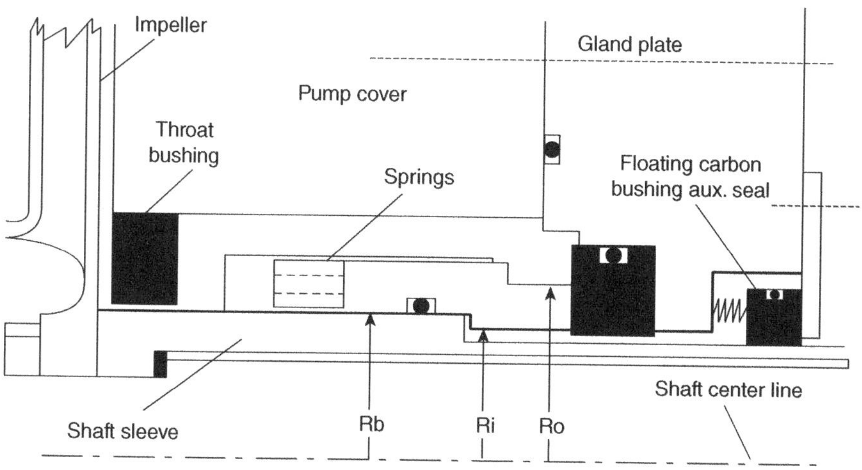

$$\text{Balance ratio} = \frac{Ro^2 - Rb^2}{Ro^2 - Ri^2}$$

Figure 10.10 Balance pressure mechanical seal schematic drawing.

Fluid Temperature Margin

To accommodate heat input the seal cavity fluid must be maintained at a temperature substantially below the temperature at which the fluid will vaporize. This is referred to as the "fluid temperature margin". The required fluid temperature margin will vary as the fluid characteristics vary; as a general "rule of thumb" a minimum temperature margin of 15 °C (27 °F) should be maintained.

There are two principal methods of achieving adequate temperature margin: (i) Cooling the flushing fluid before it enters the seal cavity. (ii) Increasing the pressure in the seal cavity so that the fluid vaporization temperature is raised.

Since increasing the pressure of the fluid in the seal cavity will also increase the heat generated at the seal faces a compromise is sometimes called for in non-cooled discharge flush applications, where increased cooling flush flow is balanced against a moderate increase in seal cavity pressure. If a high pressure increase is required in the seal cavity to obtain an acceptable temperature margin and the NPSH margin is very low, the preferred approach would normally be to increase the rate of discharge flush to the seal cavity. Where NPSH margin is more moderate and an adequate temperature margin can be achieved with a more moderate increase in seal cavity pressure, the preferred approach would normally be to increase the fluid pressure in the seal cavity. Where the situation is less clear than this the other variables help in the overall decision. Increasing the pressure of the fluid in the seal cavity is normally achieved through the installation of a close fitting throat bushing.

Pressure-Velocity limitation- As the sealing planes move relative to each other, they are affected by the actual face pressure and rotational speed. The product of the two, pressure times velocity, is referred to as PV and is defined as the power per unit area with a coefficient of friction of unity:

The PV for a given seal installation can be compared with values developed by seal manufacturers as a measure of adhesive wear.

Contacting- Seal Operating **Envelope**- Every seal has an operating envelope. The basic envelope for a contacting seal is shown in Figure 10.11. The upper limits are defined by wear

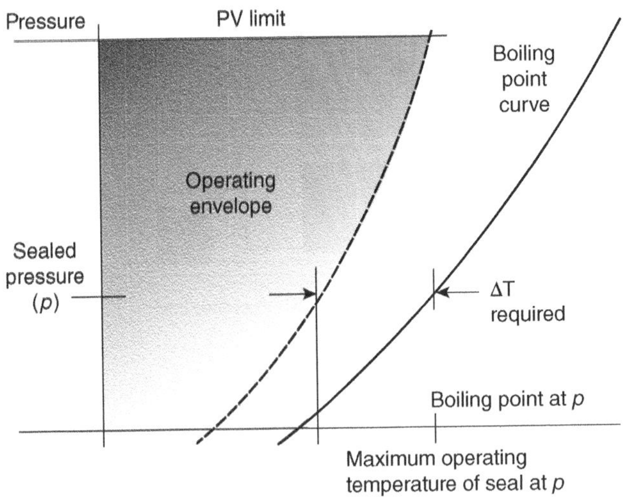

Figure 10.11 Operating envelope for a contacting seal.

of the seal faces, usually defined by a pressure-velocity limit. The fluid being sealed should be cooled so the liquid at the seal faces does not flash. Operating within the envelope will result in excellent seal performance.

Environmental Control

Apart from the temperature of the sealed fluid, considerable heat is generated at the seal interface when running and, as temperatures increase, it is necessary to consider the use of more expensive seal materials. Any increase in temperature of the fluid in the area around the seal was associated with a shorter seal life. The next step in controlling temperature is to circulate cool product fluid, usually from the pump discharge line, to the seal cavity, ensuring a continuous flow of fluid past the seal and back into the pump body. Actual circulation will depend on the seal cavity pressure relative to suction and discharge pressures, and the condition of the pump wear rings, etc. Circulation may also be used to increase pressure at the seal to prevent vaporisation of volatile fluids.

Where further sophistication is necessary, a cooler may be added to the circulation line. Other auxiliary devices, such as filters or cyclone separators, may also be included to remove solids from the circulation flow and ensure a cleaner fluid in the seal area. For more arduous duties, a completely separate, cool flushing fluid may be supplied to the seal.

Off-Design Pump Operation and Seal Performance

Proper operation means avoiding running the pump dry or with the suction or discharge valve closed, or while the pump is cavitating. If the correct size pump has been applied and installed, it will be running close to its best efficiency point (BEP). Centrifugal pumps are designed to minimize radial loading on the impeller at the BEP; however, radial forces increase as the pump is operated farther away from its BEP (more or less flow, Figure 10.12).

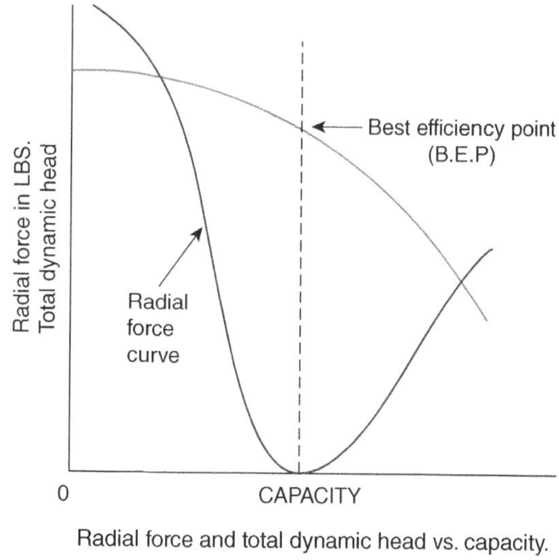

Figure 10.12 Radial force and total dynamic head vs. capacity.

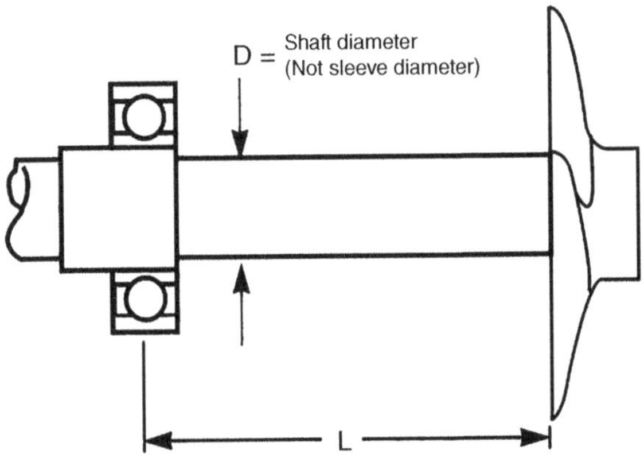

Shaft length (L) and shaft diameter (D) defined for the ratio L^3/D^4 for pump service evaluation.

Figure 10.13 Shaft length – L3/D4.

These forces can run into the thousands of pounds as the pump approaches zero flow (shut-off or deadheading). High radial loads can cause increased shaft deflection.

Many articles outline the use of the ratio L3/D4 to evaluate and compare pumps for specific service. This ratio has been called the shaft flexibility factor because the higher the number, the more flexible the shaft and the greater the deflection. It is used primarily for pumps with overhung impellers (such as ANSI pumps and most other process pumps). L is the distance from the center of the impeller to the center of the first bearing, and D is the diameter of the shaft in the seal area (Figure 10.13). If the shaft has been cut down and a sleeve used, D is the diameter of the shaft under the sleeve (not the sleeve). Obviously, the ratio will be smaller for a solid shaft, and the smaller the ratio the less the shaft deflection will be for a specific set of conditions. Although there is not an ANSI spec for the maximum allowable ratio, a number of industry engineers have suggested that a maximum ratio of 50 is preferred to achieve longer life with most seals.

Other Design Requirements

With pump and seal manufacturers working more closely, opportunities for substantial improvements appear feasible. The uptime of centrifugal pumps, the longevity of mechanical seals, and the minimization of fugitive emissions can be further improved. Studies support the following conclusions:

1) It is imperative that careful consideration be given to the pump fluid characteristics and the heat rates of mechanical seals. Ensure that the fluid in the seal chamber is always in a liquid state and capable of providing good lubrication to the mechanical seal. Past guidelines have always stressed operating 25 °F (14 °C) away from the flash temperature of the fluid in the seal chamber. This recommendation appears to be frequently overlooked or ignored, but it is still applicable.
2) Open, enlarged-bore seal chambers are more forgiving to the survival of single mechanical seals than conventional stuffing boxes during both normal and off-design pump operation. Seal face-to product boiling point margins can be increased by as much as

40 °F (22 °C) without the aid of a bypass flush. Use double mechanical seals, with barrier fluid circulating features, to achieve adequate protection to survive off-design pump conditions including low NPSH cavitation and dry-running operation independent of the seal chamber design.

3) Low NPSH cavitation conditions can cause significant damage to single mechanical seals when there is a vacuum in the seal chamber. Apply a bypass flush to reduce the seal face temperatures considerably in this situation. However, the best method is to use a close-clearance throat bushing in combination with a bypass flush to ensure a positive pressure in the seal chamber.
4) Dry-running conditions can also cause significant damage to single mechanical seals. Operating with a conventional stuffing box is less severe during dry running than with an enlarged tapered bore seal chamber. There is no known "fix" for keeping a single mechanical seal cool during dry-running operation; the only alternative is to use an external flush with a close-clearance throat bushing or use a double mechanical seal.
5) Off-design operation of typical ANSI pumps lead to vibration that may damage the seal and reduce overall long-term reliability of the equipment.
6) Metal bellows seals equipped with close-clearance vibration dampeners can tolerate vibrations from off-design pump operation many times longer than those not so equipped.
7) Dual mechanical seals can tolerate low NPSH and dry-running off-design conditions with less damage than single mechanical seals. In addition, these seals equipped with circulating features will achieve adequate protection to survive operation during low NPSH and dry-running, off-design conditions.
8) Maintaining a positive pressure in the seal chamber per API Plan 11 has beneficial effects on the performance of single mechanical seals during off-design pump operations.
9) These points, evaluating the effects of off-design pump operations on the performance of mechanical seal designs, have resulted in a seal selection guide you can use to optimize mechanical seal life. Proper application includes using an appropriate seal configuration (single, double, tandem), materials (metal parts, faces, elastomers, or other secondary seal), and environmental controls (flush, recirculation, barrier fluid system, heating/cooling) installed in a pump suitable for the service conditions (pump type, size, materials, shaft strength, bearing protection, and other appropriate options).

Train Your Operators!

Operators need training because they are the people causing the majority of your seal and bearing failures. Maintenance people are not the ones wrecking the production equipment. Untrained operators regularly run pumps dry, start them with closed suction or discharge valves, and fail to report increased noise or vibrations from the equipment they run. They do not do this to sabotage the plant; they do it from ignorance. Production "owns" the equipment; maintenance just "borrows it" when it needs to be repaired. Stop cursing your maintenance budget and start training your operators. You'll see how quickly maintenance can become a profit center for your plant instead of the ugly step-child it is often regarded as in plant budget and planning sessions.

Seal Installation, What to See

Although mechanical seal installation appears to be quite simple, a variety of factors can contribute to failure. Paying attention to details will increase the probability of a successful installation. And, if their benefits are recognized, proven procedures will become routine for the mechanics.

A successful seal installation requires the pump parameters within the manufacturer's specification. Relative movement between the seal parts or shaft sleeve usually indicates that mechanical motion has been transmitted to the seal parts from misalignment (angular or parallel), endplay, or radial run out of the pump.

Angular misalignment results when the mating ring is not square with the shaft and will cause excessive movement of internal seal parts as the primary ring follows the out of- square mating ring. This movement will fret the sleeve or seal hardware on pusher type seal designs. Angular misalignment may also occur from a seal chamber that has been distorted by piping strain developed at operating temperatures. Damage in the wearing rings can also be found here if the pump seal chamber has been distorted.

Parallel misalignment results when the seal chamber is not properly aligned with the rest of the pump. No seal problems will occur unless the shaft strikes the inside diameter of the mating ring. If damage has occurred, there will also be damage to the bushing at the bottom of the seal chamber at the same location as the mating ring.

Excessive axial endplay can damage the seal surfaces and cause fretting. If the seal is continually being loaded and unloaded, abrasives can penetrate the seal faces and cause premature wear of the primary and mating rings. Thermal damage in the form of heat checking in the seal faces because of excessive endplay can occur if the seal is operated below working height.

Radial run out in excess of limits established by the pump manufacturer could cause excessive vibration at the seal. This vibration, coupled with small amounts of the other types of motion, will shorten seal life.

To assemble a mechanical seal to a pump, a spacer coupling is required. If the pump is packed but may later be converted to mechanical seals, a spacer coupling should be included in the pump design.

Mechanical Seal Installation Procedure

Installation of mechanical seals is a seemingly simple topic that actually involves many issues, each of which must be addressed to insure success. All of these points, however, should be part of a written procedure for plant mechanics. Before a seal is installed into a pump, the mechanic must inspect the pump to ensure that it is in satisfactory condition and seal design is correct for the pump.

Step 1 – Obtain Seal Assembly Drawing: The seal assembly drawing is the document that captures the most critical aspects of the seal including the basic seal design, installation instructions, port locations, and piping plans. This document is required to develop the specific procedures that the mechanic will use to install the seal. The mechanic must obtain the seal assembly drawing before attempting to install the seal in the equipment.

Step 2 – Check Critical Dimensions: A seal OEM will do the seal selection on the information that they received as part of the order. Some of the most critical aspects of this information are the pump seal chamber dimensions and specific pump features around the seal chamber. The seal assembly drawing displays the dimensions and features that were used as the basis for the seal design.

Arguably, the most critical and precise dimensions required for a seal installation are the shaft outer diameter (OD) and the pilot diameter for the seal gland. Other critical dimensions include the seal chamber bore, the seal chamber depth, and the distance to the first obstruction. There is a tendency to get this dimensional information from the original pump

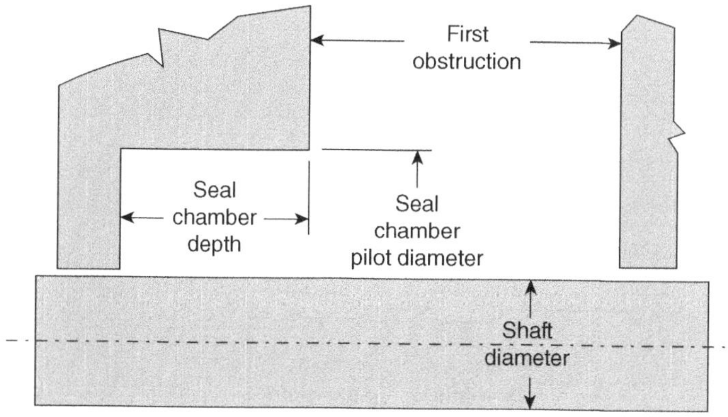

Figure 10.14 Critical pump dimensions.

OEM drawings, references to "identical" pumps, or from older seal drawings. While these sources are usually reliable, there may be applications where the original pump design was modified to accommodate a larger seal (e.g. increase seal chamber bore), include new pump features (e.g. add a port in the seal chamber), or repair damage (e.g. re-machine the seal chamber face). Using the seal assembly drawing as a guide, the mechanic must confirm that all of the dimensions on the pump match the dimensions on the seal assembly drawing.

Step 3 – Confirm the Location of any Steps, Grooves, Keyways, Internal Ports, or Passageways: The critical features of the seal chamber and surrounding area are not limited to simple dimensions. Many pumps are designed with features which are required for the pump operation or which provide additional benefits for the seal. Many pumps are designed with steps on the OD of the shaft. These steps are often used to provide specific dimensions for interfacing with other pump components. These steps may occur in the area of the seal chamber and the seal OEM must design the seal to either accommodate the step or avoid it. Steps may also occur in the bore of the seal chamber. These may have been added to accommodate earlier seal designs or repair seal chamber damage.

Step 4 – Confirm Coupling Space In some pump designs: (e.g. vertically suspended pumps and between bearing pumps) the seal can be installed into the pump without the need to remove the pump or driver (motor). This greatly simplifies the effort required to service the pump and minimizes the need to realign the pump and driver shafts after maintenance. In these cases, the seal assembly must be designed to fit between the ends of the motor shaft and the pump shaft after the coupling has been removed. This spacing is referred to as the coupling space or distance between shaft ends (DBSE). Prior to attempting to install the seal, the mechanic should confirm that the overall length of the seal is less than the DBSE.

Step 5 – Confirm Location of Brackets and Windows: The seal assembly contains a number of ports which connect the seal gland to piping or tubing from the seal piping plan. The pump however also has a number of brackets or braces to connect the fluid end to the hydraulic end of the pump. The seal assembly must be designed so the seal piping does not interfere with the pump brackets. The mechanic can confirm the location of the port and brackets from the seal assembly drawing.

Step 6 – Measure the Equipment Condition: One of the fundamental assumptions of good pump reliability is that the pump provides a good foundation for the seal. A pump that operates with minimal shaft movement and deflections and with proper alignment will

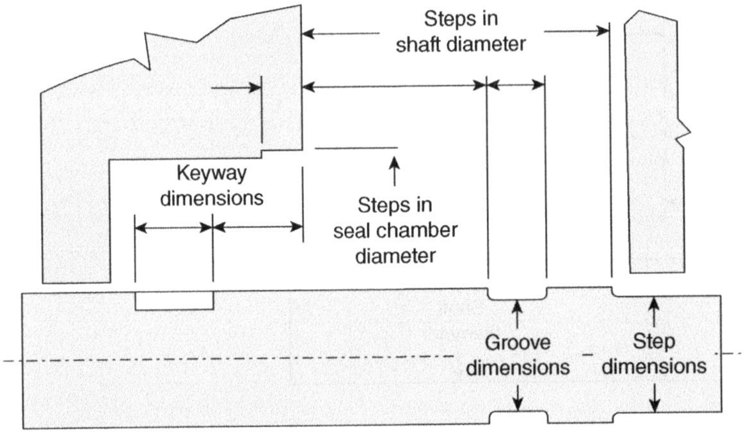

Figure 10.15 Critical pump features.

minimize the motion in the mechanical seal. This will result in lower seal wear, lower leakage, and longer seal life. These alignments and runouts can be broken into several categories.

Concentric runout: While the pump shaft is generally located in the center of the seal chamber, it cannot be assumed that the shaft is perfectly concentric with the seal chamber bore or pilot diameter. A pump is an assembly of many components and the inevitable tolerances, clearances, and assembly variations virtually ensures that the shaft and seal chamber will not be perfectly concentric.

Pump and seal OEMs recognize this and have developed standards for acceptable concentricity. API 682, API 610 and ASME B73 require that the concentricity must be less than 0.005″ (0.125 mm) TIR. This is measured by placing the dial indicator on the shaft and measuring the runout on the OD or ID pilot on the seal chamber of the pump.

Perpendicular runout – Perpendicular runout is the perpendicular deviation of the seal chamber face relative to the centerline of the rotating shaft. Like concentric runout, perpendicular runout is the result of multiple components, manufacturing tolerances, and clearances between the assembled pump components. Perpendicular runout however can have a greater impact on seal performance since some seal designs must continually compensate for this runout through continuous axial motion of the seal components. This will result in high wear on seal components as well as erratic and increased leakage. Pump and seal OEMs again recognize the need to limit this type of runout to a minimum practical value. API 682

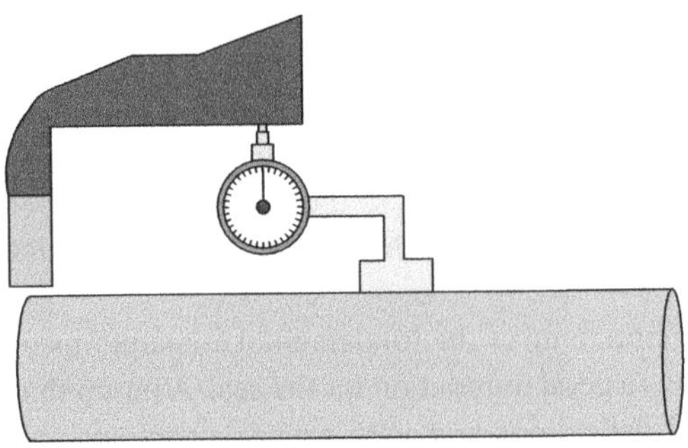

Figure 10.16 Concentric runout.

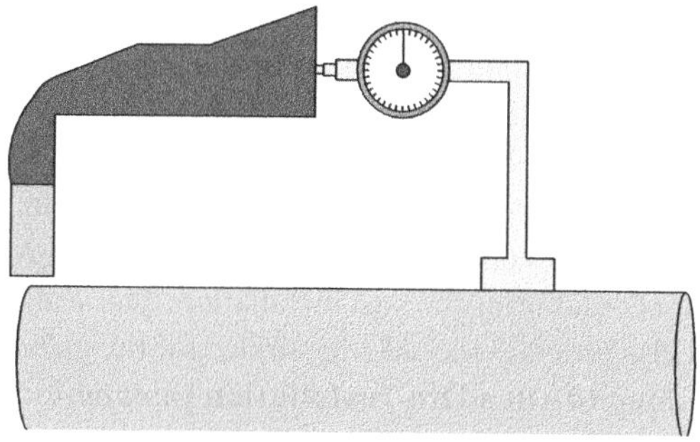

Figure 10.17 Perpendicular runout.

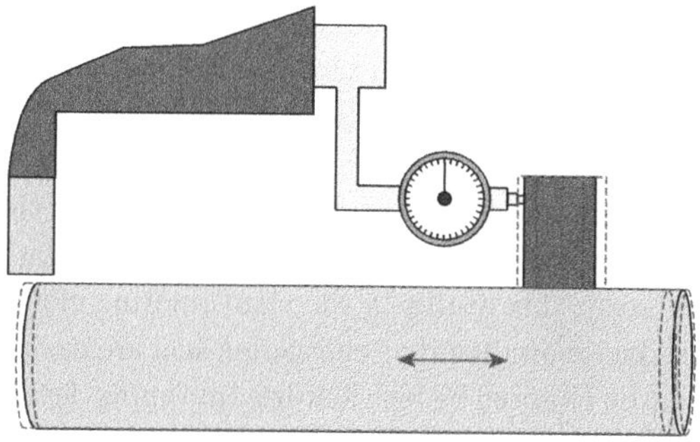

Figure 10.18 Shaft end float.

and API 610 require the perpendicular runout to be less than 0.0005″ per inch of seal chamber bore (0.5 μm/mm) TIR. This is measured by placing the dial indicator on the shaft and measuring the runout on the face of the seal chamber.

Bent shaft runout – Bent shaft runout refers to an out-of-concentric condition where the deviation is constant relative to a point on the OD of the shaft. This can be measured by placing the dial indicator on the seal chamber and measuring the TIR on the OD of the shaft during one full revolution. Different pump designs and standards address this type of runout. The end user should refer to the equipment OEM for details on allowable bent shaft runout.

End float or axial play – All pump shafts are located axially in the pump by means of a thrust bearing. The axial location is required to ensure that the impeller remains in the correct location in the pump casing. Axial motion can also have a negative impact on seal and coupling performance. In most pump designs, the pump hydraulics create a net axial thrust in one direction so shaft motion only incurs as the pump ramps up to steady-state operation. If, however, a pump does not create a net thrust or the direction of the thrust changes, the shaft can move continuously back and forth resulting in additional wear on the seals and other pump components.

Step 7 – Inspection of the Seal Chamber: Dimensions and runouts are not the only considerations when installing a seal. The areas around the seal chamber must also be inspected and prepared for the seal installation. Some of these areas are obvious while others may be unique to only specific pump designs.

Step 8 – Check the Shaft for Burrs or Sharp Edges: Even a shaft which has been machined to the correct dimensions may not be ready for the seal installation. The seal sleeve is intentionally designed with a very small clearance between the shaft OD and the sleeve ID. This close clearance fit is required to minimize seal runout and improve seal performance. It also requires that the surface of the shaft is free from any defects or burrs which may cause the sleeve to hang-up during installation.

Step 9 – Check the Seal Chamber Face for Damage: Any scratches, gouges, or other damage which results in a raised surface will prevent the seal gland from properly contacting the seal chamber during installation. This can result in a cocked seal gland with the same impact as an out-of-perpendicular alignment. Any raised surfaces or damaged areas must be stoned flat prior the seal installation. The seal chamber face not only locates the seal gland, it also provides the sealing surface for the gland gasket.

Step 10 – Read the Installation Instructions and Specific Installation Requirements from the Seal Assembly Drawing. Many mechanics have significant experience with pumps and seals in their plants. Many of these mechanics are confident that they do not need to review the instructions or documentation that comes with the seal. In some cases, this practice has led to seal failures that could have been easily avoided. The installation instruction can provide specific details which are impossible to check without looking at the drawing and instructions.

Ensure that the direction of rotation of the shaft matches the direction of rotation of the seal – Between bearing pumps will have seals at both ends of the pump casing. When facing the seals, this results in the shaft rotating clockwise on one end of the pump and counter clockwise on the other end. Some seal are designed for a specific direction of rotation. This may be required by face features, pumping devices, or seal porting. Ensure that the direction of the seal matches the direction of rotation of the end of the pump where it will be installed.

Conclusion

There are many different pump designs and a wide variety of seal options which may be applied in the field. In all of these cases the mechanical seal reliability depends on pump design, pump operation seal selection, installation etc. Since the mechanical seal vendor or OEM, who may be asked to replace or repair the seal, is seldom cognizant of the potential conditions that prevailed preceding failure, it is asking much of them to provide a conclusive diagnosis of failure. Much of this analysis, therefore, rests with the pump user, who may work with the seal vendor/repairer in determining the root cause.

11

Centrifugal Pump Failure

In a centrifugal pump, fluid enters into the casing, falls on the impeller blades at the eye of the impeller, and is whirled tangentially and radially outward until it leaves the impeller into the diffuser part of the casing. While passing through the impeller, the fluid is gaining both velocity and pressure. The two main parts that are responsible for the conversion of energy in a centrifugal pump are the impeller and the casing. The impeller is the rotating part of the pump and the casing is the airtight passage which surrounds the impeller. The other components in a pump are shaft, coupling, bearing, seals and driver.

Inability to deliver the desired flow and head is just one of the most common trouble encountered during operation. There are other many conditions in which a pump, despite suffering no loss in flow or head, is considered to have failed and has to be pulled out of service as soon as possible. These include seal related problems (leakages, loss of flushing, cooling, quenching systems, etc.), pump and motor bearings related problems (loss of lubrication, cooling, contamination of oil, abnormal noise, etc.), leakages from pump casing, very high noise and vibration levels, or driver (motor or turbine) related problems.

Many a time the root causes of failure are the same but the symptoms are different. A little care when first symptoms of a problem appear can save the pumps from permanent failures. Thus the most important task in such situations is to find out whether the pump has failed mechanically or if there is some process deficiency, or both. Thus the decision to pull a pump out of service for maintenance/repair should be made after a detailed analysis of the symptoms and root causes of the pump failure. Many a time the trouble get resolved just by changing operating parameters. Hence a thorough understanding of pump operating data, maintenance data and selection plays important role in reliability of pumps.

Pump Failure Causes

Pump reliability largely depend upon the selection process, how far it operate from the best efficiency point (BEP), how its net positive suction head required (NPSHR) compares to the worst case suction head available, the design of the piping hydraulics in the suction and discharge piping close to the pump, and the manner in which piping nozzle loads will be accommodated etc. Whether or not the ideal pump has been selected, installation must be performed in a manner that avoids hot misalignment, soft foot issues, and thermal bowing of the casing. Once the pump is installed as part of an overall process or system, it is shown how proper start up and steady operation procedures will avoid binding the pump rotor due

Root Cause Failure Analysis: A Guide to Improve Plant Reliability, First Edition. Trinath Sahoo.

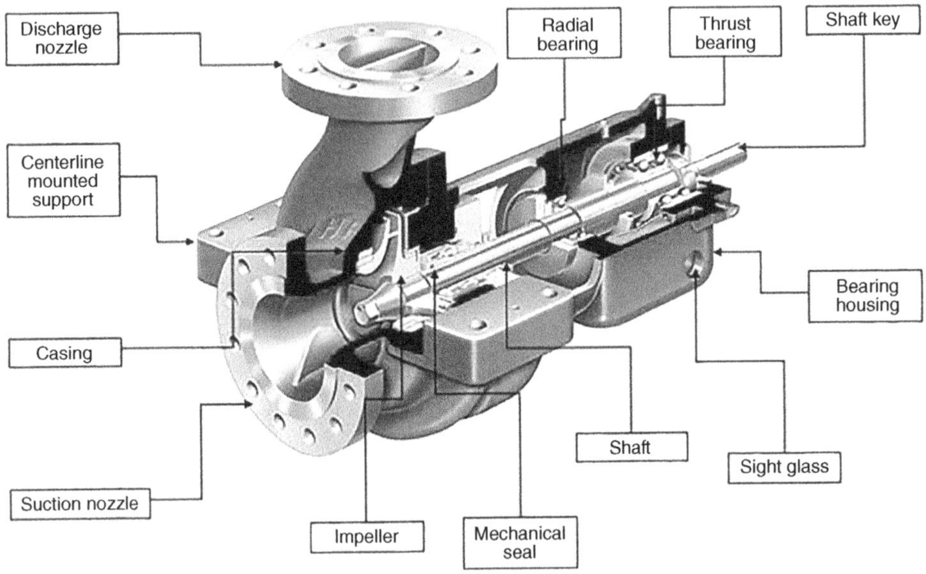

Figure 11.1 Cross section view of centrifugal pump.

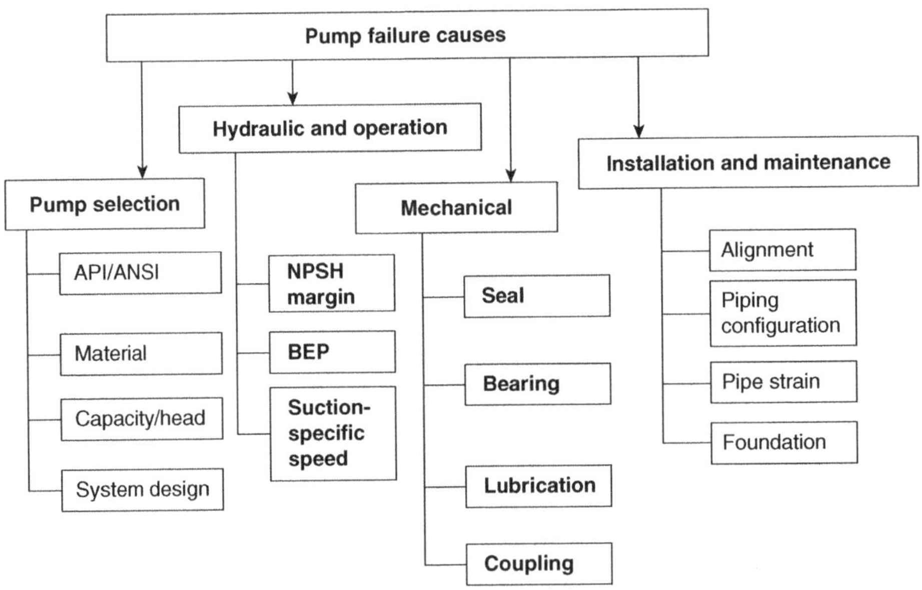

Figure 11.2 Pump failure causes.

to temperature differential within the rotor or casing. Figure 11.2 summarizes the many factors that may cause pump failure.

Pump Selection

The selection of a pump start with development of a good specification for buying the pump. It is the job of the machinery engineer to review the standard data sheets for accuracy and completeness. The various API specification clauses stipulate minimum requirements and even at that the American Petroleum Institute (API) informs industry that these are not mandatory compliance items. The machinery engineer should review and use his previous experience with equipment in similar service under similar operating conditions to have a better specification for buying the pump.

Pump selection starts with a basic knowledge of system operating conditions, fluid properties, pressures, temperatures, and system layout. The other important factors that affect the performance of a centrifugal pump are Net Positive Suction Head (NPSH) and Cavitation, Vapour pressure of the working fluid. Also before choosing a pump first decide the capacity and head you will need. You must also consider if this capacity will change with the operation of your process or not. You're going to calculate how much pressure will be needed to deliver different capacities to the location where you'll need them. Calculate friction resistance in the lines, fittings and any valves or hardware that might be in the system. As an example: high-pressure nozzles can be tricky, especially if they clog up. This resistance is called the friction head.

The viscosity of the fluid is another consideration because it'll affect the head, capacity, efficiency and power requirement of the pump. There are some viscosity corrections you can make to the pump curve when you pump viscous fluids.

Unfortunately, oversizing a pump increases the cost of operating and maintaining a pumping system and creates a different set of operating problems – including excess flow noise, inefficient pump operation, and pipe vibrations. The energy cost alone of using an oversized pump is substantial.

A key to improving system performance and reliability is to fully understand system requirements (peak demand, average demand, and the variability of demand). Problems with oversized pumps often develop because the system is designed for peak loads, while normal operating loads are much smaller. Excess flow energy is then forced into the system. In addition to increasing operating costs, this excess flow energy creates unnecessary wear on components such as valves, piping, and piping supports. The previous experience also makes it easier to review special component requirements, seal types, metallurgy, construction features, etc.

Remember that with a centrifugal pump if you change its capacity you change the pressure (head) also. But a boiler feed pump is an example of an application that needs a constant pressure with varying capacities to meet a changing steam demand. The demand for boiler water is regulated by opening and closing a control valve on the discharge side of the pump with a discharge re-circulation line returning the unneeded portion back to a convenient storage place, or the suction side of the pump. How are you going to vary the pump's capacity? Are you going to open and close a valve or maybe you'll be using a variable speed drive, or maybe a gasoline or diesel engine. Will the regulating valve open and close automatically like a boiler feed valve, or will it be operated manually? The variable speed motor might be an alternative if the major part of the system head is friction head rather than static or pressure head.

Material

The pump manufacturer will try to choose pump metal components that are chemically compatible with what you're pumping, as well as any cleaners or solvents that might be flushed through the lines. If the temperature of the pumpage changes the corrosion rate can change also. The choice of materials could have a serious effect on your spare parts inventory. Will he be selecting universal and easily obtainable materials? Unless you have a great deal of experience with the product you're pumping, do not select the metal components by using a compatibility chart. There is no material that will with-stand attack from all combinations of liquids and solids found in pumped solutions.

Some of the corrosion problems that can occur in pumps, often the result of a service misunderstanding that could be avoided if the proper information is accurately determined in the initial stages of material selection. Some pump material selection guidelines developed by OEM.

Shaft Slenderness Ratio

The ratio L3/D4 of the shaft (L- length, D- diameter) is very important for reliability. This ratio will have a major effect on the operating window of the pump and its initial cost. The lower the number the better, but any thing below 60 (2 in the metric system) is acceptable when you're using mechanical seals. A low L3/D4 can be costly in a standard long shaft pump design because it dictates a large diameter shaft that is usually found only on expensive heavy-duty pumps. We often run into L3/D4 problems when you specify, or the pump supplier sells you a low cost, corrosion resistant sleeve, mounted on a steel shaft rather than a more expensive solid, corrosion resistant shaft.

Hydraulic Requirement

Pump life improvements can be achieved through proper pump hydraulic selection. There are four basic hydraulic selection factors that can have a significant effect on pump reliability. These are pump speed, percent of best efficiency flow, suction energy and NPSH margin ratio.

The "Mean Time Between Repair" (MTBR) and Life Cycle Cost of most centrifugal pumps can be improved if slower pump speeds are used, and pumps are selected to operate in their preferred operating range (70–120% of BEP flow rate).

The flow remains well controlled within a range of rates of flow designed as the preferred operating region (POR). Within this region, the service life of the pump will not be significantly affected by the hydraulic loads, vibration, or flow separation. The POR for most centrifugal pumps is between 70 and 120% of BEP. Design characteristics for both performance and service life are optimized around a rate of flow near the best efficiency point (BEP). At around BEP, the hydraulic efficiency is maximum, and the liquid enters the impeller vanes, casing diffuser (discharge nozzle) or vaned diffuser in a shock less manner. Flow through the impeller and diffuser vanes (if so equipped) is uniform and free of separation, and is well controlled.

Speed affects pump component wear and NPSH requirements, along with the head, capacity, and the pump size. High speed pumps cost less initially, but the maintenance costs can be staggering. Speed is especially critical if you're going to be specifying a slurry pump.

Further, the mean time between repair of high and very high suction energy pumps can be increased by keeping the NPSH margin above certain minimum levels and/or by reducing the suction energy level. The easiest way to lower the suction energy and increase the NPSH margin of a pump application is by lowering the speed of the pump.

NPSH Requirement

The pump manufacturer requires a certain amount of net positive suction head required (NPSHR) to prevent the pump from cavitating. He shows that number on his pump curve. When you look at the curve you'll also note that the net positive suction head required

(NPSHR) increases with any increase in the pump's capacity. You'll also be calculating the net positive suction head available (NPSHA) to be sure that the pump you select will not cavitate. Cavitation is caused by cavities or bubbles in the fluid collapsing on the impeller and volute. In the pump there are different types of cavitation.

- Vaporization cavitation.
- Internal recirculation cavitation.
- Flow turbulence cavitation.
- Vane Passing cavitation.

In some cases you can reduce the NPSH required. This is especially true if you're pumping hot water or mixed hydrocarbons. You may have to install an inducer on the pump, add a booster pump, or go to a double suction pump design if you don't have enough net positive suction head available (NPSHA). This need to be given lot of thought during selection process.

NPSH Margin- High and very high suction energy pumps that operate with the minimum NPSH margin values may be susceptible to elevated noise levels and erosive damage to the impeller. Additional NPSH margin may be needed to cover uncertainties in the NPSHa to the pump or operating flow point. If a pump runs farther out on the curve than expected (which is very common), the NPSHa of the system will be lower than expected and the NPSHr for the pump will be higher, thus giving a smaller (or possibly negative) actual NPSH margin. All pumping systems must be designed to have adequate positive margin throughout the full range of operation.

Impeller Selection

There are multiple decisions to be made about the impeller selection. The impeller shape or specific speed number will dictate the shape of the pump curve, the NPSH required and influence the efficiency of the pump. The suction specific speed number of the impeller will often predict if you're going to experience a cavitation problem.

The impeller material must be chosen for both chemical compatibility and wear resistance. The selection will also be influenced by how far you'll venture off the best efficiency point (BEP) on the capacity side of the pump curve.

Suction Specific Speed

The concept of suction specific speed (*Ss*) must be considered by the pump designer, pump application engineer, and the system designer to ensure a cavitation-free pump with high reliability and the ability to operate over a wide flow range.

$$Ss = \frac{N \times Q0.5}{(\text{NPSHR})0.75}$$

where

N = pump rpm

Q = flow rate in gpm at the best efficiency point

NPSHR = NPSHR at Q with the maximum impeller diameter

The system designer should also calculate the system suction specific speed by substituting design flow rate and the system designer's NPSH. The higher the value of *Ss*, the sooner the pump will cavitate when operating at flows below the BEP'. Therefore, before an acceptable value of *Ss* be determined, the process system and pumped liquid characteristics must be defined. For a low-maintenance pump system, designers and most user specifications require, or prefer, *Ss* values below 10 000–12 000.

As a general rule, the higher the suction specific speed, the higher the minimum stable flow capacity will be. If a pump is always operated at its best efficiency point, a high value of *Ss* will not create problems. However, if the pump is to be operated at reduced flow, then the *Ss* value must be given careful consideration.

ANSI vs. API – Which Specification to Follow

It is generally easier to maintain alignment of API pumps since their supports are surrounded by the typically moderate-temperature ambient environment. The decision on API vs. ANSI construction is experience based. Unlike API pump bearings, which petrochemical companies often specify for an L10 life of 40 000 h, ANSI pump bearings are selected on the basis of an expected 24 000-hour life.

Compared to an API pump, the typical ANSI pump exhibits following features which affect pump reliability.

- A thinner casing, i.e. less corrosion allowance.
- Reduced permissible nozzle loads. It is even more sensitive to pipe-induced stresses than the API pump.
- Smaller stuffing box size. Unless a large bore option is chosen, an ANSI pump may not be able to accommodate the optimum mechanical seal for a given service.
- Impellers designed and manufactured without wear rings.
- Foot mounting, whereas the API pump will be centreline mounted,
- The average ANSI-design baseplate does not always provide adequate structural integrity.
- And load-bearing capability.
- Shaft deflection is often excessive.
- Bearing life is shorter than it could be.

Failure Due to Operational Reasons

Hydraulic Disturbance Due to Capacity

Centrifugal pumps should be operated as much of the time as possible at or near the manufacturer's design rated conditions called the best efficiency (BEP). Each centrifugal pump has a best efficiency point (BEP) at which its operating efficiency is highest and its radial bearing loads are lowest. At BEP, a pump operates most cost-effectively in terms of both energy efficiency and maintenance. In reality, continuously operating a pump at its BEP is difficult because systems usually have changing demands. Any pump operated at excess capacity, beyond BEP and lower head, will vibrate, causing bearings and shaft troubles and require excess power. When operating at reduced capacity, less than BEP and higher head, the incorrect vane angles cause eddy flows in the impeller and casing and will result in erosion of the impeller, casing and wearing rings. The radial load on the rotor increases, causing higher shaft stresses, increased shaft deflection, bearing troubles and accelerated deterioration of wearing rings.

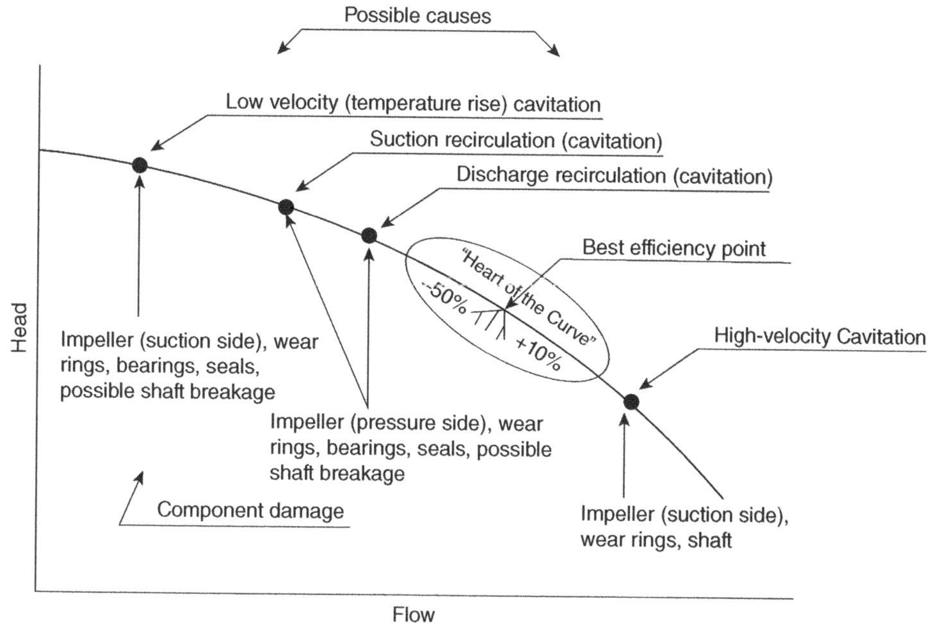

Figure 11.3 Centrifugal pump component damage and causes as a function of operating point.

Centrifugal Pump Hydraulic Disturbances

The Limits of the Centrifugal Pump Curve

The centrifugal pump curve has high flow and low flow limits which can cause significant mechanical damage to the pump if not avoided. At the low flow end of the curve, flow recirculation can damage a pump while at the high flow end excessive NPSHR, horsepower and choke flow can result in mechanical damage to impellers, casing, shaft, bearings and seals. Each of these factors is discussed below.

Low Flow Operation

As we examine these factors we can see that oversizing a centrifugal pump will result in low flows through the impeller at part loads. A portion of the flow will reverse itself and set up turbulence as it re-enters the impeller. The abrupt change in direction and very high acceleration can result in cavitation on the back side of the impeller vane (refer to Figure 11.4). As a result of oversizing an impeller, significant effects in performance and mechanical reliability can be experienced.

Pumps are designed to operate at minimum radial thrust loads at best efficiency point. Low flow operation results in high radial loads which can cause premature bearing failures unless bearings are selected to accept these higher loads in anticipation of operation at low flows. Pressure surges and flashing of the liquid can also occur at low flows. This can cause loading and unloading of the mechanical seal faces which can result in a seal failure. Operating at low flow can also lead to shaft deflection which can cause bearing and seal damage. Depending on the fluid being pumped, low flow operation can result in a high temperature rise through the pump because the amount of energy absorbed by the liquid is low compared to that absorbed by friction losses.

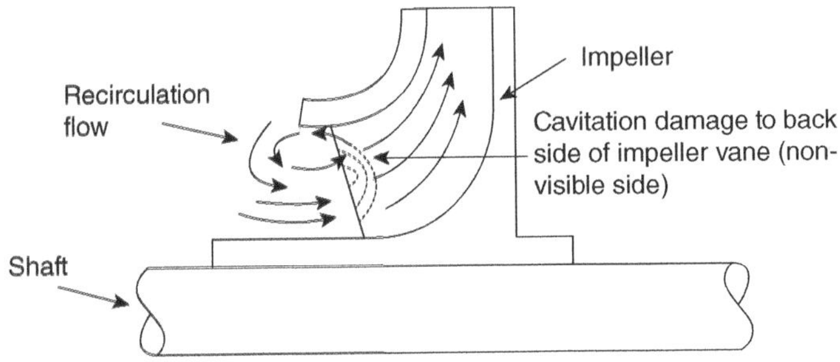

Figure 11.4 Recirculation flow pattern in impeller at low flows.

The following broadly classed factors have been recognized as the major contributors to low flow pump problems.

Hydraulic Instabilities- When the pump is operating significantly below the BEP, flow streamlines (that is, patterns) within the pump change considerably from the rated design streamlines. Fluid eddies are most likely to develop at the inlet and discharge of the impeller resulting in flashing, cavitation, and shock waves that often produce vibration and serious component erosion. This phenomenon is classically known as internal recirculation. It can occur at the pump inlet (suction) and discharge.

Recirculation- Recirculation occurs at reduced flows and is the reversal of a portion of the flow back through the impeller. Recirculation at the inlet of the impeller is known as suction recirculation. Recirculation at the outlet of the impeller is discharge recirculation. Suction and discharge recirculation can be very damaging to pump operation and should be avoided for the continuous operation of pumps of significant energy level or pressure rise per stage.

Every pump has a point where recirculation begins Recirculation is caused by oversized flow channels that allow liquid to turn around or reverse flow while pumping is going on. This reversal causes a vortex that attaches itself to the pressure side of the vane. If there is enough energy available and the velocities are high enough, damage will occur. Suction recirculation will produce the loud crackling noise in and around the suction of the pump. Recirculation noise is of greater intensity than the noise from low-NPSH cavitation and is a random knocking sound.

Suction recirculation is reduced by lowering the peripheral velocity, which in turn increases NPSH. To avoid this it is better to recognize the problem in the design stage and opt for a lower-speed pump, two smaller pumps, or an increase in NPSHA.

Discharge recirculation is caused by flow reversal and high velocities producing damaging vortices on the pressure side of the vane at the outlet. The solution to this problem lies in the impeller design. Discharge recirculation will produce the same characteristic sound as suction recirculation except that the highest intensity is in the discharge volute or diffuser.

The onset flow of recirculation increases with increasing suction specific speed. The higher the value of *Ss*, the sooner the pump will cavitate when operating at flows below the BEP'. Therefore, before an acceptable value of *Ss* be determined, the process system and pumped liquid characteristics must be defined.

Mechanical Loads- As the flow rate through the pump decreases, steady state loads increase and superimposed dynamic cyclic loads appear radially and axially on the impeller and shaft. The dynamic cyclic component increases significantly when recirculation within the pump occurs. Bearing damage, shaft and impeller breakage, and rubbing wear on casing, impeller and wear rings can occur. A bypass system may be necessary not only to reduce component loading and stress but also to prevent motor overload.

Thermal Factors- At low flow conditions, centrifugal pumps are very inefficient and a significant amount of input energy is lost and heats the liquid and the pump assembly. Temperature rise comes from energy imparted to the liquid through hydraulic and mechanical losses within the pump. These losses are converted to heat, which can be assumed to be entirely absorbed by the liquid pumped. The temperature where the vapour pressure equals the suction pressure plus the NPSHR is the maximum allowable temperature. The difference between the allowable temperature and the temperature at the pump inlet is the maximum allowable temperature rise.

Internal Recirculation- Internal recirculation is another performance-degrading effect that damages pumps in much the same way that cavitation does. Internal recirculation tends to occur at low flow rates when fluid leaving the impeller forms damaging vortices. To avoid this problem, manufacturers list the minimum flow rates for their pumps. Operators should be aware of this minimum flow requirement and avoid overly restricting pump output.

High Flow Operation

Selecting a pump to operate far to the right of best efficiency point can also result in potential pump problems. Operation at high flows can result in overloading horsepower with reduced system resistance, significant changes in head with no change in flows, Higher NPSH required than available, Recirculation cavitation at impeller tips etc.

Hydraulic Stability Guidelines

One measure of the hydraulic stability of a pump and its ability to operate away from design conditions is called "suction specific speed."

There are two specific speeds: discharge specific speed and suction specific speed.

"Discharge specific speed" is a non-dimensional design index, to classify pump impellers as to their type and proportions. It is defined as the speed in revolutions per minute at which a geometrically similar impeller would operate, if it were of such a size as to deliver one gallon per minute against one foot head.

Unlike discharge specific speed, suction specific speed is not a "type number," but a criterion of a pump's performance with regard to cavitation. The formula for suction specific speed is:

$$\frac{N\sqrt{Q}}{(H)^{3/4}} \tag{11.1}$$

where

N = Pump speed in rpm

Q = Capacity in gpm

H = Total head per stage at the *best efficiency point* (BEP)

Note: For double suction impellers, the total flow should be divided by two, in calculating the specific speed.

The specific speed determines the general shape or class of the impeller. As the specific speed increases, the ratio of the impeller outlet diameter, to the inlet or eye diameter decreases. This ratio becomes 1.0 for a true axial flow impeller.

The pumps with high suction specific speeds (above 11 000) had a failure rate of approximately double that of the ones with lower ratings. The 11 000 limit appears to be a desirable maximum.

Suction specific speeds above 10 000 should be avoided whenever possible. While higher suction specific speed impellers can be operated effectively, the operating range is narrowed to about 85–100% of BEP flow. This is too tight for industry operating practices. In other words, while it is technically correct that high suction specific speed pumps can be made to work, the loss of operating flexibility that goes with it is not acceptable unless an elaborate bypass system and other compensations are made to maintain flow.

Hydraulic Disturbance Due to Cavitation and NPSH

Cavitation

In centrifugal pumps, the acceleration of fluid into the impeller causes the fluid pressure to drop. If this pressure drop is sufficient, the liquid vaporizes, forming tiny bubbles that are unstable and prone to violent collapse. These violent bubble collapses can throw tiny, destructive water jets onto impeller surfaces. Crackling and popping noises are indications of cavitation. Not selecting the right pump – or operating the system at either higher than-design temperatures or lower-than-design suction pressure – can be a cause of cavitation. Cavitation usually occurs at high flow rates, when a pump is operating far to the right along its performance curve; however, under certain conditions, cavitation damage can occur at low flow rates as well. Cavitation damage can also result when the pump suction is starved because of the formation of air pockets or fouling of pipes. The most important effects of sustained cavitation are reductions in pump performance and erosion of the pump impeller. Cavitation degrades pump performance because the vapor in the pump restricts flow and lowers the generated head cavitation, also result in accelerated bearing and seal wear and poor system performance.

If cavitation causes enough loss of material in the impellers, they can become unbalanced, creating alternating bearing loads that accelerate bearing wear. Because it dramatically shortens pump life, cavitation is a serious threat to system reliability. Cavitation also increases other maintenance requirements by inducing vibrations that stress pump foundations and connected piping. Cavitation-like damage can also occur as a result of internal recirculation. Operating the system at low flow rates can establish damaging flow patterns in either the suction or discharge regions of an impeller. For applications in which cavitation is to some extent unavoidable, high-tensile-strength materials should be specified for the impeller. Tougher materials can withstand higher energy cavitation.

To prevent cavitation, centrifugal pumps must operate with a certain minimum amount of pressure at the inlet.

Net Positive Suction Head

When designing a pumping system and selecting a pump, one must thoroughly evaluate net positive suction head (NPSH) to prevent cavitation. A proper analysis involves both the net positive suction heads available in the system (NPSHA) and the net positive suction head required by the pump (NPSHR). NPSHA is the measurement or calculation of the absolute pressure above the vapor pressure at the pump suction flange. Since friction in the suction pipe is a common negative component of NPSHA, the value of NPSHA will always decrease with flow. NPSHA must be calculated to a stated reference elevation, such as the foundation on which the pump is to be mounted. NPSHR is always referenced to the pump impeller centre line.

Turbulence loss is extremely high at low flow and then decreases with flow to the best efficiency point. Friction loss increases with increased flow. As a result, the internal pump losses will be high at low flow, dropping at generally 20–30% of the best efficiency flow, then increasing with flow.

Figure 11.5 shows the pressure profile across a typical pump at a fixed flow condition. The pressure decrease from point B to point D is the NPSHR for the pump at the stated flow. The pump manufacturer determines the actual NPSHR for each pump over its complete operating range by a series of tests. For most applications the NPSHA will exceed the NPSHR by a significant amount, and the NPSH margin is not a consideration. For those applications where the NPSHA is close to the NPSHR (2–3 ft), users should consult the pump manufacturer and the two should agree on a suitable NPSH margin. In these deliberations, factors such as liquid characteristic, minimum and normal NPSHA, and normal operating flow must be considered.

Guidelines for selecting pumps free of hydraulic disturbances

Action to be taken to see that NPSHA is greater than NPSHR at maximum operating flow.

Increase NPSHA by:

- Increasing suction drum level
- Decreasing pumping temperature
- Decreasing suction line losses
- Reselect pump (if possible)
- Select canned pump

Suction System Design Can Affect Hydraulic Disturbance

A majority of the hydraulic troubles that occur with centrifugal process pumps are associated with problems on the suction side because of poor design, manufacture, or application of the pumping system itself. A frequently overlooked source of trouble is the suction piping

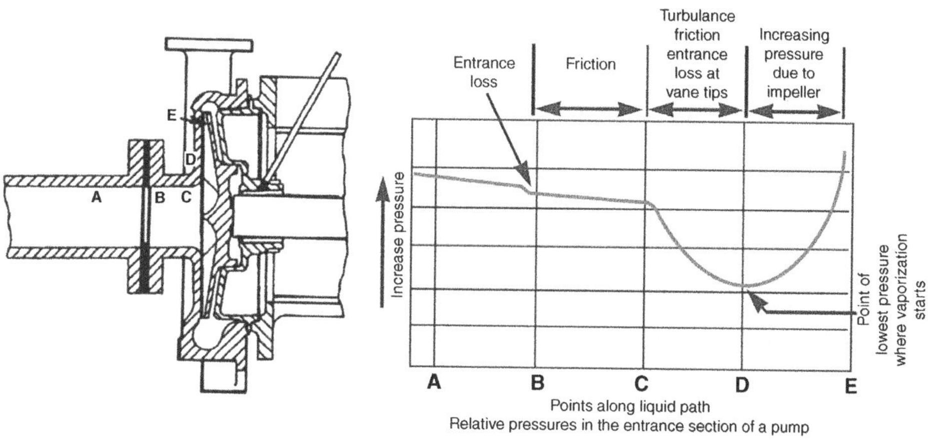

Figure 11.5 The pressure profile across a typical pump at a fixed flow condition.

design. In most cases, there is an extensive suction piping system between the tower or drum and the pump. The basis of much piping design is pipe rack layout convenience, not the assurance of a steady flow of liquid to the pump suction to

1) Meet or exceed NPSHR of the impeller pattern;
2) Have a uniform velocity profile; and
3) Have a minimum rotation velocity component of the entering liquid.

Suction Piping

The pump piping should provide a unobstructed flow of liquid to and from a pump, without adversely affecting the performance or reliability of the pump. The suction pipe should be straight, coming directly to the pump, with no turns or flow disturbing fittings close to the pump. The diameter of suction piping should be equal or larger than the pump suction nozzle. It should be sized in such a manner that the maximum liquid velocity at any point in the inlet piping does not exceed 2.5 m/s. Wrong design of suction piping lead to cavitation damage, load on bearing and seal, noisy operation.

It is important that mechanical loads are not excessive, as applied to the nozzle by hydraulic pressure (for unrestrained nozzle connections across expansion joints) and by piping end forces and moments (where expansion joints are not used). Avoid the use of flow components such as certain reducers, elbows, or strainers in the suction that might decrease suction pressure excessively compared to the vapor pressure at which the liquid will cavitate, or that will cause the flow velocity to be mal-distributed or unexpectedly swirling as it enters the nozzle.

When a pump is taking its suction from a tank, it should be located as close to the tank as possible in order to reduce the effect of friction losses on the NPSH Available. Yet the pump must be far enough away from the tank to ensure that correct piping practice can be followed. Pipe friction can usually be reduced by using a larger diameter line to limit the linear velocity to a level appropriate to the particular liquid being pumped. The suction line should have eight pipe diameters straight length before entering pump suction.

Discharge Piping

Discharge piping flow characteristics normally do not affect the performance and reliability of a centrifugal pump, but the maximum recommended velocity at any point in the pump discharge piping should not be more than 4.5 m/s.

Vortexing

Most process pumps take suction on a liquid source above them such as the side or bottom of a tower or drum. It is very important that there is adequate submergence level over the inlet to the suction pipe, to prevent vortexing. Sometimes a vortex may arise because the liquid level in the process vessel becomes very low or the velocity in the piping becomes very high.

A vortex is a swirling or funnelling action in a liquid. When this occurs between the liquid's surface and the draw off nozzle in a vessel or basin, air or vapour can be drawn into a pump. The formation of a vortex can be very damaging, because vortexing effectively reduces the NPSH available. With the suction head of a pump close to a minimum value, air or vapour entrainment is likely to occur. Only 2% air entrainment can result in a 10% loss in pump capacity. In these cases, a term frequently called "submergence" comes into play; Submergence and NPSH are not the same. It is possible to have adequate submergence and insufficient NPSH, or the reverse. A proposed installation must be checked for both adequate submergence and for NPSH equal to or greater than that required by the pump.

To avoid vortex formation, different precautions must be taken according to pump design.

As the suction piping layout at the pump becomes more complicated, the liquid velocities must be reduced. If this is not possible, the submergence must be increased. More submergence for hydrocarbons than water is recommended by many experts. Mechanical vortex breakers, at the suction can also reduce rotation velocities, but are not effective against vapour entrainment.

Vertical Pump Trouble

A submerged vertical pump has no suction piping. Manufacturers of such a pump require a minimum submergence to present vortex formation. Greater submergence is needed if the pump has no inlet strainer as the ·strainer acts as a vortex breaker. Submerged pumps should have an adequately sized suction sump. Baffles are needed if two or more pumps are in a row. The "Hydraulic Institute Standards" provide recommended practices for sump design, yet many designers continue to ignore this need the barrel of a vertical "canned" pump serves as an annular suction pipe. With this configuration, vortex forming is unlikely.

Cooling Water Pump Trouble

Most petroleum refinery cooling water pumps are non-API designs, but the problems typified with their installations are frequently encountered in process systems. The water level in the cooling tower basin determines the basin's height and cost. Consequently, a minimum water level is preferred, but the level must be sufficient to meet the pump's NPSH. Design of the suction piping is critical since submergence is a greater factor than NPSHR in good pump performance for this situation. To prevent air entrainment and overtaxing, a baffle or spoon bill is needed. We are accustomed to spoon bills in storage tanks, but we don't think to use them in cooling towers. The eccentric reducer is placed in the line with the flat side up in order to avoid an air pocket. A butterfly valve is frequently used because of little resistance and no air pockets in the housing. The vortex breaker is essential. It can be as simple as the flat plate in the sketch.

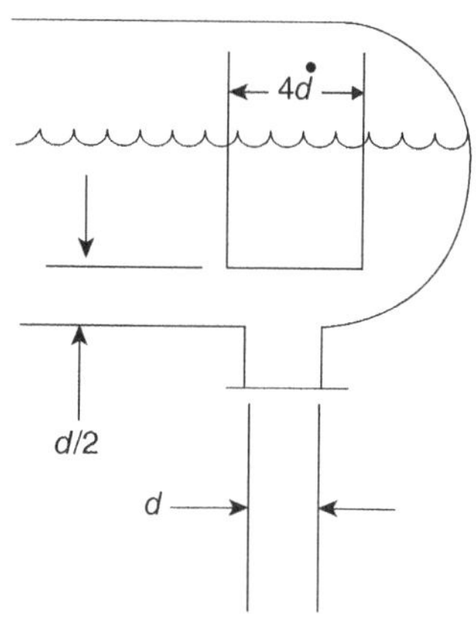

Figure 11.6 Vortex breaker in cooling tower pump inlet.

Other Mechanical Consideration

Installation and Maintenance

While design and procurement are important aspects for any application, the installation of the equipment plays a very significant role. A proper installation involves many facets; good foundation design, no pipe strain, proper alignment, just to name a few.

Piping Stress

Stress imposed on the pump casing by the piping reduces the probability of satisfactory performance. Under certain conditions the pump manufacturer may identify some maximum levels of forces and moments which may be acceptable on the pump flanges. In high temperature applications, some piping misalignment is inevitable owing to thermal growth during the operating cycle. Under these conditions, thermal expansion joints are often introduced to avoid transmitting piping strains to the pump.

The pump and piping flange faces should be parallel within less than 0.001 inch per inch of pipe flange outer diameter, or 0.010″, whichever is greater. Flange face separation should be within the gasket spacing, plus or minus 1/16 in. (1.5 mm).

Misalignment

Shaft and coupling misalignment is the primary contributing factor in over 50% of pump problems. Where misalignment exists in coupled shafts, the shafts are forced to rotate in an eccentric manner which increases power consumption and creates vibrations and stresses on the shafts, couplings and machines themselves which ultimately result in breakdowns. Shaft movement can result in fatigue or failure of the shaft, and also reduces mechanical seal life.

Misalignment in piping can occur as a result of incorrect engineering practices. Connecting fixed pipe-work to smaller pumps can increase the chance of misalignment due to a reduced resilience to excessive nozzle loading. Also, sagging from the intrinsic self weight of piping plus the weight of the fluids inside can result in misalignment if the pipe-work is not sufficiently supported.

High temperatures in a pump system can cause pipe-work to expand, the expansion being equal to the coefficient of thermal expansion of the piping material used. This thermal growth can force joints of piping apart, leading to both misalignment and increased nozzle loading.

Coupling misalignment can result in a build up of heat in bearings from increased friction and vibrations. Consequently bearing life is reduced and in the case of carbon journal bearings lubricated by the pumped liquid; wear can occur elliptically, rapidly reducing functionality. In severe instances, coupling misalignment can lead to fractures in the pump casing, increased noise, and failure of the pump shaft.

Couplings and shafts can be quickly and easily aligned by using a Laser Shaft Alignment (LSA) system. Such systems can align to accuracies of micrometers and therefore help to reduce wear and tear of couplings, seals and bearings. By correcting misalignment LSA aids in minimising energy consumption, reducing noise levels, and lessening vibrations making it a cost-effective technique.

Any rigid pipe-work should be aligned carefully with the pump and also suitably supported throughout the pump system. Inlet piping should be free from any flow disturbing fittings for a given minimum length to reduce potential pressure spikes. Flexible piping should be used where smaller pumps are concerned as they can be particularly intolerant to excessive pipe loads.

Casing

The pump casing is known as the frame or housing of pump. Its purpose is to provide the ultimate support, relative to ground, of the rotor, and to provide a well- sealed flowpath for the liquid being pumped. Mechanical issues important to casings are their resistance to distortion and breakage under pressurization and piping nozzle loading, the tendency for multistage casings to take on a "bow" on the order of the internal clearances if warmed up too fast or if one end of the pump is not free to expand (refer to the analysis of this later in this tutorial under "Casing Thermal Bowing"), the occurrence of leaks at the gaskets or split flange bolts, and the difficulty of certain types of casings of avoiding loss of interior gasket preload, and associated internal leakage.

Casing Nozzle Loading

Common causes of casing nozzle loads are piping length changes or bend curvature changes due to thermal growth, pressurization strain, unrestrained discharge lines, installation misalignment (mating up of flanges with the help of a come-along), driver/pump misalignment, and soft-foot problems. In turn, excessive pump nozzle loads sourced in the piping can induce excessive nozzle neck stress, casing internal misalignment and rubbing, and most importantly driver/pump misalignment, with possible high vibration and bearing or seal wear. API 610 provides a widely used table of allowable loads for various flange sizes.

Manufacturers typically will quote the maximum nozzle loads that they allow at each pump flange. The loads are in the form of direct forces (vertical, axial, or horizontal) at the flange face, and moments (torques) with the twisting axis about one of those three directions.

Casing Thermal Bowing

Casing problems can occur from other than pressurization effects. One of the most important of these other issues is thermal growth and/or bowing. A heated rod calculation checks out the maximum change in dimension of the casing between room temperature and the operating temperature. Sudden immersion of the rotor and casing at operating temperature will result in this growth of the rotor, while the greater bulk of the casing causes it to warm up and expand more slowly, possibly causing a severe axial rub. As a rule-of-thumb, binding becomes possible when the upper versus lower casing temperature differential exceeds about 100 °F, with rubbing beginning at about half this value. This is the reason why many users of boiler feed pumps and hydrocracking pumps put their pumps on slow roll when the pump is taken offline, why warmup cycles are often carefully specified and followed, why casings may be insulated (besides the energy cost savings), and why excess seal injection water has sometimes led to rubbing and fatigue problems in shafts.

Shaft Displacement

Bearings in centrifugal pumps support shaft, the mass of the impeller, hydraulic loads imposed on the impeller, and loads caused by couplings and drive systems. They keep axial and radial shaft deflections within acceptable limits for the impeller and shaft seal. The more the shaft deflects from the center of the stuffing box, the more likely the lapped faces of seals are to separate. Rotating seals (the spring loaded face rotates with the shaft) are very sensitive to this type of shaft displacement or any other form of misalignment between the stationary and rotating faces. Bearings are also affected by the shaft deflection and can potentially reduce their service life and reliability.

Causes of shaft deflection are

1) The pump is operating off the best efficiency point (bep).
2) The pump is cavitating.
3) The pump is vibrating.
4) Pipe strain caused by either mechanical or thermal expansion.
5) Misalignment between the pump and driver.
6) Start up thrust.
7) Water hammer.
8) High L3/D4 number.
9) Thermal growth, both axial and radial.

Wear rings

sometimes called impeller neck rings, are the annular cylindrical surfaces that form a tight clearance gap between a "covered" impeller's rotating shroud and the surrounding stationary casing piece. Their main purpose is to provide a non contacting partial seal that prevents discharge fluid in the vicinity of the impeller exit from leaking back into the lower pressure suction area. A secondary function of wear rings is to provide hydrostatic rotor support through a phenomenon known as "Lomakin effect," which allows the wear rings to act not only as seals, but also as weak but strategically placed bearings. Also, as wear rings increase their radial clearance due to rubbing wear or erosion, Lomakin effect is lost in a ratio roughly proportional to the clearance increase. This can cause rotor critical speeds to shift down as elements wear out, possibly causing unexpectedly high resonant vibration at some point before the pump would normally be refurbished due to decreased hydraulic performance.

Filters, Suction Strainers

Filters and other methods of straining foreign particles from the flow are generally needed more for some reason in the process than for protection of the pump. The primary pump issue in this regard is being certain the suction pressure does not decrease below the net positive suction head required (NPSHR) due to filter clogging, to avoid head loss and erosion damage due to cavitation. A good practice is to have a maintained gauge at the pump suction flange that continuously displays pump suction pressure.

Base Plates and Grounting

Insufficient baseplate stiffness will reduce reliability of pumps. A weak, or improperly supported baseplate will make it impossible for centrifugal pumps to operate successfully over long periods of time. The installation of new pumps is often overlooked by civil engineers and mechanical contractors when they design and construct a new pump installation. Generally, the engineer or contractor pay attention to the structural strength of the floor being adequate to support the weight of the equipment. But there are many other factors which need to be looked in to.

Grouting a baseplate or skid to a foundation requires careful attention to many details. A successful grout job will provide a mounting surface for the equipment that is flat, level, very rigid, and completely bonded to the foundation system. The most prominent installation problems involve voids and distortion of the mounting surfaces. Another field installation problem with costly implications is distortion of the baseplate machine surfaces. Distortion can either be induced prior to grouting due to poor field levelling techniques, or the distortion can be generated by the grout itself. Baseplates with sturdy cross braces are not affected by the slight volume change of the grout. For less rigid designs, the bond strength of the epoxy grout can be stronger than the baseplate itself.

It is also important to pay attention to vertical strength which impact the bending stiffness. The greater the distance of the pump centre of mass and/or radius of gyration (with respect to reed motion) relative to the "footprint" of the pump baseplate, the more important this effect is likely to be.

Bearing Failure

Pump bearings (normally antifriction for small pumps) support the hydraulic loads imposed on the impeller, the mass of impeller and shaft. The bearings keep the shaft axial end movement and lateral deflection within acceptable limits for the impeller and shaft seal. The lateral deflection is most influenced by the shaft stiffness and bearing clearance. Both magnitude and direction of the axial force may change during the pump startup process, due to varying flow conditions in the side spaces between the impeller shrouds and casing walls. The changes in flow conditions and the resulting changes in pressure distributions on the impeller shrouds cause the axial load to fluctuate. The magnitude and direction of the axial load can change from its design value if pump-out vane clearance changes due to wear, or is not set within tolerance, or if balance holes become plugged with debris.

Also there is a hydraulic radial load experienced due to the unequal velocity of the fluid flowing through the casing. The unequal fluid velocity results in a non-uniform distribution of pressure acting on the circumference of the impeller.

The different causes of bearing failure are described below.

Proper Handling and Storage

Bearing damage can often be traced back to improper storage. Bearings should never for any reason be opened until they are ready for installation. Handling bearings by opening boxes and tearing wrappings prematurely can let in dirt and expose bearings to corrosive elements. Improper storage exposes bearings to dampness and dust. Storing bearings in excessively high temperatures can also degrade a grease's shelf life, so always check with the grease manufacturer for storage specifications.

Figure 11.7 Bearing cross section view.

Improper Lubrication

According to a recent study, up to 80% of bearing failures are caused by improper lubrication. Improper lubrication is the major cause of bearing failure. Issues include: too much lubricant; too little lubricant; using the wrong lubricant; mixing incompatible lubricants; incorrect lubrication intervals; using old, deteriorated grease or oil; and water contamination. Use the appropriate type and correct amount of lubricant, avoid grease loss, and follow appropriate relubrication intervals. Contamination is caused by foreign substances getting into bearing lubricants. These include dirt, abrasive grit, dust, steel chips from contaminated work areas and dirty hands or tools. Filter the lubricant and clean work areas, tools, fixtures and hands to reduce the risk of contamination.

Overload

Bearings are overloaded when they operate beyond their rated design limits for load, speed or temperature. The greater the overload, the more it shortens bearing life, which may cause the bearing components to fracture. Some overloads may be caused by inappropriate or unanticipated operations due to changes in production requirements or operator error. Other considerations should be taken into account when encountering high operating temperatures for bearings: high ambient or process temperatures; extreme loads (radial or thrust); improper bearing clearance adjustment on tapered bore bearings; improper preloading of angular contact thrust bearings; oversized shafting; and heat transfer from machinery, through the shaft orhousing mount.

Improper Mounting

In most instances, bearings should be mounted with a press fit on the rotating ring. These include Loose shaft fits, loose housing fits, excessively tight fits, out-of-round housings and a poor finish on the bearing seat. To avoid follow proper mounting instructions and provide training to ensure all employees understand the difference between a properly and improperly installed mounting.

Misalignment

Bent shafts, out-of-square shaft shoulders, out-of-square spacers, out-of-square clamping nuts and improper installation due to loose fits can cause misalignment, which may result in overheating and failure.

Corrosion

Moisture, acid, low-quality or broken-down grease, poor wrappings and condensation from excessive temperature reversals can cause corrosion that is abrasive to the finely finished surfaces of ball and roller bearings. To overcome this Divert corrosive fluids away from bearing areas. Select integrally sealed bearings and consider external seals for particularly hostile environments.

Fit

A tight fit can be caused by excessive loading of the rolling element when interference fits exceed the radial clearance at operating temperatures. Micro-motion between fitted parts where the fits are too loose in relation to the acting forces may result in a loose fit.

In a tight fit, there may be a heavy rolling-element wear path in the bottom of the raceway, or overheating or an inner-ring axial crack. For a loose fit, fretting (generation of fine metal particles), may occur which leaves a distinctive brown color.

Fatigue

Fatigue can be indicated by the fracture of running surfaces and subsequent removal of small, discrete particles of material from the inner ring, outer ring or rolling elements. This is often the result of overloading, an excessive preload, tight inner-ring fits and using the bearing beyond its calculated fatigue life. Fatigue is progressive and will spread with continued operation. It is always accompanied by a noticeable increase in vibration and noise.

Seal Failure

The most common cause of pump downtime is a mechanical seal failure. To avoid this, it's essential to apply the right seal for the desired, and appropriate, function. The seal failures are also the result of an unfavourable seal environment such as improper heat dissipation (cooling), poor lubrication of seal faces, or seals operating in liquids containing solids, air or vapors. To achieve maximum reliability of a seal application, proper choices of seal housings with stuffing box, large tapered bore seal chamber, and seal environmental controls and API seal flush plans must be made.

Many a time the seal can be exposed to a wide variety of operating conditions, which can cause issues down the line. Sometimes operating conditions can become very different from conditions that the seal was intended for. Let's look at some of the common reasons why mechanical seals fail – and help you protect your seals and them and make them last longer.

1) **Letting the pump or machine run dry:** A mechanical seal can easily leak or become damaged if you let your pump or machine run dry. If the seals run dry, the friction will cause heat to accumulate, and cause damage – causing thermal shock. Your seal can disintegrate just 30 seconds after heating up.!
2) **Too much vibration:** Pump vibration can be caused by improper alignment, pump imbalance or operating the pump past its BEP (best efficiency point) – which means it's operating beyond its normal working point. These vibrations damage the machine and lead to reduced seal life
3) **Hammering couplings**: Many machines contain components that have a hammering action, sometimes so intense that it produces a lot of vibration and movement. This can cause problems not only for the machine but also for the seals.
4) **Human error**: Simple human mistakes can cause the seals to fail. For example, not starting the machine up properly can cause the motor to trip and the shaft to twist, causing movement that results in internal parts coming into contact, seal failure, and even reduce the life of bearings. Other human errors that can cause seals to fail include improper installation of the seal, failing to clean the areas surrounding the seal, and lack of proper maintenance. Oil, dirt and even fingerprints can damage the seal, so it's important to exercise due diligence when it comes to seal care.!
5) **Using the wrong seal**: It's important to find the right seal for the job, which means selecting the appropriate seal for the machine that you're using. However, even seasoned

engineers can get it wrong, which leads to errors during installations. Different machines or pumps require different types of seals and are built to handle different materials. Using a abrasive material can also reduce the lifespan of a seal.

6) **Improper use of, or completely omitting mechanical seal flush plans**: It's critical that a mechanical seal flush plan is properly in place. No mechanical seal recommendation is complete without a coordinating flush plan. Be Aware: if no flush plan exists, dewatered product and contaminates can build up, causing excessive heat or erosion on the seal, resulting in shortened seal life.

Other factors that affect seal reliability are

1) The pump is operating off of its best efficiency point (B.E.P.) causing the shaft to bend.
2) The rotating assembly is out of dynamic balance.
3) The shaft is bent.
4) There is misalignment between the motor and the pump.
5) Pipe strain is twisting the pump stuffing box.
6) Heat causes expansion and that always opens the possibility for rubbing or wear.
7) Cavitation, slip stick, harmonic vibration, bad bearings or some other form of vibration is causing excessive movement of the shaft.
8) An environmental control has failed.

Coupling Problem

The function of the coupling is to connect the two rotating shafts, for the transfer of rotary motion and torque. For a coupling to work at its optimum efficiency, it must match all required conditions, including performance, environmental, use and service factors. If all of these factors have been taken into consideration when selecting a coupling the coupling should have no failure issues over its lifetime. However, if just one of these factors is not met, a coupling can prematurely fail. Many pump couplings are designed to be a fuse that fails before major damage occurs to expensive pump systems. Premature wear or failure is typically the first symptom of a problem.

The most frequently encountered causes of coupling problems are:

- Misalignment;
- Improper fit or assembly;
- Overload;
- Abnormal environmental conditions; and
- Torsional Vibrations.

Lubrication Related Problem in Pump

Selection of the best oil for the application is an important factor in optimizing equipment lubrication. The quality and quantity are also most critical elements of lubrication in pumps.

Quantity

If the level of oil in the sump reaches critically low or high points, damaging conditions may occur. In a low-level operating condition, the bearing will not receive enough lubricant for

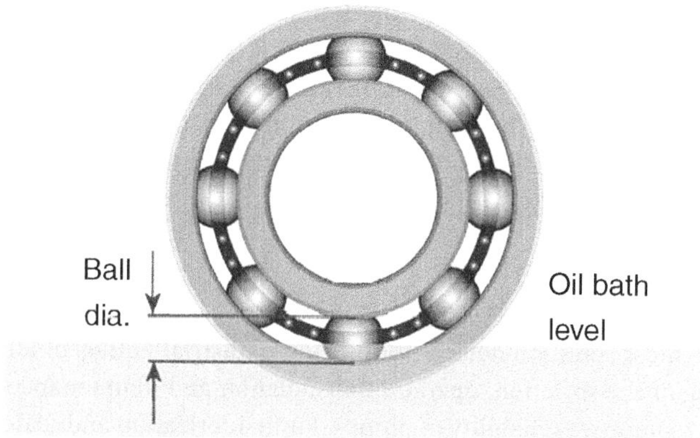

Figure 11.8 Oil bath lubrication showing a typical oil level.

proper film strength – a precursor of surface contact, skidding, and possibly catastrophic failure. In a high-level oil operating condition, churning of the lubricant will occur, accelerating the oxidation rate as a result of excessive air and elevated temperatures.

For smaller bearing arrangements and slower speeds, oil bath lubrication arrangements are commonly employed. In rolling-element bearings where oil bath lubrication is used, the normal oil level is set at around one-third to one-half the diameter of the rolling-element ball (or roller), as shown in Figure 11.8.

One of the most widely used methods of maintaining the proper level lubricant in a bearing housing is the constant-level oiler.

Quality

In basic terms, quality of lubrication can be looked at two ways.

How the lubricant can become contaminated, and
How the lubricant can degrade.

The leading causes of contamination are particulate matter, moisture, incompatible fluids, and air entrainment. The leading causes of degradation are oxidation, heat, and use. Although contamination is widely recognized for its effect on the quality of oil, degradation can be just as damaging to equipment. The leading causes of contamination are particulate matter, moisture, incompatible fluids, and air entrainment.

Water is one of the main culprits that shorten bearing life or contaminate oil. Also, thermal cycling caused by temperature changes tends to draw airborne contaminants into the bearing housing. Damaged or worn seals allow product to enter the housing and create severe problems. Water contamination of oil can cause several problems relative to oil contamination or degradation. Water in oil will dramatically decrease the effectiveness of the lubricant and therefore reduce service life.

The presence of contamination shortens bearing service life in two ways. Every time a rolling element passes over a dent, contact pressure increases at the edge of the dent. Higher stresses result in shorter fatigue life. The second mechanism is wear. Although the balls do roll in a ball bearing, because of the curvature of the balls and races, some sliding occurs as well. The sliding portions of the contact, when contamination particles are present, can result in wear of the surfaces.

The life of a lubricant is significantly reduced when the lubricant is exposed to high operating temperatures. The oxidation rate of oil doubles every 18 °F (7.8 °C) This can be significant when considering that pump operating temperatures are frequently near or above 60 °C (140 °F). By simply lowering the operating temperature of the oil to 50 °C (22 °F), a 50% reduction in the rate of oxidation can be realized – doubling the effective life of the oil.

Conclusion

Centrifugal pumps are by far the most common devices for moving industrial liquids of all types. A thorough understanding of the selection, operation, installation and maintenance of centrifugal pumps is essential to improve reliability of pumps. Pump lubrication and sealing systems significantly affect pump operation and reliability also. Often, repairing a pump doesn't solve the underlying problem and the failure reoccurs. In these cases, a truly effective solution requires correct root-cause analysis.

12

Reciprocating Pumps Failure

Working Principle

Reciprocating pump is a positive displacement pump, which causes a fluid to move by trapping a fixed amount of it then displacing the trapped volume into the discharge pipe. Figure 12.1 shows the schematic view of a reciprocating pump. The fluid enters a pumping chamber via an inlet valve and is pushed out via an outlet valve by the action of the piston or diaphragm. During the suction stroke, the piston create vacuum in the cylinder. This vacuum causes the suction valve to open and fluid enters the cylinder. During the delivery stroke, the piston compresses the fluid and increases pressure in the cylinder. This increasing pressure in the cylinder causes the suction valve to close and delivery to open, and water is forced in the delivery pipe. Reciprocating pumps are self-priming and are suitable for very high heads at low flows. They deliver reliable discharge flows and are often used for metering duties because of constancy of flow rate. The flow rate is changed only by adjusting the rpm of the driver.

Though a centrifugal pump delivers variable flow rate based on discharge pressure (head), a reciprocating pump develops a constant pulsating flow of fluid. The magnitude of these flow pulsations is dependent upon the number of plungers in the pump and will decrease as the number of plungers increases. If a smooth flow is required, then the discharge flow system has to include additional features such as accumulators. The operating pressure of a reciprocating pump is strictly a function of system resistance. System pressure is generated by the friction losses through piping and/or the pressure requirements for flow through restrictions, into pressurized vessels. A reciprocating pumps do not have a shutoff head similar to that occurring in centrifugal pumps. Instead, a reciprocating pump discharging into a closed system will continue to deliver fluid, by constantly increasing system pressure, until a relief valve opens, the driver is overloaded, the pipe ruptures, or the pump fails mechanically.

Power Pump Operation and Construction

The reciprocating pump has two distinct sides, power ends, and liquid end. The drive end of a power pump is called a power end. Its function is to convert rotary motion from a driver into reciprocating motion for the liquid end. The main component of the power end is the power frame, which supports all the power end parts and, the liquid end. The pump's liquid end consists of a fluid cylinder, valves, suction, and discharge manifold, and the plungers or pistons which convert the reciprocating motion into hydraulic horsepower. The second

Root Cause Failure Analysis: A Guide to Improve Plant Reliability, First Edition. Trinath Sahoo.

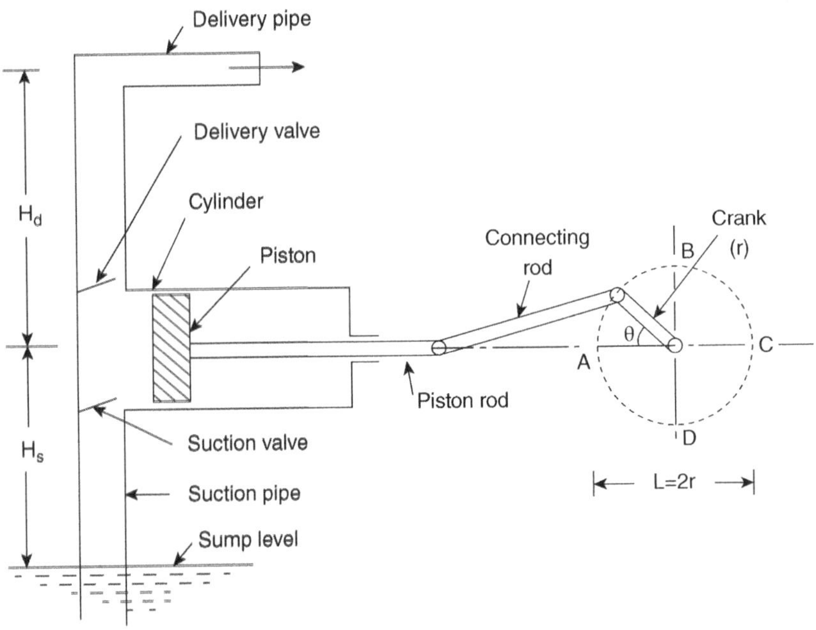

Figure 12.1 Schematic view of reciprocating pump.

major component of the power end is the crankshaft (sometimes a camshaft). The crankshaft transmits power from the driver to the connecting rods. The connecting rod is driven by a throw (journal) of the crankshaft on one end and drives a crosshead on the other. The crankshaft moves in pure rotating motion, the crosshead in pure reciprocating motion. The connecting rod is the link between the two. Main bearings support the shaft in the power frame. The crosshead transmits the force from the connecting rod into the pumping element. It is fastened directly to a plunger or piston rod. The other end of this rod is fastened to the pumping element. Power pump overall efficiencies normally range from 85 to 94%, higher than any other type of pump.

Figure 12.2 shows different components of a reciprocating pump.

Uses and Applications for Reciprocating Pumps

Reciprocating pumps are best suited for high-pressure/low-capacity services. Such services include high-pressure water-jet cleaning and cutting; amine, and lean oil charge, in gas processing; oil-field saltwater injection and disposal, steel and aluminum mill hydraulic systems, hydrostatic testing, metering; food and chemical homogenizing; well-drilling mud etc.

The Failure Causes in Reciprocating Pump

1) Inadequate NPSH, cavitation, water hammer.
2) Pump pulsations.
3) piping system.
4) Piping vibration.

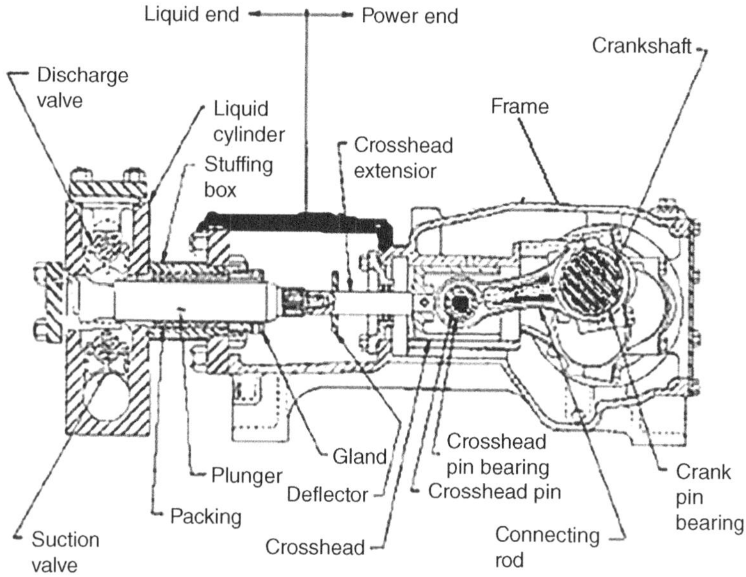

Figure 12.2 Schematic view of reciprocating pump parts.

5) Liquid end component failure.
6) Temperature extremes, thermal cycling.

Although any component in a pump system may be defective, most of the operational problems are caused by liquid transient interaction of the piping system and pump system or by purely mechanical interaction of the pump, drive system. We will discuss hydraulic and mechanical detail of reciprocating pump for determining the sources of these problems.

Net Positive Suction Head

NPSH for positive displacement reciprocating pumps is normally expressed in pressure units since a significant portion of pump NPSHR is the pressure requirement needed to push the suction valve from its seat, to overcome friction losses and acceleration head within the fluid cylinder rather than an energy per unit mass requirement. NPSH is divided into two important aspects: What is available from the suction vessel and piping (NPSH A), and what is required by the pump (NPSHR).

NPSHA

Suction system NPSHA is the pressure provided above fluid vapor pressure to ensure the pump is provided with the NPSH it requires, plus an adequate amount of pressure to overcome piping friction losses and acceleration head generated by the pump. To determine the necessary minimum, total NPSHA must be equal to or greater than pump NPSHR + pipe friction losses + acceleration head. If suction system design cannot achieve this requirement, NPSHA can be increased by one or more of the following remedies.

1) Increase suction pipe diameter to make a large volume of fluid available adjacent to the pump suction connection and reduce suction piping flow velocity, which will reduce acceleration head and piping friction losses.

2) Shorten suction piping length by providing a more direct route between the pump and suction vessel; relocate the pump closer to the suction vessel, or some combination of both. Again, rescuing acceleration head and piping friction losses.
3) Install a suction pulsation dampener or stabilizer adjacent to the pump suction connection. This will rescue the pulsating fluid mass inertial effect by changing the suction piping length portion in the acceleration head equation to approximately 20–25 times the suction pipe diameter for bladder types and to approximately eight to 10 times the suction pipe diameter for large volume flow through types.
4) Increase available static head by elevating the suction vessel, raising the fluid level in the suction vessel, or possibly lowering the pump location.
5) Cooling the fluid being pumped after it leaves the suction vessel to lower the Vapor pressure, now making this pressure reduction a portion of the NPSHA.

NPSHR

NPSHR for a reciprocating is important because it provide a user guidance on the minimum suction head that the user need to supply in the system. When the adjusted suction head at the pump inlet exceeds NPSHR the assumption that one can operate the pump properly without performance loss, vibration, cavitation, and noise. Hydraulic institute standard states that NPSHr is the suction head measured at the pump inlet manifold at which pump volumetric efficiency fall by 3% from a value measured at relatively high suction head. The primary difference between NPSH A and NPSH R is that NPSH A is a static pressure concept, and NPSH R is a dynamic pressure concept. To reduce NPSH R, one need to find a way to minimize pressure pulsation.

Problem of Inadequate NPSH

Though net positive suction head available (NPSHA) is the static head plus atmospheric head minus lift loss, frictional loss, vapor pressure, and acceleration head available at the centerline of suction connection, the acceleration head can be the highest factor of NPSHA. In some cases, it is 10 times the total of all the other losses. Data from both the pump and the suction system are required to determine acceleration head; its value cannot be calculated until these data have been established. Inadequate NPSH can cause cavitation, the rapid collapse of vapor bubbles, which can result in a variety of pump problems, including noise, vibration, loss of head and capacity, and severe erosion of the valves and surfaces in the adjacent inlet areas. To avoid cavitation of liquid in the pump or piping, the absolute liquid static pressure at pumping temperatures must always exceed the vapor pressure of the liquid. The pressure at the pump suction should include sufficient margin to allow for the presence of pulsations as well as pressure losses due to flow.

The suction pressure at the reciprocating pump inlet influences performance in the matter of volumetric efficiency and hydraulic noise, both of which affect pump life. With the reciprocating pump speed used in recent years, it has become necessary for pump manufacturers to define the minimum pressure at which a given pump will operate satisfactorily.

Test Criteria for NPSH

Power pump NPSH tests are performed by holding the pump speed and discharge pressure constant, and varying the NPSH available (NPSHA) in the system. Capacity remains constant for all NPSHA values above a certain point. Below certain NPSHA value, capacity begins to fall. NPSHR is defined as the NPSH available when the capacity has dropped 3%.

Typical Suction System Problems

The pulsating flow characteristics of a reciprocating pump cause variable feed or inlet requirements in the suction system. This pulsating flow results in pulsating pressure, which can generate cavitations, water hammer, valve knock, or possibly damage to pump components.

Water hammer is the generation of a pressure spike due to the collapsing of voids at the plunger/fluid interface. These voids are formed by the separation of the plunger from the fluid, due to the inertia of the fluid and its inability to accelerate as quickly as the plunger during the start of the pump suction stroke. Cavitations, on the other hand, results in the generation of pressure pulses because of the collapse of vapor pockets formed when system pressure drops below the fluid vapor pressure. If system suction pressure is not high enough to compensate for the fluid inertia and overcome the fluid vapor pressure, water hammer and /or cavitations will occur.

Each of these phenomena produces pressure spikes or shocks. These shocks can then lead to valve knock or more serious damage to pump components, valve knock usually results from inadequate valve response due to abnormal dynamic pressure conditions causing valve lag, reverse flow through valves, and slamming of the valve on its seat. In either case, all of these problems can be eliminated by properly designing suction systems with adequate NPSHA.

Other factors that affect the suction requirements of a reciprocating pump are as follows:

- **Rotating speed:** One of the most important parameters for reciprocating pumps is the maximum allowable speed. The velocity variation in the piping for a reciprocating pump having multiple cylinders is 25%, regardless of pump size or speed. For the same capacity, a smaller pump, running faster, will produce the same maximum and minimum flowrates and more pulses per second. Since the acceleration head increases proportionally with frequency, doubling the pump speed doubles the acceleration head, thereby reducing NPSHa from the system. Also, doubling the speed requires a stiffer valve spring, thereby increasing NPSHr by the pump. If the NPSHa drops below the NPSHr, cavitations will occur.
- **Valve spring factor:** At slow speed (100 rpm or less), suction valve spring characteristics are of little consequence, and a very low suction pressure can be tolerated. Many slow-speed pumps operate without the valve spring. The pump suction pressure requirement depends on the pump speed and the valve spring rate. Pump volumetric efficiency is greater when the pump has low rpm and the valves have higher spring rate. A higher spring rate increases efficiency at the expense of higher suction pressure. Lower suction pressure diminishes volumetric efficiency. Higher valve spring rates increase efficiency.

When stiffer springs are added to the suction valves, the spring force and the valve weight must be overcome to open the suction valve, and NPSHr will increase. These springs reposition the valves on to their seats more quickly; operation is thus smooth at higher speeds.

- **Plunger size:** Generally, in reciprocating pumps, the plunger size can be replaced with other sizes. The larger clearance resulting from using a smaller diameter plunger lowers volumetric efficiency. But, assuming the same rpm and service conditions, if plunger size increases, the required NPSH will increase.

- **Entrained gas:** Reciprocating pumps will compress gas, but the pressure developed is limited by the typically large clearance volume. Unless the pump can discharge some gas during each stroke, it will not clear itself. A second and potentially more serious problem arises if the pump is able to raise the pressure sufficiently to cause the gas to liquefy. Should this happen, the sudden change in volume can produce destructive pressure pulsation.

When the liquid contains air or is exposed to gas other than the liquid's own vapor, vapor pressure of the solution increases. Hydraulic Institute Standards recommend an NPSH margin of 3 psi for reciprocating pumps in systems where the pumpage has been exposed to a gas other than the liquid's own vapor. To minimize the problem of dissolved air, the NPSH tests are performed with water near its boiling point in the suction vessel.

Again, the stuffing box packing region could be the source of air ingress when the suction pressure drops below atmospheric. In that case, air is often drawn through the stuffing box packing to the pumping chamber on the suction stroke. Air ingress will cause a drop in capacity and noisy pump operation.

- **Differential pressure:** When the pump operates with low differential pressure, the effect of cavitation is often difficult to detect. Conversely, when the pump handles liquid with high discharge pressure, the effect of cavitation is generally more pronounced.
- **Liquid properties**: When reciprocating pumps are used in viscous service, their maximum allowable speed may have to be lowered for long-term satisfactory operation. A decrease in speed reduces the volumetric flow; this limits the feasible flow range.

Increasing viscosity has two effects. First, valve motion is impeded. At constant speed, this leads to greater valve leakage and, hence, lower volumetric efficiency. Alternatively, for a given volumetric efficiency, pump speed must be reduced as viscosity increases. The second effect is increased head loss through the suction valve and its porting, resulting in higher NPSHr. The extent of this effect increases with pump design pressure.

Pumps with high fluid temperature must be reduced in speed for proper operation. It is generally recommended to consult the pump manufacturer for relevant application parameters in both viscous and high-temperature services.

- **Suction line:** The suction line is typically sized to limit fluid velocity to 1–3 ft/sec. This low velocity reduces system acceleration head, maximizing pump NPSHa. A long radius elbow reduces pressure drop and maximizes NPSHa. An eccentric reducer is used to decrease the line size from the larger suction pipe diameter down to the pump inlet size. Orientation of this reducer is always such that an air pocket cannot form in the pipe.
- **Suction stabilizer:** The suction stabilizer usually contains internals such as a bladder, diaphragms, or baffles. Stabilizers are used in front of reciprocating pumps to reduce pressure pulsation in the pipe and acceleration head in the system. Energy from the pressure pulses is partially absorbed, and the peaks on the pulsation curves are lower. Generally, stabilizers are pressurized vessels or containers with a gas–liquid interface. The interface could be direct, or it could use an elastomeric sheet to keep the media separated.

Using a pulsation dampener reduces acceleration head and friction losses. This minimizes NPSHa when installed as a suction stabilizer in the suction piping and reduces pressure fluctuations in both piping system and pump.

Acceleration Head

It is necessary to calculate the total acceleration head (Ha) in the suction line. Because the velocities in the suction and discharge piping are not constant, the pumpage must accelerate a number of times for each revolution of the crankshaft. Since the liquid has mass and, therefore, inertia, energy is required to produce the acceleration. This energy is returned to the system upon deceleration. Sufficient pressure, however, must be provided to accelerate the liquid on the suction side of the pump to prevent cavitation in the suction pipe and/or the pumping chambers. The drop in pressure, below the average, on the suction side, caused by this acceleration, is called acceleration head.

In other words energy required to keep the fluid into the suction pipe from falling below vapor pressure is called acceleration head. The key influencing factors of the pump are pump speed, plunger size valve spring load and spring rate, valve lift, valve passage area, cylinder passage configuration, suction manifold configuration. While the key influencing factors for the suction piping are actual suction pipe length, mean flow velocity in suction line, and fluid being pumped. If sufficient energy is not available, problems such as fluid flashing, cavitation, piping vibration, noisy operation, reduced capacity, and reduced pump life can occur. To calculate the Ha required overcoming this phenomenon, use the following empirical equation.

$$H_a = \frac{LVNC}{gk} \tag{3}$$

where H_a = Head in feet of liquid pumped to produce required acceleration; L = Actual suction pipe length in feet not equivalent length; V = Mean flow velocity in suction line in feet per second; N = Pump speed in rpm; and C = Pump constant factor of

0.200 for simplex double acting
0.200 for duplex single acting
0.400 for simplex single acting
0.115 for duplex double acting
0.066 For triplex single or double acting
g = Acceleration of gravity = 32.2 ft/s^2 (9.8 m/s^2)
k = Liquid factorfor highly compressible hydrocarbons
2.0 For most hydrocarbons
1.5 For water, amine, glycol

As pump speed is increased, mean flow velocity also increases. Therefore, acceleration head varies as the square of pump speed. Acceleration head varies directly with actual suction pipe length (L). Because this equation does not adequately compensate for pumpage and system elasticity, it is recommended only for short, rigid suction lines. Some thought that a smaller pump, running faster to achieve the same capacity, would produce less pulsation in the suction and discharge pipes. Unfortunately, the opposite is true. The velocity variation in the piping for a triplex pump is 25%, regardless of the pump size or speed. For the same capacity, a smaller pump, running faster, will produce the same maximum and minimum flow rates and more pulses per second. Since the acceleration head is proportional to frequency, doubling the pump speed doubles the acceleration head, thereby reducing the NPSH available from the system. Also, doubling the speed requires a much stiffer valve spring, thereby significantly increasing the NPSHR of the pump. If the NPSHA drops below the NPSHR, cavitation and pounding will occur.

As Pressure pulsations are a product of the acceleration created by the pump and the mass of the pumpage in the system. To reduce acceleration head in a suction line, a bottle or suction stabilizer may be installed in the pipe, adjacent to the pump.

Maximum recommended speeds for plunger-type power pumps in continuous-duty services are adopted by API 674, along with the Hydraulic Institute basic speeds. Lower speeds may be dictated by factors such as pump construction, system design, low NPSHA, entrained gas, high temperature, entrained solids, high viscosity, and requirements for a low sound level.

Typical Discharge System Problems

Typically, discharge system piping problems include excessive piping vibration, piping fatigue failures, valve hammer, and possibly destruction of pump components. These are generated by the pressure pulsation in the discharge piping. The intermittent flow of a liquid through pump internal valves generates liquid pulsations at integral multiples of the pump operating speed. For example, a 120-rpm triplex pump generates pulsations at all multiples of pump speed (2 Hz, 4 Hz, and so on); however, the most significant components will usually be multiples of the number of plungers (6 Hz, 12 Hz, 18 Hz, and so on). Resultant pulsation pressures in the piping system are determined by the interaction of the generated pulsation spectrum from the pump and the acoustic length resonances of liquid in the piping. The pulsating pressure in the discharge line causes a pulsating forces resulting in vibration. Increased pressure causes the magnitude of the shaking forces to increase, resulting in higher vibrations, thus, vibrations are more of a problem in discharge systems

Pulsation Control

The primary purpose of pulsation control for reciprocating pumps is to attenuate or filter out any pump generated pulsating pressures which create forces in the piping system that cause vibration and noise. Another benefit of pulsation control is the reduction of fatigue of the pump liquid end and pump expendable parts. Reducing the pressure peaks seen by the pump piston will reduce the power end peak loading.

Therefore, the best method of solving these problems is to modify the acoustic characteristics of the system. This change can be accomplished by the installation of correctly sized pulsation dampeners, or by changing the operating parameters of the system, the effects of changing pump speed on pulsation amplitudes. However, system design requirements usually prohibit any modification to system operating parameters. If a flow rate of 1500 gpm is required, pump speed cannot be reduced, thereby reducing pulsations, because system flow rate will also decrease. In the majority of cases, the installation of properly sized pulsation dampeners will reduce pulsations for even the worst piping systems.

Pulsation Dampener Design

The use of pulsation dampeners in reciprocating positive displacement pumps systems can significantly smooth the flow of the pumped liquid and extend both pump and system life. Pumps can have any number of pistons. Simplex has 1, duplex (2), triplex (3), quadruplex (4), and quintuplex (5). Simplex and duplex generate very lumpy flows, while the triplex and quintuplex designs are considerably smoother, as shown in Figure 12.3.

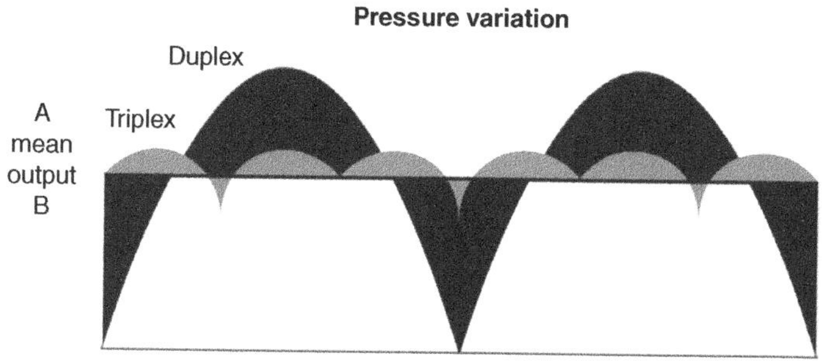

	Duplex	Triplex
A variation above mean	60%	6.1%
B variation below mean	100%	16.9%
Total variation	160%	23%

Figure 12.3 The five wave forms with the five pressure variations.

Many pulsation dampener designs are available, but the design that is used most often is the bladder type, which is simple and cost effective. Figure 12.4 shows across-section of a typical pulsation dampener. The most common design of a bladder-type pulsation dampener has a single liquid connection at one end of the metal housing and a gas-charging valve at the other end with the bladder between. The gas volume acts as a spring, compressing and expanding to meet liquid system pressure changes. To work effectively, the gas must be dry to follow the gas law: (pressure times volume remains constant). When the pressure is doubled, the volume is halved. Any moisture can change the state and vapor pressure and alter this relationship. The more gas is compressed, the more potential energy it contains to give back when needed. Dry nitrogen is the gas of choice. It is readily available and, being oxygen

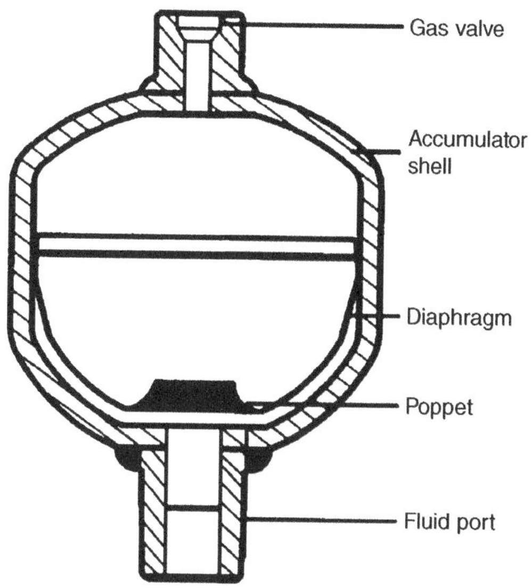

Figure 12.4 Cross-sectional view of a dampener.

free, it prolongs bladder life. The pulsation dampener housing must be made of a material that meets the corrosion requirements of the pumped liquid and the installed environment. The bladder must be an elastomer that is compatible with the liquid and permits the necessary expansion and retraction over time.

Servicing the Dampener

The dampener must be precharged before being installed, and the precharge must be checked periodically for optimum performance. The charging must be made with atmospheric pressure at the liquid port. The precharge should be set between 50 and 70% of the system pressure. When running at set pressure, the average volume of gas will be 50–30%, respectively, of the pulsation dampener volume. The manufacturer will advise the appropriate precharge to match the pump type, accumulator size, bladder, and liquid characteristics.

Problem with Piping System

One significant factor in the successful of any pump is an appropriately designed piping system; a poorly designed system can cause problems such as:

- Fluid flashing – Entrained gases in the fluid coming out when pressure in piping or pump falls below fluid vapor pressure
- Cavitation-free gases in a fluid being forced back into the fluid. These implosions cause severe pressure spike that pit and damage pump internal parts.
- Piping vibration can result from improper piping support, cavitation, or normal reciprocating pump hydraulic pulses
- Noisy operation most present when pump is cavitating
- Reduced capacity – can result from fluid flashing. if it is, this is an indication that the pumping chambers are filling up with gases or vapors.

These factors contribute to reduced pump life, and are a potential hazard to personnel and associated equipment. It is possible to fracture a fluid cylinder and or piping and damage the pump drive end internals with high-pressure surges occurring when fluid is flashing or cavitating.

The following basic piping guidelines represent a combination of hydraulic recommendations and criteria established by experienced designers of system containing reciprocating pump.

Pump Suction Piping System

Piping should be laid out so no high points occur where vapor pockets may from. Vapor pockets reduce the effective flow area of the pipe and consequently make pump priming and operation difficult. Vent any unavoidable high points and provide gauge and drain connections adjacent to pump.

Line Size

Many pump problems result from a suction line that is too small in diameter, or too long. Suction piping should be as follows to provide a smooth transition of fluid flow and result in reduced piping friction losses.

Be Short and Direct

- Be one to two sizes larger than pump suction connection. Use eccentric type pipe reducers at pump with flat side up to avoid a possible vapor pocket.
- Contain a minimum member of turns. Accomplish necessary turns with long radius elbows or laterals.

Line Velocity

The suction line velocity is based on an accelerating head of 0.7 feet per foot of suction line length and thc acceleration head equation from the hydraulic institute standard. For discharge line, a velocity not exceeding three times the suction line velocity is considered good practice.

Pump Discharge Piping System

Listed below are the fundamental requirements for a discharge piping system.

Piping should not be smaller than pump discharge connection, and should:

- Be as short and direct as possible.
- Be one to two sizes large than pump discharge connection with increasers used at pump.
- Contain a minimum number of turns. Accomplish any necessary turns with long radius elbows or laterals.
- Have a fluid velocity not exceeding three times maximum suction line fluid
- Be provided with gauge and drain connections adjacent to pump.

All positive displacement reciprocating pumps deliver fluid and build pressure until action is taken to control and stabilize the pump's work or a failure occurs. To protect pump, piping, and personnel from hazards a "positive displacement" pump must have a safety rellef valve.

The safety relief valve should be sized to pass the entire pump capacity, and the cracking pressure should be set at 10% over the specified working discharge pressure and have an accumulation pressure not exceeding 110% of cracking pressure.

The safety relief valve outlet connection should ideally be piped back to the suction supply vessel. Piping back to the suction pipe can cause discontinuities in the suction pipe flow that can result in poor pump operation and damage. Should it become necessary to pipe the safety relief valve back to the suction piping, the connection into the suction pipe should be a minimum of 10 suction pipe diameters in length back toward the suction supply vessel away from the pump suction connetion. This will allow any flow discontinuity created by the relief valve flow into the suction pipe to be smoothed out by time and viscous effect.

A discharge bypass line from pump discharge piping back to the suction supply vessel permits lubrication to reach critical pump and drive parts during start-up without subjecting them to high loads and allows all fluid cylinder pumping chambers to become fully primed. A bypass line with a shut off valve should be installed in discharge piping between pump and check valve back to suction supply source, not into the pump suction line to prevent flow discontinuity.

Install a discharge check valve beyond the bypass connection to protect pump from discharge system pressure during pump idle period and start-up.

Discharge piping "dead ends" is to be avoided or provided with dampening device. This type of feature can be responsible for undesirable piping harmonics and can contribute to elevated levels of vibration and noise.

For some services, the natural pump pressure or flow fluctuations may not be appropriate. In these cases, it is prudent to use a pulsation dampener for the installation. For maximum effectiveness, the dampener should be mounted adjacent to the pump fluid cylinder. Recommendations for dampener size and type can be obtained from dampener manufacturers based on details of pump type and size, service conditions, and piping system.

Install flanges or unions as close to the pump as practical to allow for fluid cylinder removal during maintenance.

Shut off valves are required in both suction and discharge lines to isolate pump when maintenance is required. They should be of full opening design, such as a gate valve.

When connecting two or more pumps to a common suction and discharge line exercise care to prevent a mutually reinforcing pressure wave from occurring during operation. This can be achieved by adding the capacities of all pumps that will operate simultaneously to determine line velocities for sizing pipe and calculating the acceleration head. The best way to avoid a mutually reinforcing pressure wave is to install independent suction and discharge lines to each pump.

Figure 12.5 gives an example of the recommendations outlined in the previous section for an appropriate pump piping system.

Piping Vibrations

When mechanical resonances are excited by pulsations, vibrations in the pump and piping can sometimes be 20 times higher than under off-resonant conditions. When the mechanical resonances coincide with the acoustic resonances, an additional amplification factor as high

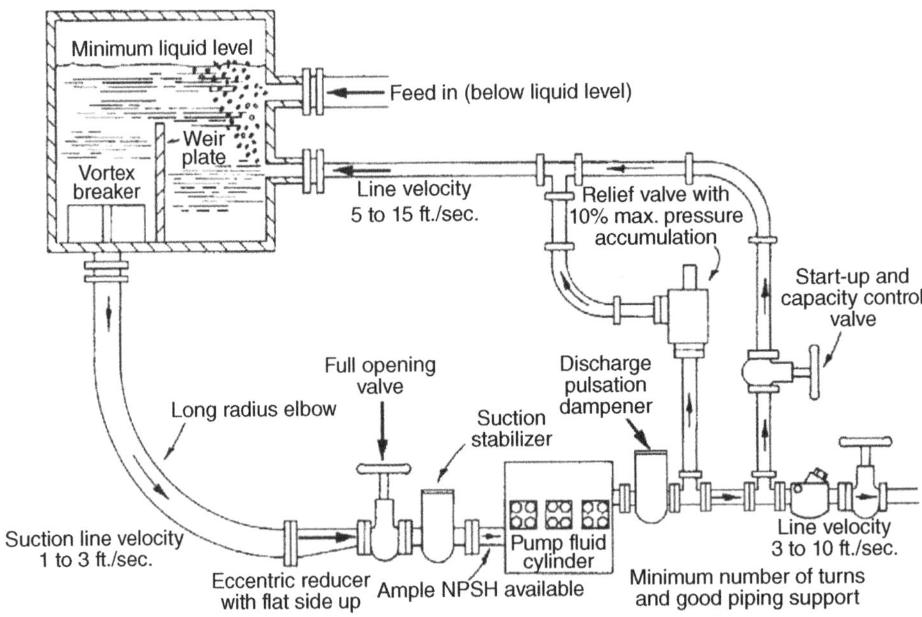

Figure 12.5 Correct piping design.

as 300 can be encountered. Piping system mechanical natural frequencies can be calculated using simplified design procedures to provide effective detuning from known excitation sources. To minimize piping vibration problems, all unnecessary bends (considering routing and thermal flexibility) should be eliminated because they provide a strong coupling point between pulsation excitation forces and the mechanical system. When bends are needed, use the largest enclosed angle possible and locate restraints near each bend. Piping should also have supports near all piping size reductions and at large masses (valves, accumulators, flanges, and so on). Small auxiliary piping connections (vents, drains, pressure test connections, and so on) should be designed so the mass of the valve and flange is effectively tied back to the main piping, thus eliminating relative vibration.

Problems with Liquid End Components

All reciprocating pumps contain one or more pumping elements (pistons, plungers, or diaphragms) that reciprocate into and out of the pumping chambers. The liquid cylinder is the major pressure-retaining part of the liquid end and forms the major portion of the pumping chamber. Each pumping chamber contains at least one suction and one discharge valve. Sealing between the pumping chamber and atmosphere is accomplished by a stuffing box. The stuffing box contains rings of packing that conform to and seal against the stuffing box bore and the rod (or plunger). All the components used in the liquid-end come across some or the other problems. Care must be exercised to enhance their reliability.

Packing

The component normally requiring the most maintenance on reciprocating pumps is packing. Figure 12.6 Reciprocating pump stuffing box (nonlubricated) showing pressure gradient

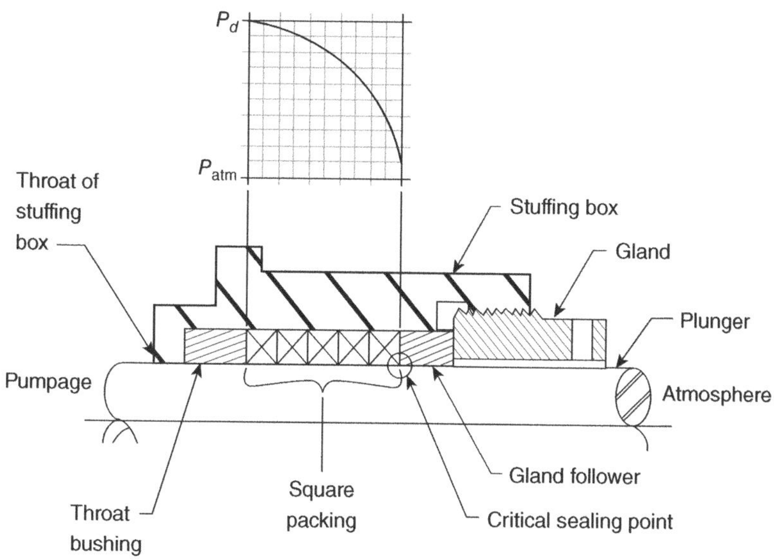

Figure 12.6 Reciprocating pump stuffing box (nonlubricated) showing pressure gradient during discharge stroke.

during discharge stroke. Although the life of packing in a power pump is typically varies from 3 months to 2 years. Short packing life can result from any of the following conditions:

- misalignment of plunger with stuffing box;
- worn plunger, rod, stuffing box bushings, or stuffing box bore;
- improper packing for the application;
- insufficient or excessive lubrication;
- packing gland too tight or too loose;
- excessive speed or pressure;
- high or low temperature of pumpage;
- excessive friction (too much packing);
- packing running dry (pumping chamber gasbound);
- shock conditions caused by entrained gas or cavitation, broken or weak valve springs, or system problems;
- solids from the pumpage, environment, or lubricant;
- improper packing installation break-in; and
- icing caused by volatile liquids that refrigerate and form ice crystals on leakage to atmosphere, or by pumpage at temperatures below 32°F (0°C).

As is evident from this list, short packing life can indicate problems elsewhere in the pump or system

To achieve a low leakage rate, the clearance between the plunger (or rod) and packing must be essentially zero. This requires that the sealing rings be relatively soft and pliant. Because the packing is pliant, it tends to flow into the stuffing box clearances, especially between the plunger and follower bushing. If this bushing does not provide an effective barrier, the packing will extrude, and leakage will increase.

The last ring of packing, adjacent to the gland-follower bushing, will experience the largest axial loading, resulting in greater deformation, tighter sealing and, therefore, the largest pressure drop. The gap between the plunger and the follower must be small enough to prevent packing extrusion. Most packing failures originate at this critical sealing point. Because this last ring of packing is the most critical, does the most sealing and generates the most friction, it requires more lubrication than the others.

Therefore, to maximize packing life, the overall stack height of the packing should not exceed the stroke length of the pump. Because the last ring of packing requires more lubrication than the others, lubrication of the packing from the atmospheric side is more effective than injecting oil into a lantern ring located in the center of the packing. Care must be exercised to get the lubricant onto the plunger surface and close enough to the last ring, so that the stroke of the plunger will carry the lubricant under the ring.

Plungers

Next to packing, the plunger is the component of a power pump that typically requires most frequent replacement. The high speed of the plunger and the friction load of the packing tend to wear the plunger surface. For longer life, plungers are sometimes hardened, although it is more common to apply a hard coating. Such coatings are of chrome, various ceramics, nickel-based alloys, or cobalt-based alloys. Desired characteristics of the coatings include hardness, smoothness, high bond strength, low porosity, corrosion resistance, and low cost. No one coating optimizes all characteristics. Ceramic coatings are harder than the metals, but are brittle, porous, and sometimes lower in bond strength. Porosity contributes to shorter packing life. Mixing of hard particles, such as tungsten carbide, into the less-hard nickel or cobalt alloys has resulted in longer plunger life at the expense of shorter packing life.

Stuffing Boxes

In stuffing box designs, various lubrication and bleed-off arrangements are provided to minimize leakage and extend packing life. The most significant advance in packing arrangements in recent years has been spring loading. Spring loading of packing has the advantage that it requires no adjustment of the gland. The gland is tightened until it bottoms and then is locked. This removes one of the biggest variables in packing life: the technician skill.

Spring loading is applied most to V-ring (chevron) packing but also works well with square packing. The spring must be located on the pressure side of the packing. Springs of various types can be used, including single-coil, multiple coil, wave-washer, belleville, and thick-rubber-washer. The force provided by the spring is small compared to the force imposed on the packing by discharge pressure. The major functions of the spring are to provide a small preload to help set the packing and to hold all bushings and packing in place during operation.

Valve

The valves in a reciprocating pump are simply check valves which are opened by the liquid differential pressure. Most valves are spring loaded. They have a variety of shapes, including spheres, hemispheres, and disks. Figure 12.7 depicts typical Reciprocating pump disk valve.

In plunger pumps, valves open and close sequentially: suction valves close, then discharge valves open, then discharge valves close, and then suction valves open. A suction opening event ***cannot*** happen until discharge closing has occurred; a discharge opening event ***cannot*** happen until the preceding suction closing has occurred. Valve opening events occur largely due to pressure differential across the valve; closing events occur largely due to spring force (combination of spring stiffness and preload).

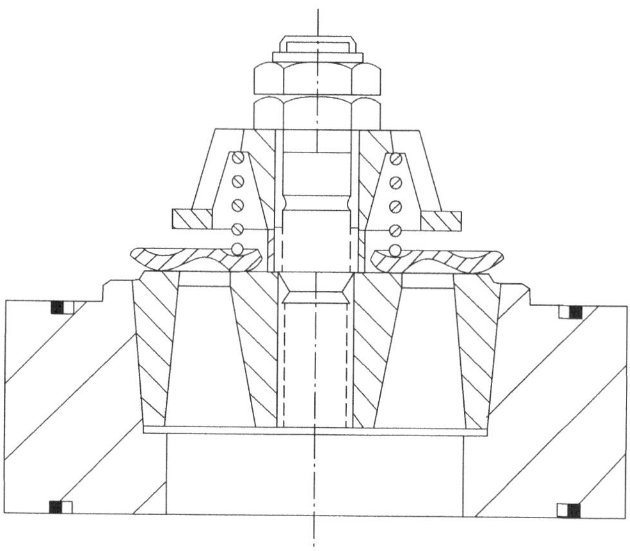

Figure 12.7 Reciprocating pump disk valve.

Good pump performance for low compressibility liquids such as water or amine is obtained when valve closing events occur no more than 7 to 10 degrees after dead center. In these cases, plunger velocity is quite low, and therefore, the pressure spike which occurs during the valve closing event is minimized. Other things being equal, vibration levels are reasonable because forces in the system are low. Similarly, if valves are performing well, opening events are at a relatively low velocity and the plates does not contact the backguard at all. Near theoretical capacity is achieved, because the plunger is pushing liquid during almost all of its stroke.

Cavitation does not occur when valves are good (given adequate suction pressure) because the valve opens before the chamber pressure goes below the vapor pressure of the fluid (cavitation happens when a liquid reaches vapor pressure), allowing bubbles to form and then implode when the pressure rebounds.

The Cause of Late Valve Closure

"What could cause the valve event to be so late?" Insufficient spring stiffness could cause the suction valve to close late because there is less contribution from the spring toward forcing the valve to close. To compensate for the lack of force from the spring, drag force from backflow is required to close the valve. The discharge valve opening is late accordingly, since it must happen after suction closing. Figure 12.8 shows the cross section view of a valve.

The Effects of Late Valve Events

Valve events that are late occur after the plunger has had significant time to accelerate, and therefore, the plunger velocity is greater. When closing occurs, the sudden change from high to zero velocity of the fluid through the valves causes shocks similar to water hammer in the liquid system. The shocks are very sharp pressure peaks that produce shaking forces in the system piping, which in turn cause piping vibration. For opening events, a similar pressure spike occurs due to the sudden change from zero to high velocity. In addition, the increased velocity causes the valve plate to move much faster and to have a greater chance of contacting the backguard. Such contact drastically increases wear and tear on the valve, and should not occur. Another effect of late valve closing is reduced pump capacity due to backflow.

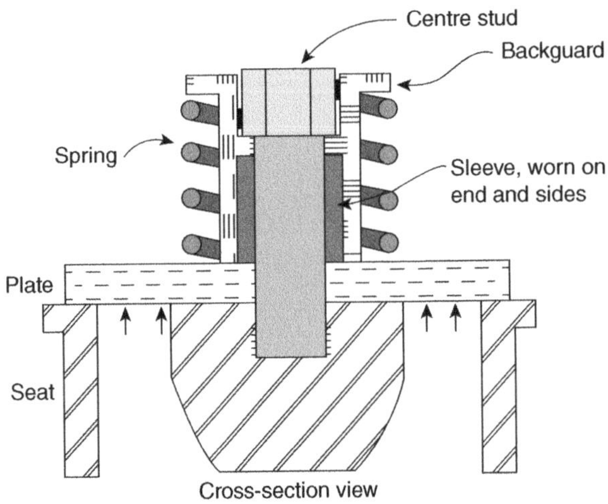

Figure 12.8 The structure and wear pattern of the valve is an important piece of the puzzle.

On the suction side, liquid gets pushed back into the suction line as long as the valve remains open after dead center. On the discharge side, liquid gets drawn back into the pump chamber as long as the valve remains open after dead center. Suction cavitation may occur with the later discharge valve closure because the chamber pressure falls faster. The inertia of the suction valve tends to prevent it from opening until after vapor pressure is reached.

Thermal Problems

In pump systems with high thermal gradients, large forces and moments on the pump case can cause misalignment of the pump and its driver as well as pump case distortion resulting in vibrations, rubbing (wear), bearing failure, seal leakage, and so on. High stresses can be imposed on the piping, resulting in local yielding or damage to the piping restraints, snubbers, or support system. Misalignment problems commonly exhibit a high second-order component of the shaft vibrations. Proximity probes can be used at the bearings to measure movement of the shaft relative to the bearing centerline.

Shaft Failures

Pump and driver shafting can experience high stresses during start-up and normal operation because of the uneven torque loading of the positive displacement pumping action. Shaft failures are strongly influenced by the torsional resonances of the system, which are the angular natural frequencies of the system. Torsional problems can usually be solved by changing the coupling stiffness between the driver and pump or by using a flywheel in an effective location. The addition of a flywheel will tend to smooth the torque oscillations. Pumps with a greater number of cylinders and equal cylinder phasing usually operate more smoothly with lower shaft stresses.

Conclusion

Reciprocating pumps have some disadvantages, the most common being pulsating flow. Because of the pulsation, special consideration must be given to system design. Guidelines are provided later in this section. In most applications, the initial and maintenance costs for a reciprocating pump will be greater than for a centrifugal or rotary pump. The packing in a typical power pump lasts about 2,500 hours, less than a mechanical seal on a rotating shaft. Most problems with reciprocating pumps can be minimized by selecting pumps to operate at conservative speeds, by carefully designing the pumping system, by careful operating procedures, and by maintenance practices which preserve the alignment of the plunger (or rod) with the stuffing box.

13

Centrifugal Compressor Failure

Centrifugal compressors are rotary continuous-flow machines in which the rapidly rotating element accelerates the fluid as it passes through the element, converting the velocity head into pressure, partially in the rotating element and partially in stationary diffusers or blades. This type of compressors is employed in numerous fields for example fertilizer plants, nuclear reactors, and chemical and petro chemical industries. Figure 13.1 shows the process diagram of a centrifugal compressor.

Characteristics of Centrifugal Compressor

The characteristics of a dynamic compressor are to increase gas energy (a function of mass and velocity) by working on the gas with blades. The positive displacement compressor characteristic results in relatively constant flow regardless of system required energy, whereas the dynamic compressor characteristics results in significantly large flow changes for changes in system resistance. As opposed to a positive displacement compressor, the dynamic compressor is self-limiting and does not require a relief valve to protect the compressor and driver. They are

- Variable volume delivery
- Fixed head capacity (for a certain flow)
- Self-limiting

Velocities and gas density plays an important part in the make-up of this compressor's characteristics.

Anything that will result in a velocity and/or density change at the tip of the dynamic compressor blade will result in a different differential pressure and a corresponding flow change.

Therefore, the dynamic compressor will be extremely sensitive to gas composition changes since these changes will produce mass and velocity changes within the compressor blades or impellers. In addition, this compressor type will also be sensitive to system resistance changes since an increased system resistance requirement will force the compressor to operate at a lower volume throughput. This is because the only way this compressor can produce higher delivered energy is at a lower velocity throughput.

Therefore, the operating point of a dynamic compressor will be much more sensitive to both changes in system resistance and the velocities inside the impeller.

Root Cause Failure Analysis: A Guide to Improve Plant Reliability, First Edition. Trinath Sahoo.

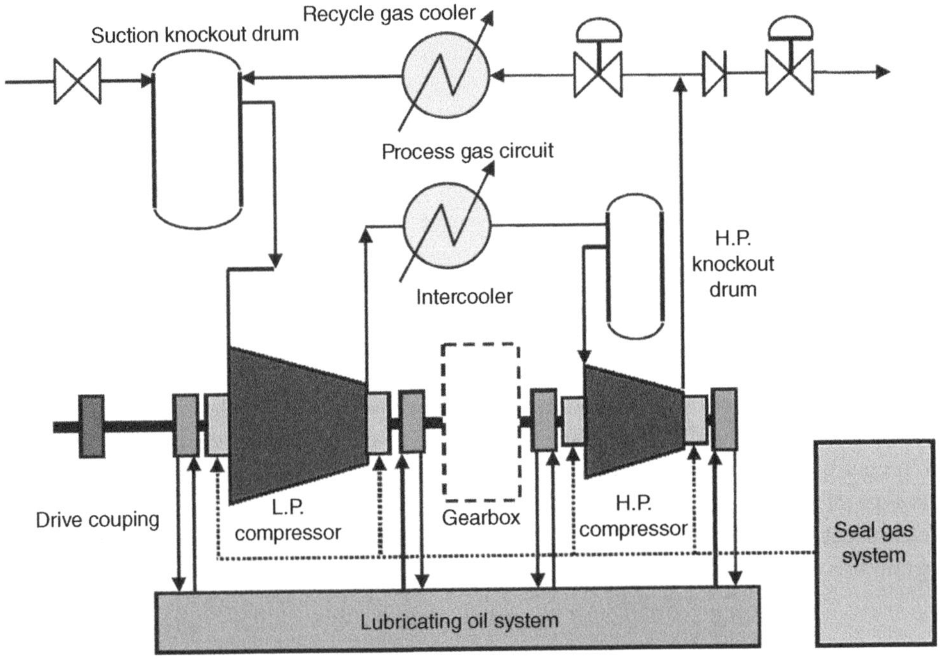

Figure 13.1 Process diagram for compressor.

Major Components of a Centrifugal Compressor

The two main components of a multistage compressor are rotating element, called the rotor assembly and the stationary parts which include casing and diaphragm. Figure 13.2 shows the major parts of centrifugal compressor.

The rotor assembly consist of impellers, sleeves, a balancing drum, a thrust disk etc.

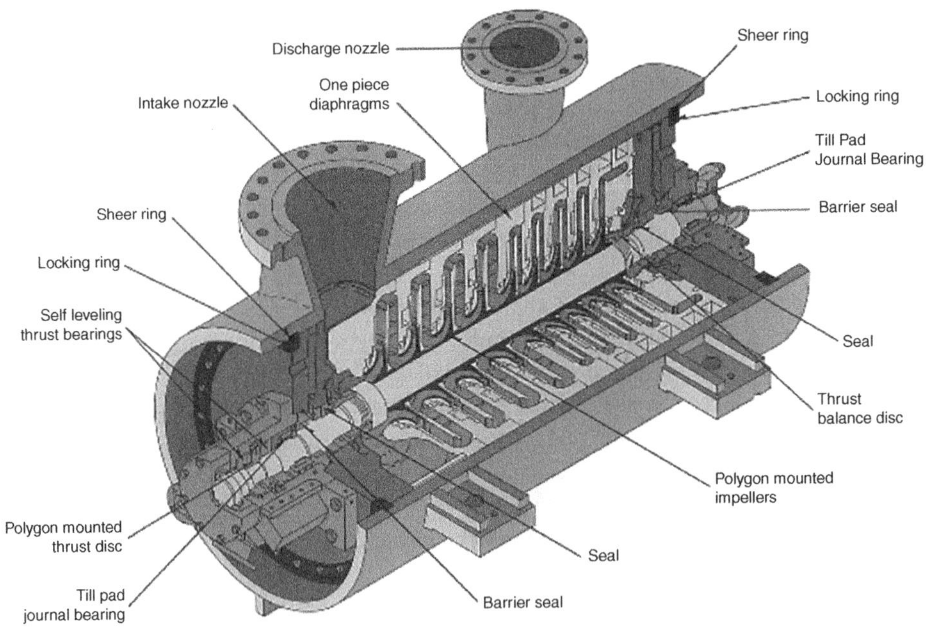

Figure 13.2 Parts of centrifugal compressor.

The sleeves are installed between adjacent impellers along the shaft. The sleeves act as wearing surfaces for interstage seals and also isolate the shaft from the process gas. The manufacturer can make the shaft from a relatively inexpensive high-strength material, giving the shaft high torque carrying capabilities.

The balancing drum or piston is a thrust equalizing component. It equalizes a large percentage of the rotor thrust. One end of the balancing drum is essentially at discharge pressure, the other end connect to a low pressure point normally the inlet end of the compressor thru a balancing drum line. The resultant pressure differential across the balancing drum cause a thrust opposite to the impeller thrust. It is common practice to balance 75% of the rotor thrust with the balancing drum.

The thrust disk fits between the thrust-bearing assemblies and along with the thrust bearing absorb the net thrust of the rotor, i.e. the portion of the rotor thrust not equalized by the balancing drum. The thrust disk positions the rotor assembly in the casing such that the impellers line up with the gas passage in the stationary components.

Rotor

Figure 13.3 shows the schematic of a compressor rotor. The major components of a centrifugal compressor rotor are:

- Shaft
- Impeller
- Balancing Drum(if required)
- Impeller spacer
- Thrust disk
- Coupling hub

Shaft- The shaft is precision machined from an alloy steel forging. This solid rotor shaft design ensures maximum parallelism of rotor components. Impellers and balance pistons are normally forged steel, SAE 4330, with stainless steel available for corrosive gas applications. Impeller spacers are typically machined from a 400 series stainless steel.
Sleeve-In the seal area, sleeve are provided to protect the shaft. Under labyrinth seal, the sleeves are stainless steel. Under oil film seals, the sleeves are often monel with a hard colmonoy or similar overlay to protect against scratches from dirt particles in the oil or gas.

Impeller

Impeller is the heart of the compressor. In which the mechanical energy (rotary motion) of the rotor is converted in to kinetic energy of the process fluid (compressing fluid). In compressor, impellers are the highly stressed component and generally become the limiting

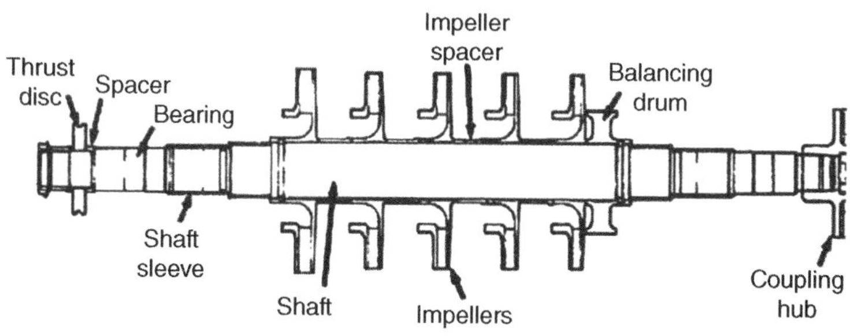

Figure 13.3 Parts of compressor rotor.

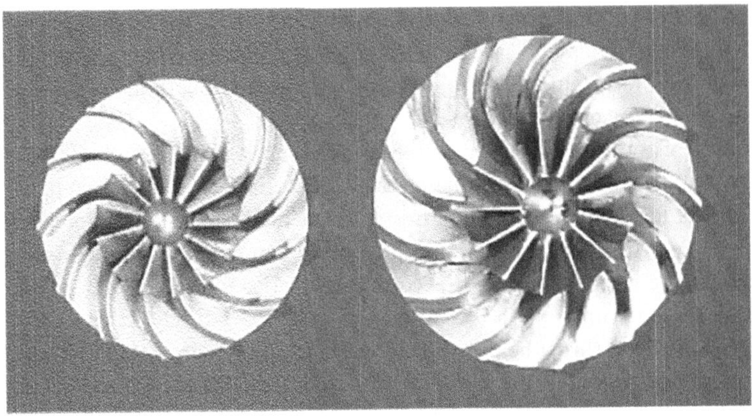

Figure 13.4 Impeller.

factor when it comes to the establishing the performance limit. Figure 13.4 shows the impeller of a compressor.

There are three types of impellers

- Open impeller
- Semi open impeller
- Closed impeller

Open impellers have the vanes positioned radial direction and have no enclosing covers on either front or backside.

Semi-open impellers have the vanes positioned in a radial or backward leaning direction and have a cover on the backside with extended to the periphery of the vanes. The radial blades semi-open impellers could develop high head and maximum flow in a single stage, even in large diameter impeller.

Closed impellers have enclosed cover on both front and backside. This is the most common type in large processing compressors. The blades are usually backward leaning and in some special cases radial curved vanes. Normally, backward curved impellers are used by all compressor manufactures because of their energy efficient. Due to the reaction force created by the fluid flow.

Impellers are mounted on the shaft with a shrink fit with or without keyways, depending on the frame size. Prior to rotor assembly, impellers are dynamically balanced and over speed tested. Impellers are mounted in pairs beginning at the center of the shaft, successive pairs of impellers are added one from each end until the rotor is complete. The rotor is dynamically balanced after the addition of each set of impellers. At each balancing operations, balance correction is done only on the newly added components.

Balancing drums are employed to modify or adjust the axial thrust developed by compressor rotor. These drums are typically required when all impellers are facing in the same direction. The balancing drum is mounted behind the last stage of the impeller.

Problems Related to Rotor

Common problems associated with rotors are; component unbalance, rotor unbalance caused by improper impeller assembly, excessive blade or impeller stress and rotor system natural frequencies, change in rotor position, over speed etc.

Impeller Forces and Shrink Fit

All centrifugal compressor rotors are "^built-up" type. That is, the major components are not integral with the shaft and must be shrunk on.

The forces acting on an impeller that create impeller stress and require positioning stress (shrink fit) are:

- Centrifugal force
- Gas load (force from torque)
- Thermal difference between impeller and shaft
- Thrust force

The impeller must be designed such that the shrink fit (shaft diameter – impeller bore diameter) is sufficient to maintain impeller/shaft contact at maximum conditions. Typically, impeller shrink fit values range from 0.002 to 0.003 in. shrink per inch of shaft diameter.

Single-stage rotors usually use hydraulic methods for rotor assembly/ disassembly since most designs require impeller disassembly for seal removal. Regardless of the method of rotor assembly, residual stresses remaining after assembly can deflect the rotor. Small amounts of rotor deflection (0.001" runout) can cause large amounts of rotor unbalance. If a rotor is low speed balanced in a deflected state, high amounts of vibration due to rotor unbalance will be experienced in the field. This is because the normal forces on the impeller will remove the residual shaft stresses at operating speed resulting in an unbalanced rotor. In order to prevent this occurrence, all compressor vendors and qualified repair facilities check the rotor deflection (runout) before and after the addition of each impeller to the shaft.

Rotor Position

Rotor position is the most common thrust-bearing condition parameter and provides useful information regarding the direction of thrust. It also provides an indication of thrust load but does not confirm that thrust load is high or not. All axial displacement monitors have preset (adjustable) values for alarm and trip in both thrust directions. Typically, the established procedure is to record the thrust clearance during shutdown and set the alarm and trip settings as follows:

Alarm = Clearance/2 +10 mils (each direction)

Trip = Alarm Setting + 5 mils (each direction)

The above procedure assumes the rotor is in the mid or zero position of the thrust clearance. An alternative method is to hand push the rotor to the assumed active position and add appropriate values for alarm and trip.

Condition monitoring and trending of critical thrust-bearing parameters will predict rotor position that impact turbocompressor reliability. The critical thrust-bearing condition monitoring parameters are:

- Rotor position
- Thrust pad temperature
- Balance line AP

Rotor Balancing

Rotor unbalance is a common cause of synchronous rotor vibration. The causes of unbalance can be varied with the actual causes depending on manufacturing methods and procedures, repair practices, and balance condition changes during operation. Some operational causes of unbalance include rotor fouling (dirt or other deposits on the rotor), bowing of rotors due to uneven heating or shaft damage, loss of rotor material possibly from rubs, or other causes.

The purpose of balancing rotors is to improve the mass distribution of the rotor and its components (caused by machining tolerances and nonuniform structure) so the mass centerline of the rotating parts will be in line with the centerline of the journals. To accomplish this, it is necessary to reduce the unbalanced forces in the rotor by altering the mass distribution. The process of adding or subtracting weight to obtain proper distribution is called balancing.

Balance correction is most effective when it is applied at or on the component that actually has the unbalance. In most cases the rotor is manufactured from a number of components (impellers, balance disks, thrust collars, etc.) that will each have some level of unbalance during assembly of the rotor. During operation, the unbalance state of each mounted component could change due to reasons stated above. As a result, the actual balance condition of an assembled rotor is never fully known prior to, during, or after a successful balance procedure is executed either in the shop or field.

Rated Speed Balancing is for high-speed rotating components where standard, "low speed" balancing (less than 2000 rpm) is not sufficient. Typically, High-speed balancing is required when additional vibration modes (critical speeds) or radial growth of components are encountered at operational speed.

Properly balanced rotating components minimize vibration and noise, extend the life of bearings, and help to ensure a component intended levels of safety, reliability and performance. As the machine operates, the balance condition of the rotor or components on the rotor can be evaluated at least in part by measuring and assessing the vibration characteristics of the machine. This force rotates about the shaft that is phased to the shaft which results in vibration at 1 × RPM.

Rotor Dynamic Considerations

Rotor Instabilities

Different bearing geometries, internal clearances, oil film characteristics, and their controlled interactions are all needed to avoid rotor instabilities. Rotors can become unstable, or prone to vibrate excessively, if supported by the wrong bearings. This tendency to become unstable is a function of rotor geometry, gas conditions, and gas properties, but also bearing curvature, pivot location, and axial width.

Critical Speed

In the case of turbocompressor rotors, their natural frequency must be excited by some external force to produce a response that will result in increased amplitude of vibration. The excitation force that could produce this result at the speed is termed as "critical speeds." Every turbocompressor that is designed must have the rotor system critical speeds determined prior to manufacture. Remember, changing of any value of support stiffness will change the critical speed. The primary components of support stiffness which influence critical speed are:

- Oil support stiffness
- Bearing pad or shell
- Bearing housing
- Bearing bracket
- Casing support foot
- Baseplate
- Foundation

Manufacturers generally establish their own minimum margin requirements between operating speeds and critical speeds. These requirements basically follow those set forth in API 670. The majority of compressors operate with flexible rotors, running between their first and second critical speeds. The location of the critical speeds plays a large part in the design of a compressor. A manufacturer has to ensure that the compressor will be able to operate over the full speed range intended; this insurance comes from rigorous computer analysis of the critical speeds, unbalance sensitivities, and stability of a system. It is generally in the best interest of overall reliability to make the system as stiff as possible (i.e., to minimize the distance between the journal bearings).

Centrifugal compressors are generally designed to operate above their first critical speeds. Critical speed is somewhat analogous to the resonant frequency at which a rotor assembly would vibrate if struck by a hammer.

The location of the second critical speed can be greatly influenced by overhung weight, which is the weight outboard of the journal bearings at each end. This typically includes the thrust disk and coupling(s). Relocating the thrust-bearing inboard of the journal bearings on a drive-through unit increases the second critical speed. This permits a higher operating speed or increased bearing span, resulting in operating speeds farther removed from the first critical speed.

Over speed Protection

Over speed protection is a key requirement in industries where failure of the turbine- or pump-speed governing system can lead to rotor damage, with expensive consequences. In the event of coupling failure, this control is lost along with the drive. Any turbocompressor can reach a dangerous over speed in reverse if the discharge nonreturn valve does not close. For this reason, twin nonreturn valves are often used. It is good practice to use one simple nonreturn valve (self-closing, not a tight shut-off) in series with a power-operated trip valve (this maybe the remote isolation valve). It is important that the volume of gas between the compressor discharge and the nonreturn valves is small. This minimizes the energy available to create reverse flow.

Independent speed monitoring and protection is therefore an essential requirement to minimize these risks as far as practicable.

Failure of Stationary Components

The Casing

Casings in centrifugal compressor can be either horizontally or vertically (barrel type) split. Generally, barrel type casings are used above 600 Psi (40 Bar) pressure or for mixed gases with a high hydrogen content. The major problem connected with casing design is distortion of the casing caused by excessive piping or foundation forces. During construction or whenever piping is removed from the casing, dial indicators mounted external to the casing should be positioned on the shaft in the horizontal and vertical planes. Movement of more than 0.002 in. (0.05mm) when piping is bolted up to the casing should require correction to the piping system (pipe flange alignment or spring support modification). The only way for the turbocompressor shaft to move is if the piping or foundation cause excessive casing strain. A check for excessive casing strain caused by the foundation is commonly known as "soft foot". To check for soft foot, a dial indicator is mounted external to each casing support. If the support moves more than 0.002 in. (0.05 mm) when the support foot is unbolted, soft foot exists. Soft foot should be checked with process piping disconnected.

Excessive or unexpected movements can distort the case, cause tilting of a head that can lead to excessive run out between the thrust-bearing and thrust disk, and cause gas leaks both internal to the machine (recycle) and/or through a case seal. This concern leads to strain gauge and dial indicator measurements of various cases and heads during pressure testing. These data are compared with predictions in a continuing effort to update and improve the analytical tools.

Horizontal Split Casing

Centrifugal compressors with horizontal split casings typically permit internal of around 700 psi (70 bars) with flow rate of around 300 000 Nm^3/h at low pressure. It varies with respect to manufacture and requirements.

Figure 13.5 shows an eight stage horizontal split centrifugal compressor. In this type of compressors are easy erection and maintenances. For inspecting the bearings by removing

Figure 13.5 Cross section view of horizontal split compressor.

top half of the bearing housing and no need for removing the top casing there by the duration of maintenance is reduced.

Barrel Casing

Figure 13.6 shows the Barrel type casing of a compressor.

This is a one-piece pressure retaining housing in the form of a cylinder with nozzles for process connections. It will normally be centerline supported. This gives a very robust design with very predictable stress levels, and can be of fabricated or cast construction. Two end covers, carrying the mechanical seals, complete the containment. Covers may be bolted on, or in some cases retained by proprietary ring joints.

It is a mandatory design for high pressures or for compressing gases rich in hydrogen. The cylindrical casing ensures good stress distribution and extremely good gas tightness. Above 700 psi horizontal split casing lose their sealing capability, resulting in an increased in potential for the escape of process gas. Hence vertical casing is more preferred for high pressure application.

Diaphragms

A diaphragm consists of a stationary element which forms half of the diffuser wall of the former stage, part of the return bend, the return channel, and half of the diffuser wall of the later stage. Due to the pressure rise generated, the diaphragm is a structural as well as an aerodynamic device. For the last stage or for a single-stage compressor, the flow leaving the diffuser enters the discharge volute. Diaphragms can be manufactured of either cast iron or cast steel. Both materials are acceptable. However, in the event of severe axial rubs caused by excessive continuous surge, cast iron diaphragms can crack and will require replacement. Since centrifugal diaphragms are rarely spared, failure will cause long periods of downtime and loss of product revenue.

Figure 13.6 Barrel type casing. *Source*: MauriceStewart - Surface Production Operations Volume IV - Pump and Compressor Systems: Mechanical Design and Specification 2019, Pages 457–525. Doi: https://doi.org/10.1016/B978-0-12-809895-0.00007-7

Since velocities are relatively high through the diffuser section (several hundred feet per second), surface finish/friction factor is crucial to overall efficiency of the unit. In many processes, dirt or polymer buildup on the impeller and diaphragm surfaces will give the aerodynamic surfaces a rough finish. In some cases, polymer buildup has been known to severely restrict the diffuser passage. Both conditions cause increased pressure losses and result in reduced overall efficiency of the compressor. Compressor performance is best preserved by including a wash system that includes water with detergents or hydrocarbon solvents to wet aerodynamic surfaces preventing attachment of the polymers and to help wash compressor surfaces once bonding of the polymers occurs.

Interstage Seals

Due to the pressure rise across successive compression stages, seals are required at the impeller eye and rotor shaft to prevent gas backflow from the discharge to inlet end of the casing. The condition of these seals directly affects the compressor performance. The simplest and most economical of all shaft seals is the straight labyrinth. This seal is commonly utilized between compression stages and consists of a series of thin strips or fins, which are normally part of a stationary assembly mounted in the diaphragms. A close clearance is maintained between the rotor and the tip of the fins.

A rub on an aluminum labyrinth causes the tips of the aluminum fins to mushroom out (Figure 4.18). This creates undesirable flow characteristics across the labyrinth and increases the radial clearance. These factors are detrimental to compressor efficiency due to the resulting increased leakage and will have an effect on the thrust loading of the machine. The overall efficiency improvement attainable by using abradable seals in a compressor varies with several factors, most notably the size of the compressor.

The causes of labyrinth seal wear are as follows: rotor vibration, excessive moisture and fouling. If centrifugal compressor head and efficiency fall off greater that 10%, the compressor should be inspected at the next opportunity if the compressor is known not to be fouled or seals are rubbed.

Balance Piston

A balance piston (or a center seal) is utilized to compensate for aerodynamic thrust forces imposed on the rotor due to the pressure rise through a compressor. The purpose of the balance piston is to utilize the readily available pressure differentials. To oppose and balance most of these thrust forces. This enables the selection of a smaller thrust bearing, which results in lower horsepower losses.

A certain amount of leakage occurs across the balance piston since a labyrinth seal is utilized. This parasitic flow is normally routed back to the compressor suction, thus creating a known differential pressure across the balance piston. Since the balance piston seal must seal the full compressor pressure rise, integrity of this seal is crucial to good performance. A damaged seal results in higher leakage rates, higher horsepower consumptions, and greater thrust loads. The balance piston damage was a result of surging and vibration excursions.

Impeller Thrust

Impeller thrust is generated by the differential force on the cover and hub of the wheel. These forces are the summation of the product of the pressures acting on the cover, hub, and the differential area from the shaft to the tip of the wheel. The impeller generates thrust between the eye and tip of the wheel, as well as below the eye. These forces (thrust) are caused by several different effects: Due to the pressure rise in the diffuser, the return channel pressure is greater than the pressure behind the impeller hub. Leakage therefore occurs from the return channel toward the impeller hub and outward toward the impeller tip.

Other Components

Bearings

The shaft is supported by a radial bearing at each end and a single thrust bearing, normally at the nondrive end. Thrust bearings will be double-sided tilting pad bearings. The bearing cannot carry anything like the total axial thrust generated by a compressor. The majority of the thrust is carried by internal gas pressure balance arrangements, normally a combination of impeller geometry and a thrust piston. To avoid instability, residual thrust is always in the same direction. This residual thrust is carried by the "active" thrust bearing, the thermocouples monitoring the thrust-bearing pads (usually 2 in the active bearing, 1 in the inactive) should indicate a healthy but not excessive load. This load will change as the compressor duty load changes, but should be repeatable with the load. Excessive temperatures can indicate overload or restricted oil supply. Overload or reverse loading (shown by axial shaft probes and hot ***inactive*** thrust bearing) suggest problems with the gas pressure balance system. As this is fully internal, the compressor must be stripped down. It is normal practice to fit proximity probes to monitor bearing behavior. The thrust-bearing wear can be monitored by these probes.

Bearing material options include tin/lead-base Babbitt and various copper-bearing alloys (bronzes) to suit specific applications. In each case, the compressor designer desires not to exceed 50% of the bearing manufacturer's allowable load rating.

Regardless of overall geometry, most hydrodynamic bearings with steel backing have a Babbitt lining of about 0.8 mm (0.032 in.) thickness, and clearances of 0.0015–0.002 inch per inch (mm per mm) of journal diameter. They utilize oil supply pressures in the relatively low range of 1–2 bar (~15–30 psi). Shaft rotation and bearing geometry cause higher pressures to be self-generated within the bearing. An oil film varying in thickness from 0.0001 to 0.001 in. prevents metal-to-metal contact. Since shear action on the oil produces heat, the lubricant must be cooled. Circulating-oil systems are best suited to accomplish the necessary cooling and filtering.

Thrust bearings must have the correct axial clearance, typically 0.008–0.010 in. (0.2–0.25 mm), to perform properly. During maintenance-related shutdowns and before further dismantling, the exact rotor float from its mid position toward the active and non-active sides of the thrust bearing should be checked. When the bearing is removed, the pads should be examined for wear and other surface damage. Also, the rotor free-float should again be checked before the compressor is fully assembled at site. Free-float shows the feasible axial impeller movement within the casing before the thrust bearing is being installed. Final readings are taken and recorded after installation of the thrust pads. In essence, the correct positioning of the rotor in its casing and observing thrust collar locations relative to the active and inactive thrust-bearing pads require considerable care and patience. Axial position probes (proximity probes) are used for onstream monitoring of rotor location.

Although bearing oil supply, oil drain headers, and bearing metal temperatures are typically monitored, bearing metal temperatures are the more important of the three. Alarm settings of 110°–120°C (230°–248°F) are generally used for bearing temperatures, whereas high header temperature (bulk oil) alarms are often set at 80°C (~176°F). Tin-based Babbitt is preferred over lead Babbitt due to its superior corrosion resistance and better bonding with steel backing. The tin-based material is, however, not as "forgiving" of dirt inclusions in the lubricating oil as is the lead-based version.

The bearings in gas compressors are dependent on a continuous supply of oil for lubrication and, more importantly, cooling. Generally, oil must be pressure-fed prior to machine start, and continue to flow until the machine has fully stopped. Oil quality in terms of grade cleanliness, low moisture content, temperature and adequate flow, are vital. Oil grades, even "equivalents", should not be mixed without draining the complete system.

Seals

Compressors require two nos of seals whether dry gas seal or wet seal as per design requirement and cost.

Baseplate

This will normally be common with the base frame under the driver (and gearbox if fitted).

Rigidity is vital to maintain alignment during temperature and load changes.

Lubrication System

The system will be mounted on or adjacent to the compressor baseplate, and designed to cope with start-up, normal operation, and rundown requirements.

Seal Gas System

Many modern machines have dry gas seals, the seal faces being lubricated and cooled by Nitrogen gas. This has the benefit of being simpler and taking less power than liquid supported fails.

Seal Oil System

Where liquid supported mechanical seals are used, the seal liquid is usually compatible oil.

Pressurized seal oil may be derived from the lubricating oil system, or be a self-contained system.

Suction KnockoutPots

Depending on the gas being handled, liquid water or light hydrocarbons may condense out.

Liquid droplets, or slugs in particular, can damage high-speed compressors. It is good practice to install an appropriately designed knockout pot immediately prior to the inlet of each compressor casing. Level detection and drainage are required.

Inter-Cooler/After-cooler

For efficient compression, within limits defined by material properties, it is often necessary to cool the gas within the compressor system. This may cause additional condensation of liquids, with interstage cooling requiring an integral or associated liquid knockout and drainage system.

Recycle Valve

Centrifugal compressors have a limited range of acceptable suction flow. For start-up purposes, and to control capacity, discharge gas is returned to suction through a recycle valve. When a high proportion of the flow is recycled, it is important that the flow route includes a cooler. Either the normal after-cooler or a dedicated recycle cooler may be used.

Control and Management Systems

The main control requirements will be dictated by the function that the compressor must perform. Flow control, suction pressure or discharge pressure control are common. Control may be done within a dedicated PLC or a plant DCS system. Parameters to be managed will include shaft speed, suction (or discharge) throttle valve control, and recycle valve control. Shaft speed control is achieved by a link to the gas turbine control.

Anti-surge control. When the process control calls for the compressor to operate with too low a suction flow, the anti-surge element of the control system will over-ride the process control and progressively open the recycle valve. If, due to fault conditions, this does not happen quickly enough, the compressor will enter "surge". This is characterized by violent oscillation of suction pressure and absorbed power. On detection of surge pulses, the compressor drive will trip and the recycle valve will be driven open rapidly.

Lubrication and seal service system controls. These are normally simple mechanical controls to maintain pressure, flow, and temperature of services. Loss of pressure will normally cause a compressor trip.

Troubles Related to Process

Change in Process Parameter

Since turbocompressors are designed for a chosen duty, it is important that the vendor be aware of, and understand, the full envelope of the duty. It is also important that any proposed extension of the operating envelope be explored with the vendor. Fortunately, changes in load resulting from changes in molecular weight or inlet pressure, will normally be limited by the normal machine controls. Problems will occur if changes are rapid, or if instruments are designed on the assumption of a fixed gas composition or inlet condition. The safety concerns relate to the possibility of over-pressurizing the discharge line, or overloading the bearings.

The greater the density of a gas, the greater the discharge pressure and the closer the surge point to the design point. This is because once any dynamic compressor stage is designed, the head (energy) produced for a given flow and speed is fixed. The greater the density of the gas (proportional to molecular weight), the higher the pressure produced and the lower the volume flow. Flow is inversely proportional to pressure (Boyles Law). Since surge is caused by low flow, a dynamic compressor handling a denser gas will surge sooner.

In the case of reduced gas density, the opposite effect will occur. The discharge pressure will be reduced and the surge point will move to the left, farther from the design point.

Effect of gas and temperature changes on pressure ratio

MOLECULAR WGT.	INLET TEMP	PRESSURE RATIO
INCREASES	CONSTANT	INCREASES
DECREASES	CONSTANT	DECREASES
CONSTANT	INCREASES	DECREASES
CONSTANT	DECREASES	INCREASES

Surge

Surge is a characteristic behavior of a centrifugal compressor that can occur when inlet flow is reduced such that the head developed by the compressor is insufficient to overcome the pressure at the discharge of the compressor. Once surge occurs, the output pressure of the compressor is drastically reduced, resulting in flow reversal within the compressor. When a centrifugal compressor surges, there is an actual reversal of gas flow through the compressor impeller. The surge usually starts in one stage of a multistage compressor and can occur very rapidly. Figure 13.7 shows the compressor surge diagram.

How Surge takes place in Centrifugal Compressor

Suction Throttling

Now consider the inlet flow decreases due to the suction valve throttling (they are many reasons will cause the compressor inlet flow rate decreases). Consider the compressor operating at the following conditions

Flow rate = 5500 kg/h

Discharge pressure = 20 bar.

Speed = 6000 rpm

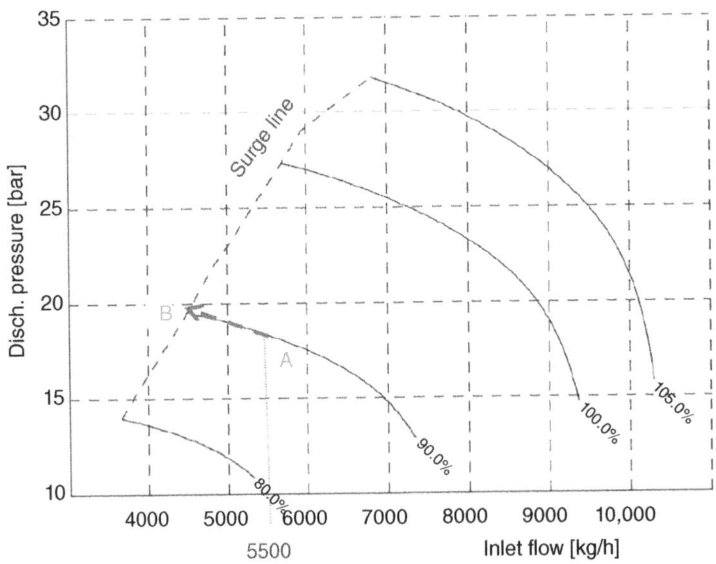

Figure 13.7 Compressor surge diagram.

Due to suction valve throttling the inlet flow rate of the compressor is decreased from operating point (A) to the new operating point (B) (Refer above figure). At the new operating point (B) the compressor flow is reduced on the other hand the discharge pressure will rise further. Due to the rise of pressure and decrease flow will cause the Surge cycle.

Discharge Valve Throttling

Similarly, if the discharge side system resistance will increase due to discharge valve throttling. The pressure developed by the compressor will increase and flow rate will start decreases. The same phenomena will happen, that is the operating point "A" shifted to the new operating point of "B". This will cause the surge in the centrifugal compressor.

Change in Speed

An increase in operating speed of the compressor also causes compressor surge. Now consider the operating speed of the compressor is 6000 rpm. If the speed increase to 7500 rpm. The Operating point of the compressor will shift from "A" to "B". (Refer below figure). Figure 13.8 shows how the change of speed affect surge in a centrifugal compressor.

At the new operating point "B", the discharge pressure of the compressor will increase. Due to the increase in pressure the point "B" fall in the surge line. Due to this, surge phenomena will occur in the compressor.

Other reasons that cause Surge in Centrifugal Compressor are.

Inlet Filter Chocking

Due to dirty particles present in inlet filter will decrease the flow rate and reduce the suction pressure. Due to the reduce flow rate, the operating point will move toward the surge line. Once the operating point touches the surge line then Surge occur in the compressor.

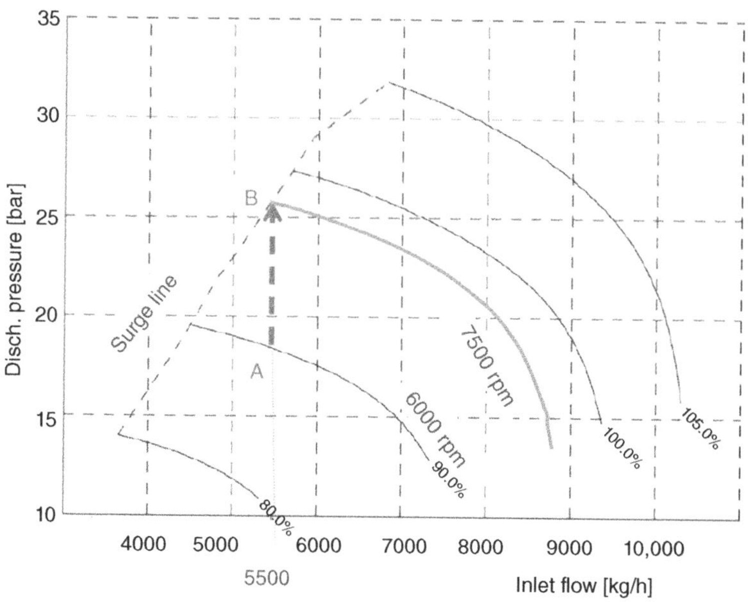

Figure 13.8 Change of speed affect surge.

Driver Input Speed

If the compressor is driven by Turbine or Variable speed drives sometimes the increase in speed may cause operating point to shift to surge limit line and surge will occur.

Change in Compressed Gas Property

The change in gas composition can cause compressor surge.

Effect of Surge in Centrifugal Compressor

As a result of the surge in the compressor, it may lead to damage of compressor or compressor system. The following are some of the resulting due to surging.

It is a high-speed phenomena, flow reversals can occur in less than 150 milliseconds. During surge, a significant mass of gas will flow in the reverse direction. As a result of a large dynamic force act on the impeller or blading within the compressor. Due to this, the components of the compressor (such as thrust bearings, journal bearing, casing) exposed to large changes in axial force on the rotor. If the surge is not controlled it may result in fatigue damage to compressor or piping components.

During surge pressure rapidly fluctuates, noise is generated, temperature increase can be very rapid and mechanical damage can occur. The intensity of surge varies from application to application and is proportional to the density of the fluid. Higher pressure and higher molecular weight applications can result in greater mechanical damage. Low density applications (hydrogen) can cause surge damage that may not be detectable until the equipment is disassembled. The cause of surge results from low velocity in the blade or impeller passages.

During surge, the reversal of flow within the compressor results in hot compressed gas returning to the compressor inlet. If the surge is not controlled, as a result the temperature at compressor inlet will increase and leads to a potential rubbing of close clearance components. The differential thermal expansion of components within the compressor can cause internal rubs of the compressor at operating speed resulting in impeller breakage, diaphragm breakage, extreme labyrinth seal wear and possible case breakage.

Avoiding Surge– The most important item for protecting the compressor from surge operation is the anti-surge control system. This control system should maintain a minimum volume of flow through the machine so that the surge condition is never encountered. This is achieved by bleeding flow from the discharge of the machine to maintain a minimum inlet flow. This flow can either be dumped to atmosphere or recirculated back into the inlet of the compressor. Provisions must be made for start-up and trip-out of the machine with sufficient through flow to prevent surging and excessive heating of the inlet gas.

Stonewall– Again, consider a constant-speed compressor with fixed suction conditions. As network resistance decreases, the operating point will move along the performance curve to the right. Eventually, a point of maximum flow and minimum polytropic head is encountered, beyond which further decreases in network resistance will not increase the flow rate. This is known as the choke point or stonewalls. The physical phenomenon is that the gas velocity has increased to the local acoustic velocity (therefore, Mach 1) at some point in the compressor. When choke occurs, the flow rate cannot increase unless conditions change in the choked region. Stonewall is not particularly damaging to single-stage centrifugal compressors but can cause serious damage to the rotors and blades of multi-stage centrifugal and axial compressors. In such situations, a suitably designed antichoke controller can be used to manipulate an antichoke control valve, thus maintaining sufficient system resistance to prevent choke.

Liquids in the System

One of the most potentially damaging occurrences for a compressor is the ingestion of liquid with the process gas stream. If there is liquid anywhere in the system, it is possible that some may carry over into the compressor. Knockout drums and demister pads do not always work the way they should. This liquid carryover will give erroneous results on the performance test. Another liquid problem is liquid in pressure tap lines. Care should be taken to ensure that all lines are properly sloped and drained. If lines are too small (less than 1/2 in.), capillary action will hold liquid in the lines. Ensure to open drain valves at low spots in process piping and instrument lines before, during, and after test.

The worst design is a long horizontal suction line, with a rising bend to the compressor. Liquid can collect in the line at low flow rates; this is swept up into a wave when flow is increased. A suction knockout pot as close as possible to the compressor, with drains, level detection, alarm and trip, is the normal solution. If liquid is rarely found, the drain valve may be blanked off.

Fouling

Compressors inherently increase the superheat in the gas being processed. Hence any free moisture in the gas will impinge on the first stages, and be evaporated. Any tars or salts will be deposited on the blades. This will cause fouling. Similarly any solids e.g. dust or catalyst particles will potentially be deposited on the blades (stator or rotor). The accumulation of debris in the impeller or blade passage causes fouling and it reduces the flow area and the surface finish. The distribution of the foulant on the impeller or blade row is nonuniform and usually changes with time. Flow patterns within the impeller or blade cause unequal distribution. In addition, the forces exerted on the foulant cause it to chip off with time as it becomes dry and brittle. This results in a change in rotor balance and a change in performance (head and efficiency).Also these build up /wash or wear off evenly, they might cause some loss in performance but no mechanical problems. Heavy build-ups that break off unevenly can throw the compressor out of balance. The unbalance, which gradually builds until the unit exceeds its allowable vibration limit and has to be shut down to correct the problem.

In addition, operation with significant rotor unbalance can lead to fatigue loading and a possible reduction in component life.

However, the stage head produced by the impeller at any flow rate is reduced. Therefore, for the same process system to maintain the head required, the impeller flow rate will be reduced thus forcing the operating point closer to the surge line. Note that the surge margin actually increases slightly in the fouled condition. This is because the cause of surge is low gas velocity. Since the area of the flow passage is reduced, the gas velocity increases thus increasing the surge margin.

Fouling affects efficiency through three basic loss mechanisms:

1) Friction losses
2) Flow area reductions
3) Random changes of pressure distribution on the blade

These mechanisms affect both the stationary flow path and the rotating element.

The Causes of Fouling

The phenomenon of fouling occurs as a result of air borne debris, Suction separator carryover or Process gas characteristics.

Preventing and Correcting Fouling

Some of the Preventive measures against fouling are

- Process control
- Proper inlet air filter selection and maintenance
- Condition monitoring (gas pressure, temperature and vibration)
- Online solvent injection
- Impeller or blade coatings

Corrective measures

- Slow roll or stationary wash
- Online random washing

Cleaning Centrifugal Compressors

Sometimes dirt, polymer build up, or other substances can clog the compressor internally and seriously degrade performance. Very small amounts of dirt on axial blades alter the blade profile and degrade performance. Cleaning a compressor may be all that is required to regain "like new" performance. A centrifugal compressor can be easily cleaned during normal operation (design speed) by using mild abrasives such as cooked rice or walnut shells. More common is the use of liquid cleaning agents sprayed into the process and into the return channel areas of the compressor.

Mechanical Related Troubles

Alignment

A gas compressor whose shafts are well aligned will run smoothly with minimum bearing loads. Thermal expansion of the compressor, the driver and the baseplate cause the alignment to change. In addition, thermal and gas pressure loads are applied by the process pipe work. The normal design arrangement is for the compressor to be supported at its centerline, with the pipe work supported to achieve flexibility. To allow for the machine movement without high stresses, a set of sliding keys is normally used, between the support posts and the casing. Jamming of these keys can cause misalignment and seal rubs. Similarly, jamming of any moving pipe supports can cause problems. Some designs use intentionally flexible supports. Provided that the thermal expansions are repeatable, the "cold" alignment is made with offsets for predicted expansion. This is a highly skilled job, preferably using laser alignment tools.

Lubricating Oil Systems

Lubricating Oil Systems supplied as part of packaged equipment are a packaged sub unit of the equipment. The system will provide oil to act as lubricant to bearings and wearing surfaces which may in particular cases provide a buffer fluid for seals. Figure 13.9 shows the process schematic diagram of a compressor lubrication system.

The above diagram is a typical lubrication system for a large compressor package. It has duplicate oil pumps, filters and coolers and a rundown tank. The pumps should be powered from independent sources; e.g. separate electric supplies or one electric pump and one shaft driven pump or turbine driven.

Oil pressure is regulated to maintain the correct supply manifold pressure, and excess cooled oil is returned to the tank. Prior to machine start, the oil pump is started, which fills the Coast Down (or Run Down) Tank (if fitted). When the tank is full, excess oil overflows at a slow rate to prevent stagnation, a level switch providing a "run permit" signal to the Machine Train. In the event of complete loss of oil supply, the Machine Train must be tripped, and the Coast Down Tank then provides enough oil to supply the bearings until the train stops. Note that the Coast Down oil supply rate is not enough to supply the bearings under full process load. Small units may use a hydraulic accumulator as a substitute for the Coast Down Tank. It is common to heat the oil tank to about 40°C prior to start and to arrange for

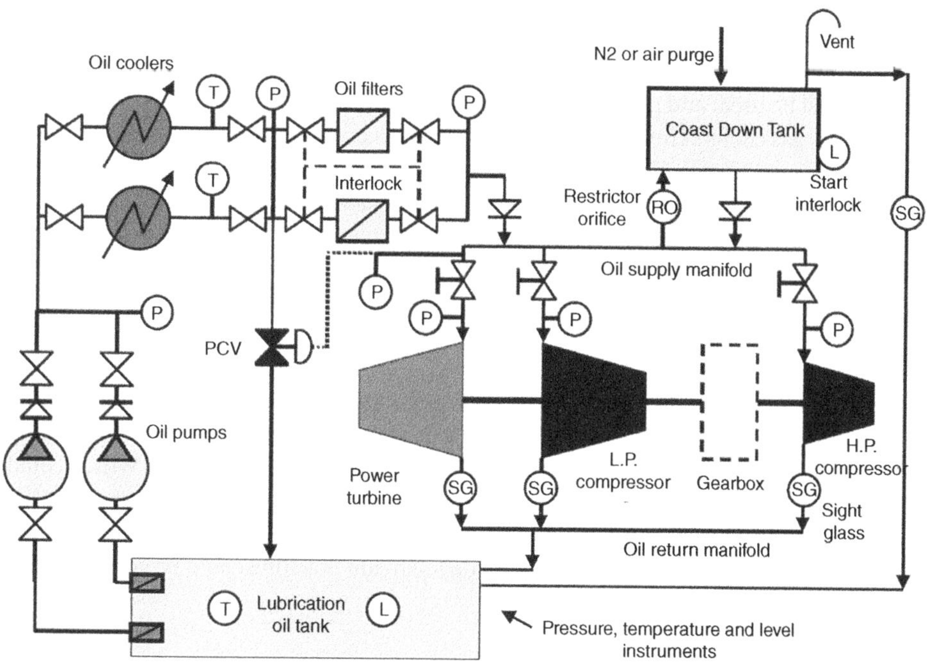

Figure 13.9 Process schematic diagram – power turbine/gas compressor lubrication.

the normal tank temperature to be 40–50°C. This keeps the oil viscosity in the correct range, and prevents moisture condensation. The oil system may have its own simple logic control, or be managed within the Package Control. Start-up and Shutdown sequences should be automatic and alarms / trips based on excursions of pressure, level and temperature. Changes to / operating out of sequence should be subject to a formal change control process.

Oil quality, temperature and pressure are all key parameters for successful machine and rotating equipment operation. This is particularly so when tilting pad bearings are running at maximum load (usually this is at maximum shaft power). Oil temperature and pressure are continuously controlled and monitored, quality is checked by periodic sampling e.g. at monthly intervals. It is appropriate to test for water, metals (steel from gears, shafts, ball bearings: white metals from plain bearings: aluminum from labyrinth seals, copper from cooler tube corrosion), chlorides (indicates cooling water leakage), and process gas contaminants. At longer intervals, it is appropriate to test the condition of the additives within the oil, and these can be topped up to maintain the condition. On a "clean" duty e.g. where the oil does not come into contact with process or combustion gases, it should seldom if ever be necessary to change the oil. Topping up should be with the same oil grade from the same supplier. Mixing of "equivalent" grades can give problems with incompatible additives.

Oil storage should be well managed; opened containers should be mounted above containment trays, and re-closed after use. Oil labeling and handling procedures should cover oil management, hygiene, spillage, and avoidance of contamination, use of correct containers.

Cooler

The main obvious cooling requirement on a gas compressor is the lubricating oil; this can be cooled against air or water. Loss of cooling effect normally results in a controlled trip with no safety implications. A cooler tube failure can be more subtle - air cooled tube failure can spray oil mist causing a fire risk, water tube failure can contaminate the oil causing rapid bearing failure. The cooler design should permit tube inspection/testing, corrosion resistant materials must be used, and gasketed water/oil joints avoided.

The process gas coolers (inter, after, recycle) provide a cooling load dependent on process duty requirements. The cooling load may be massive (of the order of the compressor drive power) thus take a considerable part of the plant's available cooling capability. Any limitation in cooling will give higher than intended gas temperatures, reduced compressor efficiency. Provided that gas cooler inlet and outlet temperatures and pressures are available, loss in cooling performance can easily be detected and creates no hazard.

Intercooler

Also any condensed liquids must be efficiently separated. Intercoolers are often used to improve compression efficiency and limit discharge temperatures. If the intercooler is ineffective, operating temperatures in the downstream machine will be excessive, Which could result in mechanical problems, such as seal rubs. The intercooler will usually condense liquid, if this is re-entrained, particularly as slugs, damage to the downstream machine blading could result. If the intercooler tubes leak, the downstream machine could be contaminated with salts from the cooling water.

Control Related Troubles

There will be a simple logic control to ensure that services like oil and seal pressure are maintained. These must be established prior to start, and loss in, e.g. oil pressure will raise an alarm and bring in line a spare pump. Loss of such services will trip the compressor. Operating the compressor with such trips defeated is unsafe. Bearing vibration, shaft position and bearing temperatures will raise alarms first, may then tripas conditions worsen.

The operators require a clear understanding of which alarms/ trips are over-ridden during start, and why. Test procedures are required to validate that the important trips are re-enabled at the correct time. Formal testing and recording of alarm / trip tests, at intervals defined by or agreed with the manufacturer, are required. If tests are done with the machine on line, it must be recognized that the trip is disabled during the test, and that a partial test only is possible. The control system will initially be configured to suit the intended duty of the machine. Should the duty change significantly, the control system may have to be re-configured to achieve effective results.

As gas flow through the compressor is reduced, the compressor will approach surge. This will be countered by altering Inlet Guide Vane geometry (where fitted) and by opening the recycle valve. Should the surge control system fail to prevent surge, the surge pulses will be detected and trip the machine. Normally the recycle or anti-surge valve is rapidly driven open and the driver tripped. Surge conditions cause rapid and severe pressure & flow fluctuations, which can damage items like bellows.

Ancillaries

Where certain ancillaries e.g. filters, oil pumps, are twinned, it is possible to service or remove one unit while the compressor is on line. The operating instructions must cover the attendant risks of operating with one such unit unavailable, and the potential consequences of an incorrect changeover. One example is putting an oil filter on line without priming it, lubricating or control oil supply may be interrupted. Instructions and, if necessary, labels, may be required if changeover valves must be operated in a particular sequence, or can readily be mal-operated.

Some compressor installations have header tanks or pressure accumulators to provide lubricating oil for rundown in the event of complete power failure. These systems only work if the tank/accumulator is full, the relevant valves are open, and the drive is tripped immediately the main oil supply is lost. The gravity oil flow is not sufficient for full speed / load operation. As live testing of such systems can cause significant machine damage, it is prudent to carry out simulated tests with the compressor stopped. If the oil supply lasts longer than the run-downtime, one can be confident of effective operation.

Monitoring during operation and preparation for planned maintenance. Online checks and inspections are the first area requiring attention. Vibration monitoring equipment and the interpretation of the vibration signature is a critical area. Important performance parameters from the compressor train and its support systems need to be collected and periodically reviewed. This includes computer recorded data as well as the machinery engineer's field observations. The daily operation and maintenance requirements must be fully understood to maintain equipment reliability.

Seal Systems (Dry Gas Seals and Liquid Seals)

Dry Gas Seal Systems

The primary purpose of a dry gas seal system is to provide clean dry seal gas to the compressors within an acceptable range of pressure, temperature, and flow. Monitoring the following will facilitate the effort.

Filters – Monitoring differential pressure across the filter in operation provides indicates the cleanliness of the element. A log should be kept of this parameter, and the filter element should be changed when recommended differential pressure is reached. The seal gas supply system must be kept free of liquids with an appropriate monitoring strategy to check for oil (from previous seal oil systems) or gas condensate. If liquid is noted on regular basis, verification of the seal gas heater operation (if equipped) should be checked.

Log Pressures/Temperatures/Flows – Following commissioning of the system, regular/expected values (seal supply, intermediate seal supply, barrier seal supply, vent flows, etc.) will be identified for the various pressures/temperatures/flows in the system. Trending this information will allow the operator to identify any changes over time.

Low Point Drains on Primary/Secondary Vents – Regular checks (timing dictated by findings) for presence of condensate or other liquids. If liquids are found on a regular basis, you need to verify what type of liquid it is(condensate, water, lube oil,) and source. This could be an indication of a future seal failure.

Heater Elements – If supplied, seal gas heater performance should be monitored and trended. The intent is to determine whether fouling of the heating element is occurring, which is indicated by increasing/excessive electrical demand by the heater.

Heat Tracing Performance – Many systems include heat tracing of the seal supply lines between the gas seal panel and the compressor to prevent liquid drop out. Recommendation is to monitor the electrical demand (as compared to baseline information post commissioning) for this system to ensure it is functional.

Liquid Seal Systems

As with dry gas seal systems, the intent is to provide a clean sealing medium (oil in most cases) to the compressor. Please note that in a sweet gas compression process, seal oil systems are often incorporated directly into the lube oil system. As such, many of the performance monitoring requirements are the same. The key difference is the presence of degassing tanks, and the necessary checks on these components.

Log Pressures/Temperatures– Following commissioning of the system, regular/expected values (sour seal oil traps, degassing tank,) will be identified for the various pressures/temperatures in the system. Trending of this information will allow the operator to identify any changes over time.

Sweet Oil Condition– Regular analysis of the sweet oil returning to the main reservoir should be performed. Intent is to verify that all the process gas has been removed from the oil.

The sour seal oil leakage Flow rate can be measured by "bucket-and-stopwatch" approach, e.g. block in the seal drainer outlet, and time the level increase in the sight glass. Large changes in sour oil flow rate can indicate a damaged or distressed seal. Trending of this data over time will indicate seal degradation.

On some systems that have sight glasses where the leakage can be seen, it provides a qualitative amount of leakage which indicates when the seals are not performing as designed.

Other Auxiliary Systems

Lube Oil Systems– The key purpose of the lube oil system is to provide clean oil at an appropriate pressure/temperature/flow rate, to the bearings, as dictated by the OEM. Monitoring the following will facilitate this effort.

Filters– Monitoring differential pressure across the oil filter during operation indicates cleanliness of the filter element. A log should be kept of this parameter, and the filter element should be changed when recommended differential pressure is reached. They should also be replaced on a time limit because they may deteriorate due to contact with the oil or process gas.

Oil Cleanliness/Quality– Regular oil analysis should be performed to ensure that there is no buildup of contaminants or water content.

Log Pressures/Temperatures/Flows– Following commissioning of the system, regular/expected values (pump output pressure, reservoir temperatures, and oil flow) will be identified for the various pressures/temperatures/flows in the system. Trending of this information will allow operator to verify any changes over time.

Coolers– Monitor/record pertinent cooling water flows/temperatures. Trend cooling water flow required in order to maintain required oil temperature levels. An upward trend could be a sign of fouling on either the oil or the water side.

Oil Level– While there are generally alarms associated with low level, it is prudent to regularly check the oil level in operation. Small changes trended over a long period may indicate loss of oil either via piping leakage, excessive misting or worst case into either the dry gas seal system (if equipped) or the process (if oil seals installed). If identified, the source of loss needs to be determined

Auxiliary or Secondary Oil Pumps – It is good practice to test the secondary pump at an outage to verify pump start-up maintains oil pressure without initiating a trip of the unit.

Accumulators – check on accumulator pressures on any systems that have them

Overhead Rundown Tank – Daily visual inspection while online of sight glass for signs of water in the oil.

Maintenance Intervals

There are two basic categories to consider when setting the maintenance strategy for the unit planned outages: operating constraints and maintenance/mechanical constraints

- Operating constraints can include: fouling/corrosion/erosion of nonmachinery operating equipment (drums, towers, reactors, etc.). There can also be government regulatory inspection interval requirements. Sometimes, these intervals are shorter than the machinery requirements.
  - Maintenance/mechanical constraints should initially be based on OEM recommendations when the unit is first commissioned. As unit experience is accumulated, the interval on certain items can be adjusted with careful review of the operational history. Where the operating constraints interval is shorter than some (or all) of the

maintenance mechanical constraints, the strategy for certain planned activities may be a mixture. For example, the machinery must be shut down for a reactor catalyst change every three years. The OEM recommends seal overhaul/replacement every five years. The strategy may be to replace the seals at the first outage after three years of operation. If the seals are found to be in excellent condition, and the risk is believed to be low enough, then the seals will be replaced again in six years (skipping a planned outage).

Items that can set turnaround intervals are:

- Steam turbine internal erosion (or fouling) of rotating or stationary components
- Compressor internal fouling, erosion, corrosion
- Compressor seal leakage rate increase (sour oil loss) due to fouling or coking of the seal
- Bearing temperature – long-term trends
- Coupling wear (for the older gear type couplings)

Importance of Engineering Rounds

Most turbomachinery tends to be the heart of large scale production facilities, and it is important to recognize the benefits beyond instrumentation that field inspection can offer. While field inspection by operators can identify many issues, a set of trained eyes that haven't been narrowed by daily walk arounds provides yet another means of ensuring production by predicting problems and improving work scope development.

Online Checks and Inspections

Vibration Monitoring System

Vibration monitoring is likely the most fundamental condition monitoring system for turbomachinery. Thus, ensuring that this system is properly designed and installed and that the data is regularly reviewed are crucial to satisfactorily monitoring the equipment.

Vibration monitoring of turbomachinery trains is typically accomplished using proximity probe systems. Gearboxes with their inherent high frequency vibration typically require the addition of accelerometers. *API 670 Machinery Protection Systems* provides a thorough overview of system design information for vibration monitoring as well as temperature and speed detection systems. During maintenance activities, probes that cannot be accessed while the machine is in service are typically replaced.

System Performance Monitoring of Auxiliary Systems

Auxiliary systems are necessary to support the rotating equipment. As such, their performance and readiness are critical to ensure trouble-free operation. Critical systems included in this discussion are the seal systems, auxiliary steam systems, and lube oil systems. Following are recommendations on how to monitor the performance of these systems.

Conclusion

Centrifugal compressor represents the most critical equipment in the process industry. Common mistakes and carelessness can cause damage to the machine and loss of production. Issues from bearing damages to compressor internal damages can be experienced. There are quite a few aspects to consider relating to the operations and maintenance of compressor trains and their auxiliary systems. Neglecting any one of these aspects can lead to decreased performance of the equipment, decreased reliability, and an increase in maintenance costs. Through careful consideration and planning, these risks can be mitigated by understanding the equipment requirements and diligent adherence to the best practices. With regular review of operational trends and physical inspections one will become more intimately familiar with the equipment. This will ultimately support the overall goals of safe operation, increased reliability and decreased maintenance costs associated with the equipment.

14

Reciprocating Compressor Failure

Reciprocating compressors are positive-displacement machines, most widely used type of compressors in process industries. They are often some of the most critical and expensive systems at a production facility, and deserve special attention. They are manufactured in a variety of different configurations, including vertically oriented, skid-mounted, lubricated and non-lubricated type. Reciprocating compressors have a wide range of applications also in oil and gas, petrochemical and power industries.

Major Components

Frame: The frame consists of a rigid casting with ribs and partitions incorporated to provide high strength at reasonable weight. In the side walls and the partitions the saddles for the high precision crank shaft bearing shells are machined. Lubrication of the frame is accomplished either by a pump driven from the crank end or by a separately mounted pump.

Crankcase: The crankcase is a U-shaped cast iron or fabricated steel frame. It serve as a reservoir for the lubricating oil. The top is left open for installation of the crankshaft. The main bearings, spaced between each throw, have removable top covers for ease of assembly and ready removal of the babbitted bearing liner shells.

Crankshaft: The crankshaft is the heart of the machine, and usually the most expensive component. It is a single piece heavy steel forging. Each throw is forged and counter weights are bolted on to balance the reciprocating mass of the crosshead and piston. The crankshaft has holes drilled from the main bearing surface through to the connecting-rod bearing face. From here, the oil passes up through a hole in the connecting rod to the wrist pin and from there, through holes to the crosshead sliding faces. Oil scraper rings in the frame end prevent oil leakage out along the piston rod. Because of this tortuous passage of oil, prepublication is required before start-up. This is accomplished with an auxiliary lube pump.

Connecting Rod: The connecting rod is a forged steel and have split big ends, in which the outer bearing shell is dowelled to the big end bearing cap to provide locking against rotation. Axial location of the connecting rods is provided by the big end bearing end faces and the crank journal shoulders or by the cross pin bush and the cross pin bosses.

Root Cause Failure Analysis: A Guide to Improve Plant Reliability, First Edition. Trinath Sahoo.

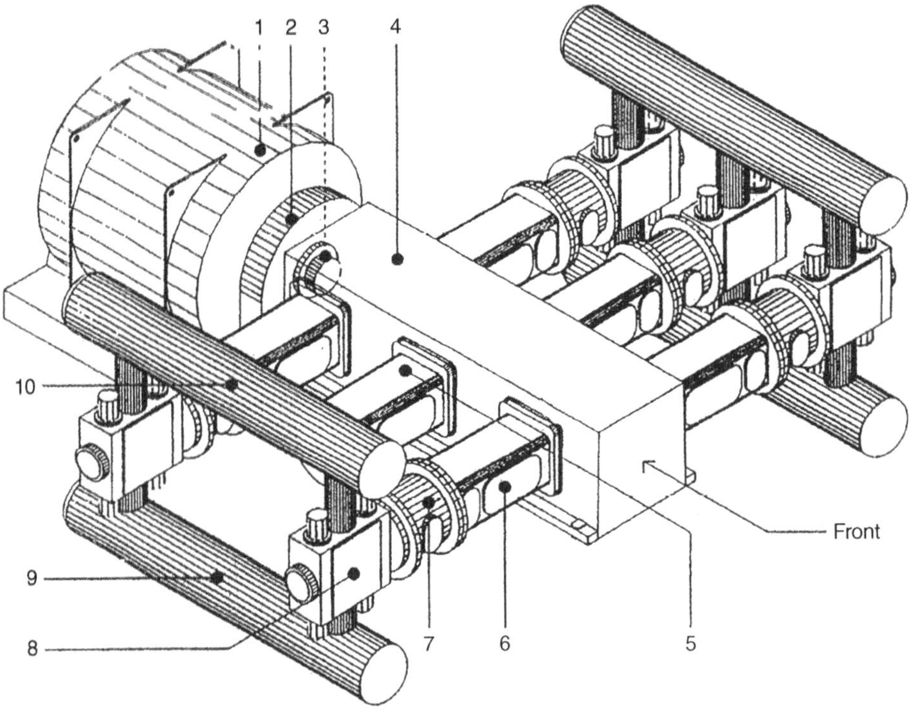

1 Driver
2 Flywheel
3 Crankshaft
4 Frame
5 Dummy cylinder
6 Crosshead section
7 Compartment
8 Cylinder
9 Discharge pulsation damper
10 Suction pulsation damper

Figure 14.1 Compressor components.

Distance Piece: The distance piece is a separable housing that connects the cylinder to the frame. Compartments in the distance piece collect and control packing leakage when the gas is toxic or flammable. The leakage can be directed to a flare or buffer inert gas is used to prevent the toxic gas leakage to atmosphere.

The closed cross head guide housing provided with inspection holes and covers, giving access to the stuffing box or to the cross head for inspection and dis-assembly. In some cases, removal of the cross head can take place through the frame top cover.

In the partition, oil wiper rings are built in cross head side, while gas sealing rings are fitted at the cylinder side. An oil slinger ring is mounted on the rod to prevent oil carryover. In those cases where absolutely no gas leakage is allowed, two separate compartments are built between the cylinder and the guide to provide total separation. In addition one compartment can be purged with inert gas.

Crosshead: All crossheads are made of cast steel and have white metal bearing surfaces. These are provided with removable shoes. The crosshead runs between two guides with about one mil per inch (0.001 per inch) of diameter clearance. It is often weighted so that the mass inertia of all reciprocating parts is sufficient to reverse the stress on the wrist pin, even when one end of the piston is under pressure. If this is not done, the wrist pin will wipe all the oil from the side under stress and will bind.

The latest development in piston rod to crosshead connection is the so called hydraulic connection. The design avoids all the problems normally encountered in piston rod assembly. During assembly the piston rod crosshead connection is preloaded with hydraulic tools to such an extent that the actual cyclic load only results in a low cyclic stress. This combined with rolled threads give maximum fatigue strength.

Cylinder: Normally cylinders of the reciprocating compressor are separable from the frame. They are attached to the frame by way of an intermediate part known as the distance piece. The cylinder will accommodate the both suction and discharge valve plates.

Cylinders are normally double acting, means on both sides of the piston gas is being compressed. Materials Up to 1000 psi, cylinders are normally made of cast iron. Above this working pressure, materials are cast steel or forged steel, at the manufacturer's discretion. Nodular iron castings are sometimes specified in preference to cast iron. API-618 specifies that all cylinders must have replaceable liners. Cylinder liners are removable type and are made of cast iron or steel alloy. Liners should be honed to a finish of 10–20 micro inches. The purpose of the liner is to provide a renewable surface to the wearing portion of the cylinder. This saves the cost of replacing a complete cylinder once the bore has been worn or scored. On the smaller frames, particularly the single-stage models, the smaller cylinder size is such that the replaceable liner is not economical and may not be available. Normally cylinders are water cooled some times with a mixture of water and glycol.

Pistons: Pistons are usually cast iron and are often hollow, to reduce weight. This space can fill with gas and is an explosive hazard when removing the piston from the rod. One should, therefore, specify that an easily removable plug must be supplied to vent this space before handling. Larger pistons are made of aluminum to reduce weight. These pistons have large clearance in the bore to allow for thermal expansion, on the order of 20 mils per inch of diameter.

Piston Rod: The piston mounts into the crosshead and locked, either by a locknut or a pin, to prevent backing off. The rod is adjusted in the crosshead to equalize the end clearance of the piston in the cylinder. This is checked by barring over the machine, crushing a piece of soft lead, and measuring the remaining thickness. This is called the bump clearance. The piston rod is normally constructed of alloy steel and must have a hardened and polished surface particularly where it passes through the cylinder packing (double-acting cylinders). Some rods are chrome plated, but problems have occurred with them, especially on high-pressure machines with a high heat buildup. This can easily cause spider web cracks in the chrome which, in turn, can flake off and destroy the packing. The best arrangement is to purchase a flame-hardened rod. As wear occurs, the rod could be plated with tungsten carbide, which should last the life of the machine. Rod loading must be kept within the limits set by the compressor vendor because overloading can cause excess run-out of the rod resulting in premature packing wear. This, in turn, leads to leakage, reduced efficiency, and increased maintenance expense.

Piston Ring and Rider Ring: The rider rings, used mainly in oil free or mini-lube compressors, is to support or guide the piston and rod assembly and prevent contact between the piston and the cylinder (risk of seizure). Nowadays rider rings are also used in lube compressors.

Piston rings, mounted on the pistons of lubricated or non-lube (oil free) compressors, are designed to ensure that the gas is compressed and to provide a seal between the piston and the cylinder.

Rod Packing: Rod packing is required to prevent the gas leakage along the piston rod where it passes through the crank end cylinder closure. Compressor packing is made up of two rings in pairs, mounted in steel or cast iron cups with the open end of the cups facing away from the pressure. The cups are bolted together and have vent, oil, and drain holes drilled in them where required. These must be correctly aligned each time the packing is opened. If Teflon is specified for packing for pressures above 500 psi, generally an additional metallic backup ring is used to prevent the Teflon from extruding out of the cups. Compressor manufacturers supply a distance piece between the cylinder and the crankcase for access to the packing. This is usually good for a three- or four-cup set. In process applications, however,

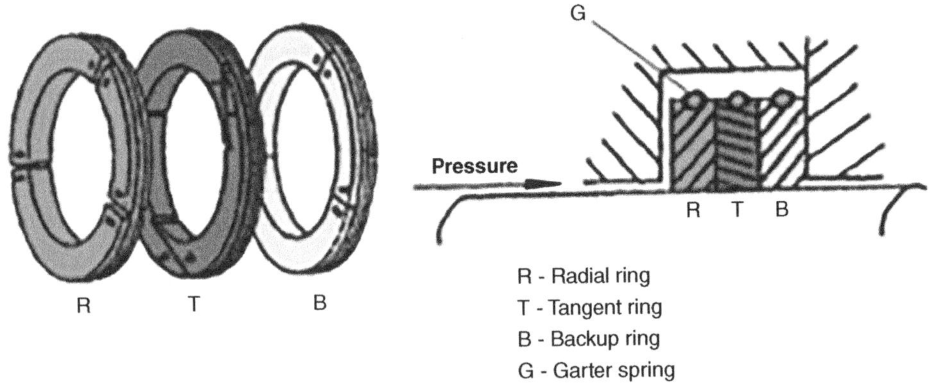

Figure 14.2 Schematic view of piston rod packing.

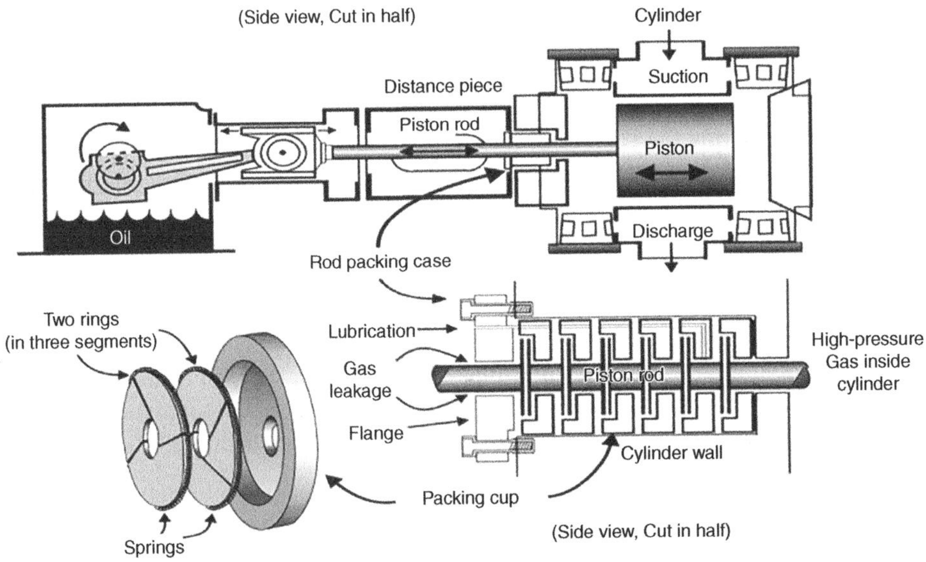

Figure 14.3 Typical compressor rod packing system positioning in the compressor.

especially above 2000 psi, packing sets run to 8 or 18 cups or more. Extra long distance pieces must be specified in order that sufficient space is available to dismantle the cups and change the packing rings.

The main packing rings types are

- Tangent Ring
- Radial Ring or Pressure Breaker Ring
- Backup ring

If cooling of packing is required, the stuffing box may be jacketed for liquid coolant.

Valves: In a reciprocating compressor cylinder, the valves have the greatest effect on the operating performance of the machine, both from an efficiency stand point and from a mechanical reliability stand point. The main component of a valve are:

- Valve seat
- Sealing element
- A stop to contain the travel of the sealing element

Valves should have the following features.

- Large passage area and good flow dynamics for low throttling effect (pressure drop).
- Low mass of the moving parts for low impact energy.
- Quick response to low differential pressure.
- Small outside dimension to allow for low clearance volume.
- Low noise level.
- High reliability.
- Ease of maintenance.

Pulsation Dampener: Since a reciprocating compressor has an intermittent delivery, pulsation in the gas flow will happen. This may be the source of vibration in the piping and equipment. In order to decrease the pulsations to a harmless level Pulsation dampners are used.

Lubrication System

Force feed type lubrication is use for medium and heavy duty compressor. The main component of the force type lubrication is the main oil pump, dual oil filter, lube oil cooler, pressure regulating valve, lube oil tank, and its associated accessories.

Rod Load

Due to gas forces on both sides of the piston, a resulting force is acting on the piston rod. This force is called the rod load.

$$A = \text{area piston in m}^2$$

$$a = \text{area of piston rod in m}^2$$

$$F_1 = A \times p_s \times 10^5 \text{ Newton}$$

$$F_2 = (A - a) \times p_d \times 10^5 \text{ Newton}$$

$$\text{The rod load is: } F_r = F_2 - F_1 \text{ Newton}$$

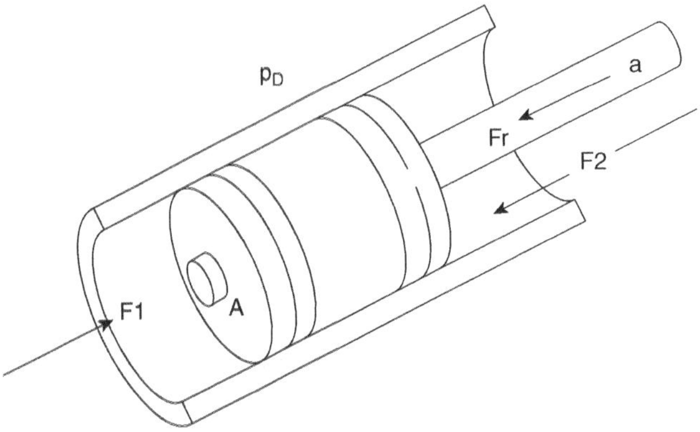

Figure 14.4 Forces acting on a compressor piston.

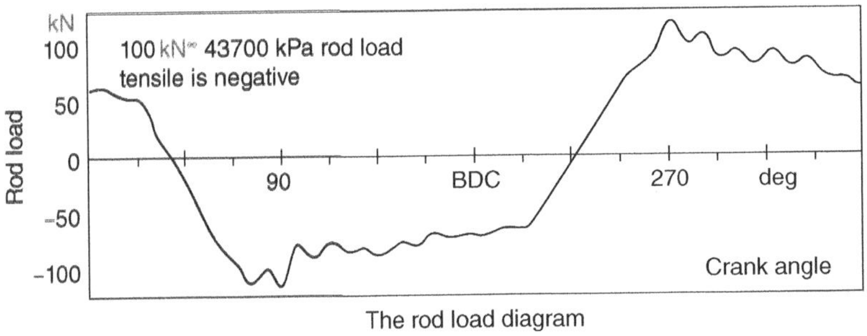

Figure 14.5 Rod load diagram of a reciprocating compressor.

The calculations shown above provide the gas load imposed on the rod (and crosshead bushing) by the compressor cylinder piston. Rod load is calculated from the crank angle pressure curves measured from the head and crank end of the compressor, and the inertial effects of the reciprocating mass. This is used to determine if the compressor is too heavily loaded either in tension or compression. The maximum tension and compression values are compared to limits provided by the compressor manufacturer to determine if a problem exists. When rod loads are exceeded, components break or have accelerated fatigue failure due to excessive stress.

To provide a reasonable crosshead pin bushing life, the rod loading at the crosshead bushing must change from compression to tension during each revolution. This is commonly referred to as "rod reversal" and allows oil to lubricate and cool one side of the bushing while load is being applied to the other side of the bushing.

When the piston moves the other way, the direction of the force will reverse. Because of the rocking movement in this bearing, normal hydrodynamic lubrication can not be established. Therefore oil can only be admitted in to this bearing by virtue of an alternating clearance which is caused by the rod load reversal. This rod reversal is essential to guarantee proper lubrication of the connecting rod small end bearing.

Rod reversal is calculated from the same information as the rod load. The rod must reverse in load from compression to tension and back for a sufficient amount of degrees of rotation in order to provide proper lubrication to the crosshead pin and bushing. A rough guide is that reversal must occur for at least 15% of rotation and have a magnitude of greater than

Table 14.1 Percentage failure due to separate compressor components.and process problems.

Compressor valves	36.0 %
Pressure packings	17.8%
Process problems	8.8%
Piston rings	7.1%
Rider Bands	6.8%
Unloaders	6.8%
Cylinder lube system	5.1%
instrumentation	5.1%
other	6.5%

three percent of the loading in the opposite direction. For more information about rod load and rod reversal, consult API 618. Alarms can be set to check for a minimum number of degrees of reversal for both directions of loading.

Reciprocating Compressor Failure Causes

In most reciprocating compressor installations the maintenance cost of valves, packings and rings account for approximately 70% of the overall maintenance budget. Table 14.1 highlight the percentage failure due to separate compressor components and process related issues.

As Table 14.1 indicates, leaking valves, packing, and rings, process problems are the primary cause of compressor failure. However, when compressor valves, rings and packing leak, the following occurs:

- Gas discharge temperatures rise;
- Percent of rod reversal may diminish; and
- Rod loads increase on other stages.

The effects of these changes can have a significant impact on how long a compressor will operate safely and efficiently. It has been shown that when gas discharge temperatures approach 300 °F, valves and rings deteriorate faster, the lubricating ability of oil is diminished, and Teflon products start to degrade. When the percent of rod reversal is minimized, the crosshead pin and bushing do not get adequate lubrication and clearances begin to open. When rod loads are exceeded, components break or have accelerated fatigue failure due to excessive stress. We will discuss the root cause of failure of valves, packings, rings and process related issues in the following section.

Compressor Valves Failure Causes

The compressor valves are the most critical component in a reciprocating compressor because of their effect on the efficiency (horsepower and capacity) and reliability of the compressor. Compressor valves are nothing more than check valves, but they are required to

Figure 14.6 Poppet valve.(take drawings to below as marked)

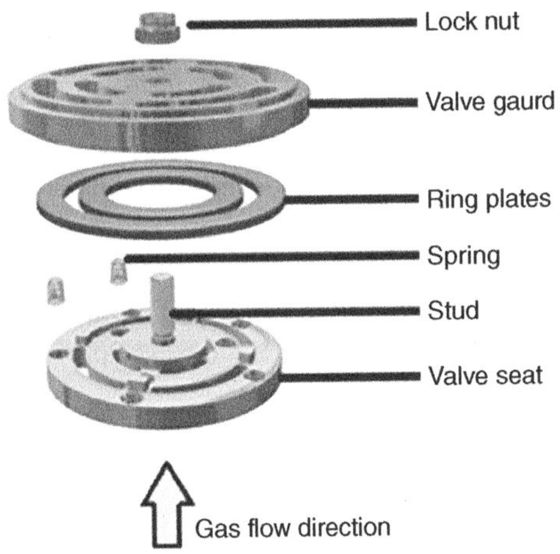

Figure 14.7 Ring valve.(take drawings to below as marked)

operate reliably for about a billion cycles, with opening and closing times measured in milliseconds, with no leakage in the reverse flow direction and with low pressure loss in the forward flow direction. To make matters worse, they are frequently expected to operate in highly corrosive, dirty gas, while covered in sticky deposits.

The commonly used valves are – poppet valves (Figure 14.5), ring valves (Figure 14.6), and plate valves (Figure 14.7). In all valves, the differential pressure across the seat must be greater than the balance spring force before gas flow through the valve. The lift characteristics, seat area, and flow areas determine the advantages of each design. In older compressor stations, channel valves were commonly used. Channel valves are now considered obsolete and used only in small air compressors.

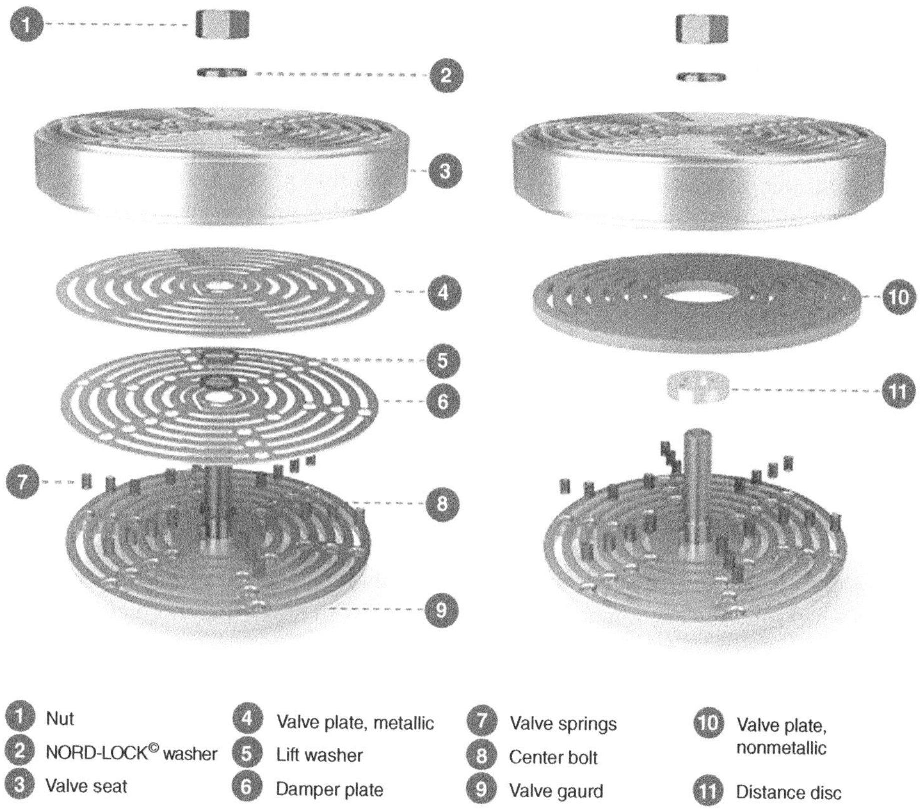

Figure 14.8 Plate valve.(take drawings to below as marked)

Poppet valves: are typically used for low compression ratio applications, such as pipeline booster compressors. As the pressure differential increases across each of the individual poppets, They lift and allow gas to pass through the flow openings in the stop plate.
Ring valves: are typically used for slow speed, high pressure process compressors. Instead of individual poppets, these valves use concentric rings, which open and close the valve ports.

Plate valves are typically used for high speed compressors.
Plate valves: are similar to ring valves with the rings connected by ribs. Instead of individual elements opening and closing, all valve ports open and close at the same time.

Valve type and size should be specified by the compressor manufacturer. Normally, the manufacturer will quote a valve velocity, which can be calculated from:

$$V = 288D/A \tag{11.1}$$

where

$V =$ average gas velocity, ft/min

$D =$ cylinder displacement, ft^3/min

$A =$ product of the actual lift, the valve opening periphery, and the number of inlet or discharge valves per cylinder, in^2.

At lower velocities the valve has less pressure drop and thus has less maintenance associated with it. Velocities calculated from this equation can be used to compare valve designs.

In addition to valve velocity, the manufacturer can furnish the effective flow area of the valve. This area is determined by measuring the pressure drop across the valve with a known flow rate and then calculating an equivalent orifice area that provides the same pressure drop. Valves with larger effective flow areas have less pressure drop and better efficiencies, The effects of the seat area, the lift area, and the flow paths are automatically included when the effective flow area is used to compare compressor valves. This in turn provides a better comparison of valve performance than just looking at valve velocity. In addition to valve efficiency, the following should be considered in valve selection: ease of maintenance, durability, and spare parts required.

Valve Failure Effect on System

1) **Higher-Than-Normal Suction Pressures**
   Irregularities are generally a sign of poor system health, including instances of valve failure, which sometimes results in uncharacteristically high suction pressure. A tell-tale sign is when surges in suction pressure are combined with lower-than-normal discharge pressure.
2) **Low Discharge Pressure**
   If the discharge from your reciprocating air compressor is weak and inefficient, the valves are a probable culprit. A compressor is designed to discharge pressurized air at specific consistencies as dictated by the system settings. When this fails to occur or when the pressure appears inadequate, the issue is likely due to obstructive valve passage. The situation is most apparent when the weak discharge pressure is coupled with excessive suction pressure because here you have correlating processes mutually failing at their respective functions.
3) **Unusually High Discharge Temperature**
   If a discharge valve fails to seal securely, an unusually low discharge pressure will result from the problem. During each upstroke, air will be discharged from the cylinder. Within seconds, the air becomes pressurized, and some of it slips back into the cylinder due to the unsecured valve seal. This short cycle through which the air is subjected results in unusually high discharge temperatures.
4) **Carbonized Valve Oil and Leakage**
   When a discharge valve is forced to cope with exceedingly high temperatures, carbonization is liable to take root in the oil along the lining, plate and backings of the valve. As the overheating spreads, the iron material that comprises the valve is liable to accelerate the chemical reaction. The valve is also likely to leak as a result of these effects.
5) **Contamination of the Valve Plate**
   Another effect of excessive valve heat is the development of contamination, which can form along the valve plate as a chemical reaction to the high temperatures. Contamination comes from a mix of broken-down oil and dust in the pressurized air, which combine to form a sludge along the valve plate.
6) **Overheated Air Compressor**
   If a reciprocating air compressor is unable to cool itself properly, the uncharacteristic heat could easily stem from an obstructed air passage within the system. As the pressure

mounts, the compressor could overheat due to its inability to pass air as programmed. Such problems are often rooted in faulty valves.

7) **Overloaded Air Compressor**
If too much pressure mounts within an air compressor, air is not flowing fluidly from one compartment to another. The process is intended to move at an even pace from the time that air first enters the system to its ultimate end use. If the pressure mounts and the system becomes overloaded, there is likely an issue of obstructed flow that could easily be rooted in a valve.

Valves Failure Effect on Capacity-

Compressor valves affect performance due to the pressure drop caused by flow through the valve; the leakage through the valve in the reverse direction.

The valves due to its inertia, does not open instantaneously, due to the springing, the valve does not stay at full lift for the full time it is open; and the valve does not close exactly at the dead center. All of these factors affect both the capacity and the power of the compressor.

The power losses caused by the valves are well known. The effects of the valves on the capacity of the compressor are less obvious, but equally important. The valves affect the capacity in three ways.

1) As the valves never close exactly at the dead center, the amount of gas trapped in the cylinder is never that predicted from simple theory. The springs in a compressor valve should be designed to close the valve at about the dead center. In practice, the exact closing angle will vary as the conditions of service vary and will depend on how strongly the moving parts of the valve adhere to their stops. This will depend on the amount and nature of gas and deposits on the valve.
Note that either too heavy a spring, which causes the valve to close early, or too light a spring which causes the valve to close late, will reduce the capacity.
2) The gas is heated by the loss associated with flow through the suction valve. This causes the gas trapped in the cylinder when the valve closes to be at a temperature higher than the suction gas temperature. Thus the density is reduced and less gas is trapped in the cylinder to be compressed.
3) The valve pressure drop, if it is large, can directly affect the capacity loss. This occurs if the valve equivalent area is so small relative to the application that the gas cannot flow in through the suction valves fast enough to fill the cylinder. The pressure at the end of the suction stroke will then be less than the suction pressure and the amount of gas compressed will be reduced.

Environmental Causes of Valve Failure

1) Corrosive contaminants
If the valves come into contact with corrosive elements, the metal will gradually wear away, and the valve will lose its ability to perform its proper function. Corrosion eats away at the surfaces of metal components and can ultimately spread to adjoining surfaces. Once corrosion takes hold in select areas, it can eat holes directly through a metal surface.
2) Foreign material
Like any component a valve can be infected with foreign elements that can dirty the inner lining and obstruct the flow of air. If a compressor is in a dusty or heavily dirt-prone

working environment, impurities could pose a threat to the cleanliness and functionality of a reciprocating compressor valve if it takes suction from atmosphere.

3) Liquid slugs
In valves that are subject to liquid flow, irregularities in the flow of liquids can have a damaging effect on the inner lining of a valve, especially over the course of years of exposure. When a mix of flowing and bubbly fluids pass through a valve – a typical occurrence in any valve where water and oil both pass through – liquid slugs can take their toll on valve metal over time.
4) Improper lubrication
If a reciprocating compressor valve comes into contact with acidic lubricant, the effects can be damaging to the metal. Though essential, lubrication can also be detrimental if allowed to linger past its expiration. If a lubricant becomes mixed with corrosive elements, it can take on acidic properties. If it hardens, it serves no useful function within an air compressor. Lubrication can also be detrimental if excessively applied to a given component.

Mechanical Causes of Valve Failure

Compressor valve failure can also be the result of mechanical factors, some of which may develop internally unknown to the operator and others which are due to user negligence. If a valve incurs excessive levels of stress due to system overload or overheating, the valve could easily wear down and lose its ability to function as expected. If an air compressor is used in a manner that goes against the advice of the manufacturer, this too could lead to valve failure. The following factors are among the more common mechanical causes of valve failure:

1) High-cycle fatigue
If a reciprocating compressor runs at high cycles over lengthy periods of time, the stress is liable to take its toll on various system components, including the valves. If you consistently run a compressor at higher-than-average settings, you should keep an eye on the valves. If you run your compressor nonstop for hours on end at high cycles, you should also inspect the valves more frequently.
2) Off-design operation
When it comes to reciprocating compressors, each model is designed for its own range of operations. If you use a compressor beyond its intended means, you are liable to run into performance issues after weeks or months of such overuse and abuse. If your compressor has been used in ways that go against the recommendations of the manufacturer, performance issues that do emerge could be the result of valve failure.
3) Impact stresses
If a compressor endures undue stress, the effects are liable to reverberate throughout the system and its various components, including the valves. Whether the impacts are physical or mechanical, a valve could easily wear down through repeated impact stress. If a compressor is accidentally dropped or shoved around within its working space, these sudden movements may ultimately lead to weakened valves.
4) Spring failure
If the valve is attached to a spring, the assembly could fail to function properly if the spring wears down from accumulated stress. Springs are designed with specific levels of

flexibility. If a spring becomes too soft or too firm for its intended function, it loses its ability to serve its corresponding components. If the valve relies on the functions of a spring, failure of the latter will have a detrimental effect on the former.

5) Pulsations
   If air travels through a compressed-air system at irregular intervals, the impact could be harmful to the valves. A reciprocating air compressor is designed to pressurize air at predictable frequencies, and each part within the system is fitted for a certain level of performance consistency. If the system suddenly functions in abrupt or violent fits and starts, the valves are liable to incur unfavourable impacts.

How to Improve Valve Reliability

1) Measure and understand the pulsation characteristics of the machine and design the valve appropriately to eliminate over travel and flutter.
2) Select the proper lift and spring factor K for the conditions (and if conditions change over time retrofit lift and spring factor K).
3) Understand the service, if there are contaminants that go through the valve, use a poppet or a moppet. If the service is clean a plate valve is fine.
4) Lubrication rate is important. Spend some time and effort to optimize lube rates.
5) Monitoring valve temperature is ok to get early indication of a failure, but won't improve reliability much. Regular PV analysis will give you the proper data to optimize valve design.
6) Liquids kill valves, make sure you have proper Knock out drum and heat tracing on the suction, and proper alarms on KO drum level to prevent carry over.
7) Contaminants in the gas do too. If contamination damage is occurring consider cyclone filters or other devices to clean up the gas.
8) Keep your cylinders cool, dry, and well lubricated.

Pressure Packing Failure

The purpose of piston rod packing is to prevent leakage of gas between cylinder and the piston rod. As the clearance between the rod and the opening increases or decreases, the amount of leakage will increase and decrease as well. The larger the passage for the gas to escape, the larger the leakage will be.

The packing consists of a series of rings mounted in a packing case, which is bolted to the cylinder. The piston rod moves in a reciprocating motion through this case. Figure 14.9 shows a typical packing case. The packing case is constructed of a number of pairs of rings, as shown in Figure 14.10.

The gas pressure is higher on one side of each ring. This compresses the rings against the sealing area. Each pair of rings consists of one radial cut ring and one tangential cut ring. The radial cut ring is installed toward the cylinder (pressure) side. Gas flows around the front face of the radial cut ring and then around the outside of both rings. Since the ring outside diameter is greater than the rod inside diameter, a squeezing force is exerted on the rod.

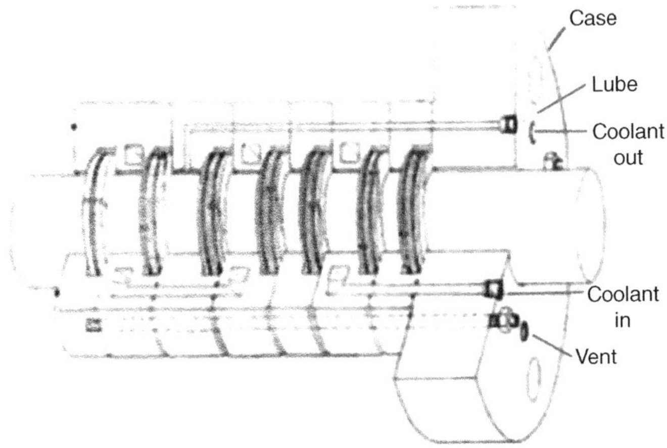

Figure 14.9 Cut-away view of packing case.

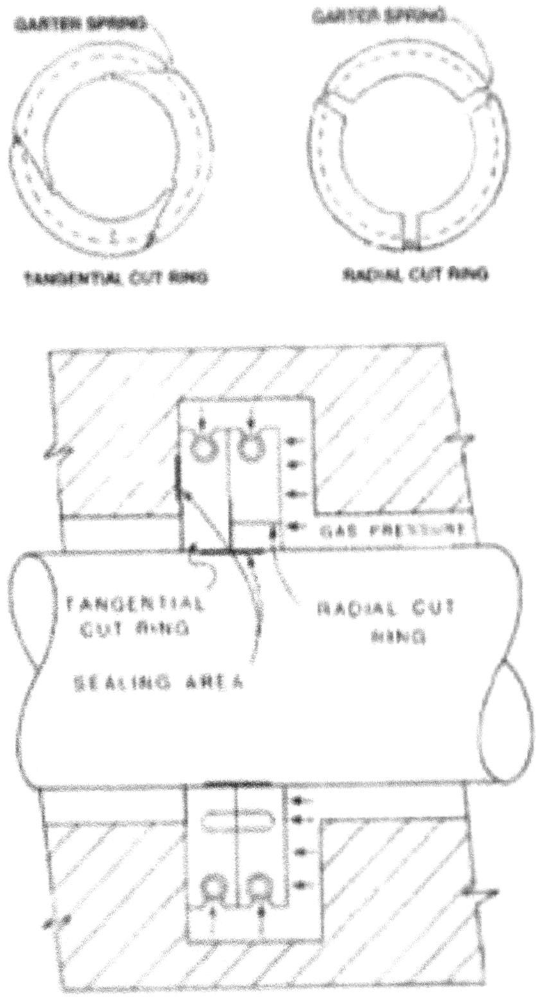

Figure 14.10 Cross section view of packing over piston rod.

This seals the path between the rings and the rod. The radial cuts are positioned in the ring assembly so that they do not stay line up with the tangential cuts. Cylinder pressure will force the ring assembly against the packing case lip, thus preventing flow around the rings.

The amount of pressure differential one set of rings can withstand is limited. Therefore, several pairs must be installed to handle typical field gas compression applications. The basic design of the packing is left up to the manufacturer.

Lubrication is needed to reduce friction and provide cooling. Lubricating oil, which must be finely filtered to prevent grit from entering the case, is generally injected in the second ring assembly. The pressure differential moves the oil along the shaft. A separate cooling system may be required for high-pressure service (5000 psi) or where high compression ratios and long packing cases are installed. The causes of ring wear are described below.

A) **Normal wear**

Packing rings must first undergo a period of wearing out, followed by a period of wearing in, then followed by the normal wearing out. This first wearing out period is caused by the fact that the rings must have a gastight fit with the piston rod and must obtain that fit by wearing out to the point of making that contact. Temperatures may vary substantially from front to back of packing and between rod and rings. This means that even though the fit between the rod and the rings is perfect at room temperature, at operating temperature, due to the gradient through the packing and the difference in expansion rates, each pair of rings must wear to a running fit to make a satisfactory seal. Having obtained this fit we should now expect the surfaces to burnish and the original high wear rate to be replaced by a much slower normal wear pattern. What then will change this pattern of normal wear, accelerating it to unacceptable limits?

B) **Causes of abnormal wear**

i) **Temperature:** Temperature can put the packing to the high wear rate. Radical changes in cylinder or packing temperature can be caused by failure of the cooling system, excessive cylinder temperatures caused by recompression, lubrication failures causing excessive frictional temperatures, and sticking of the rings causing excessive friction.

ii) **Lubrication:** If lubrication is reduced beyond the point of effectiveness by the gas or by excessive temperatures, it can permit metal contact, giving high frictional temperatures and excessive wear. Build-up of gummy or sticky carbon deposits, due to poor lubrication or reaction with the gas, can cause restriction of free movement of the rings, further contributing to excessive wear.

iii) **Blow-By:** caused by improper break-in or erosion due to leakage past the bore of the rings after a sudden temperature change, can cause excessive wear or what can be classified as wear.

iv) **Moisture**: causing pitting, erosion and corrosion, can take place on the rubbing surfaces during shutdown, giving unusual abrasive conditions.

Contamination: of the gas with foreign material, such as welding beads, sand, pipe scale, catalyst, etc., will also cause rapid wear through abrasion.

v) **Improper Break-In**

One of the large contributing factors to the ultimate life of a packing is the break-in. Break-in consists of wearing the rings to a suitable fit with the rod at operating temperature. If the rings are improperly worn in and a burnished rubbing surface is not obtained leakage will continue. If this condition exists, the rings may continue to wear without obtaining a burnished condition.

During break in, where a discrepancy may exist between the rings and the rod due to expansion, blow-by can destroy the oil film and cause excessive packing temperatures, causing further expansion and blow-by. As a result leakage will occur. The best way to overcome this is to run at the operating temperature for few hours.

vi) **Wear**– can be the most obvious cause of leakage. Complete wearing out of the rings and failure to compensate for additional wear would, of course, permit clearances and leakage. Wear on the sealing face of the cup can prevent proper sealing of the rings against the cup, causing leakage. Wear or damage of the cup may be caused by inadequate or poor lubrication, corrosion, build-up of carbon formations or abrasive material in the gas.

vii) **Rod Damage**– Scuffing or scoring of the rod can be a source of leakage. If excessive wear has occurred on the rod and rings of the original diameter are used, leakage will occur.

viii) **Improper Assembly**– Improper assembly is also a source of leakage. Rings installed improperly, that is, with the radial ring facing away from the pressure, will allow the packing to leak. If the set has been assembled in such a way that the packing cups are not perpendicular to the rod, the packing rings will not be perpendicular to the rod and can cause leakage.

Packing Care During Assembly

Assembly requires care in handling of the parts to prevent damage, observing a few simple precautions and following the assembly drawing provided with the operator's manual. The parts should all be free of nicks, burrs, scratches, etc., and should be laid out in the order in which they are to be assembled. As the rings are assembled on the rod, care should be taken that the numbered side faces the pressure, and that dirt or foreign material does not separate the joints of the rings or the rings of a pair. On lubricated packing, the rings should be liberally coated with oil. The cup joints should also be free of anything which would hold them apart and interfere with their seal. With the packing assembled and in the stuffing box, the flange studs should be tightened as any gasketed joint-applying pressure through opposite studs gradually to obtain even crushing of the gaskets. This will prevent cocking of the packing cups and ensure their being in a perpendicular plane with the piston rod. Before making oil connections to the packing, the lubricators should be hand operated to ensure that oil is reaching the point of the connection. After making the connections, the lubricator should continue to be hand operated to fill the oil holes in the case. Check the cooling system to ensure that it is functioning properly and that an unrestricted flow of coolant is obtained. Temperature control is a most important factor during break-in.

Piston Ring/Rider Ring Wear

In horizontal-cylinder compressors equipped with plastic rider bands, the piston-rod assembly drops during operation due to normal rider-band wear. Dirt, liquids, catalyst debris or insufficient lubrication may considerably and unexpectedly reduce rider ring life. Rider bands must therefore be changed periodically to prevent contact between the piston and cylinder, as this would seriously damage both components and result in expensive repairs. But replacement based solely on hours of operation is not the most efficient

Figure 14.11 Piston ring and Rider ring mounted in the piston.

method – operators often find, in retrospect, that some shutdowns were unnecessary because rider bands were still in acceptable condition.

If rings are replaced too late, the piston contacts the liner which may lead to damages causing expensive repair works and production losses. On the other hand, good operating conditions may extend ring life beyond scheduled service intervals.

Causes of Ring Wear

Cylinder

In lubricated service, the cylinder should be cleaned to remove dirt and wear debris. Carbon, gum and lacquer deposits from decomposed lubricant should also be removed. Such deposits can cause ring wear and sticking, and they impede heat transfer to the cooling jacket. Heavy deposits caused by overheating must be removed from the cylinder. Cylinder cooling jackets should be checked for blockage or deposits, which can impede heat transfer.

Dirt Generated by Process Gas

Large quantities of foreign material can cause wear of the cylinder and/or rings. Often, the presence of process gas contaminants in the form of abrasive particles is not taken into consideration during the design stage. The sign of accelerated wear due to hard particle contamination is only usually observed after an unscheduled process gas compressor shutdown.

Lubrication

With substantially less than the amount of oil required can damage the rings due to lack of proper lubrication.

Rider Ring Clearances

PTFE materials can be affected by thermal expansion. Provision has been made in sizing the ring gap and side clearance, which may make the rings compensate when operated at different temperature. At operating temperatures the clearances will change and may lead to ring wear.

Piston Ring Clearances

The suggested minimum side clearances between ring groove (side) and end (gap) clearances for piston rings of various material should be maintained. Caution should be taken to make sure that each ring has the minimum end clearance. Unless there is sufficient clearance, ring ends may butt together from thermal expansion, causing ring wear, cylinder scoring, and other damage.

Process Related Problems in Reciprocating Compressor

Gas Properties

A complete and accurate gas analysis is essential for selection of compressor. Percentage by volume of component gases, entrainment of liquids and solids, and percent water vapour and minute quantities of contaminants should be analysed before procurement done. Trace amounts of sulphur compounds and chlorides can cause corrosion or other mechanical difficulties. Even slight corrosion can produce failure of cyclically stressed parts of compressor. In a reciprocating compressor, solid particles will cause high maintenance by accelerating wear of valves, pistons, cylinders, piston rods, and packing. If at all possible, any solid particles should be removed from the gas stream before the gas reaches the compressor. Except for a change in efficiency due to valve-related losses, the reciprocating compressor is not affected by the molecular weight of the gas. Periodic changes in gas composition will have little effect on compression horsepower and pressure. On the other hand, the pressure developed by a centrifugal compressor at a particular speed is directly proportional to the gas density or molecular weight.

Liquid in Process Gas

If liquid is present in the process gas, it will vaporize because of the heat in the suction bottles. As the amount of liquids increase, the droplets will hit the suction valves and again vaporize because the valves are hot (the heat is from the temperature rise in the cylinder). As more liquids enter and hit the valves, damage will occur, as the liquids vaporize, the heat they absorb causes localized cooling of the metal and the valves will break. Some compressors have metal springs and plates that break easily, plastic materials will hold up better in these conditions.

If the liquid does get through the valves in the liquid form, they will fall (you should have suction on top, discharge on bottom) through the cylinder and they can again vaporize due to heat of compression. At this point, the valves may be broken and gas will pass back and forth through the broken valves cause an increase in average cylinder temperature. Hot suction valves indicate this type of problem. Finally, a huge amount of liquid will have a hard time getting into a cylinder because the amount of passage area through the suction valves and discharge valves is really small. But with broken suction valves, more liquid can enter than the discharge valves can let out. This large amount of liquids will fill the cylinder so on the next compression cylcle, the cylinder will stop because it cannot push the liquids out. Of course you can not stop the cylinder so some thing will break. The typical breakage point is not the cylinder walls, but the piston rod, rod pin or crank will fail and the unit should stop because of vibration, pressures errors, or other sensors.

Temperature of process gas

Temperatures of process gasare important in a reciprocating compressor. Temperatures within a conventionally lubricated cylinder should not exceed 350 °F. Theoretically, a gas with a low exponent can be compressed with much higher ratios of compression per stage, yet hold temperatures within desired limits. One drawback to this, however, is that volumetric efficiency decreases with high ratios, and a low exponent may make it uneconomical to compress to high ratios.

If the compressor discharge temperature is very high then it may result in breakdown of the lubricating oil, causing excessive wear and reduced life of the compressor valves (mainly the discharge valve). In hermetic compressors, the high discharge temperature adversely affects the motor insulation (unless the insulation is designed for high temperatures). When the temperature is high, undesirable chemical reactions may take place inside the compressor, especially in the presence of water. This may ultimately damage the compressor.

Variation in Suction Pressure

In reciprocating compressors, pressure drop takes place due to resistance to fluid flow in the suction line and in the control valves. This pressure drop can have adverse effect on compressor performance as the suction pressure at the inlet to the compressor will be lower than the design.

When the suction pressure to a reciprocating compressor is reduced the following can occur:

1) Assuming that the discharge pressure is maintained constant, the compression ratio increases; this, in turn, usually causes that stage's brake horsepower demand to decrease but the discharge temperature increases.
2) The specific volume of the suction gas increases and the mass capacity of the compressor decreases.
3) If the suction pressure is reduced to atmospheric or sub-atmospheric levels, there is a propensity to suck in atmospheric air into the process gas. If air (specifically Oxygen) can cause a hazard or upset in the downstream process, this can present serious difficulties. If the non-condensable nature of the Nitrogen and Oxygen in the air can cause downstream process upsets (such as non-condensable accumulation in refrigeration cycle condensers), then this will cause process upsets.
4) The different compression ratio resulting from a suction pressure reduction can also affect the piston loading and the piston rod reversal characteristics. This may affect the effective oil lubrication on the crosshead and the crankshaft of the machine – depending on the cylinder design limitations and the bearing design.
5) As a suction pressure decreases, the pressure ratio and density of gas decreases. This in turn reduces the volumetric efficiency, gas mass flow rate and increases work of compression.

As suction pressure increase, keeping discharge pressure constant, the pressure ratio across the compressor will reduce and because of the higher density of the fluid at the suction mass throughput will increase. But for two stage compressor the overall pressure ratio is distributed across the stages and inter stage pressure ratios dependent on the relative values of the cylinder diameters.

Effect of Leakage

In reciprocating compressor gas leakage takes place between the cylinder wall and piston, across the suction and discharge valves and across the packings. The magnitude of these losses depends upon the design of the compressor valves, pressure ratio, compressor speed and the life and condition of the compressor. Leakage losses increase as the pressure ratio increases, compressor speed decreases. Due to the leakage, some amount of gas flows out of the suction valves at the beginning of compression stroke and some amount of gas enters the cylinder through the discharge valves at the beginning of suction stroke. The net effect is to reduce the mass flow rate of gas. Even though it is possibly to minimize gas leakage across cylinder walls, eliminating leakages across valves is not possible as it is not possible to close the valves completely during the running of the compressor.

Change in Oil viscosity

In general, heavier hydrocarbon gases and higher discharge pressures cause the cylinder oil to become more diluted, resulting in decreased viscosity and thinner oil films protecting components. To avoid losing too much viscosity, heavier ISO grade mineral oils or synthetic lubricants resistant to dilution are recommended. Many a time liquid contamination (e.g. water, hydrocarbon) and oil starvation also interfere with oil film quality by decreasing oil viscosity and/or removing the oil film. Poor lubrication also results in excessive temperatures which decreases component life.

Maintenance Related Issues

Compressor Alignment

Proper cylinder alignment is critical to the longevity of the compressor components. The size and weight of the compressor rod, crosshead and piston can cause the rod runout to make cylinder bore misalignment. A heavy piston and rod can cause the crosshead or piston to tilt and results in rod runout with out-of-tolerance cylinder bore alignment. Also compressor cylinder bore misalignment can cause Crosshead shoes uneven wear, compressor piston rod failure at both the crosshead and at the piston, compressor piston rod coating failure, piston wear and compressor cylinder bore wear.

We can employ one of several methods to check the alignment of the compressor crankshaft system and the relationship of the compressor cylinders to the frame. Cylinder alignment should be checked if the bolting between the crosshead housing, distance piece, cylinder or cylinder support legs are damaged or loosened.

The tolerances the OEM has for bore alignment is small. These tolerances depend on the bore size and the operating pressure. The higher the operating pressure the smaller the tolerance. The OEM's tolerance on the cylinder alignment in most cases is 0.005″ maximum parallel offset and 0.003″ maximum angularity.

Trouble with Cylinders

Inspection of cylinders should be done at every shutdown, for example on the occasion of a valve change-out. Inspection of the cylinder bore for any sign of scuffing or scoring may be done looking through the valve ports. Cylinders measurements are taken from time to time to determine what wear has taken place and whether that wear is within limits allowing operation to continue. If measurements show that the wear is beyond acceptable limits, the measurements determine what repair procedures should be implemented. Compressor cylinder bore measurements are taken in the vertical and horizontal directions and in at least three locations along the bore, each end of ring travel, and in the center. All measurements are taken with a micrometre and should be recorded on appropriate inspection report forms.

Trouble with Cylinder Liner

Often, the presence of gas contaminants in the form of abrasive particles cause accelerated wear of the liner. The origin of such gas contamination can be materials used as adsorbents in pressure swing adsorption (PSA), unit. The lack of piping cleanliness after welding and fitting, the absence of rust inhibitors during hydro testing are potential sources for contamination also. The material selection on the gas compressor can be made in order to counteract the accelerated wear created by the foreign particle ingression. The cause/effect understanding of hard particle contamination is critical for a reliable compressor operation and for drafting of maintenance schedules.

Valve Maintenance Related Issue

If torquing of the valve jam screws not done properly that hold the valves in position, then Valve seats become damaged. If the jam screw is not torqued to proper value, the valve may become loose during operation and damage the seat. If over-torqued, distortion or cracking of the seat may occur. Valve seat damage may occur also when the valve seat gasket is not in the proper position to seal, and wire drawing might then take place. A folded or wedged gasket can cause indentations in the seating surface.

Bearing Clearances

The clearance in a babbitt bearing, or the clearance between piston and cylinder head, can be determined by opening up and inserting a soft lead wire, then measuring the thickness of the lead with a micrometer after the lead has been compressed in the bearing or between piston and cylinder head. Fuse wire is normally used, if it is soft. The most accurate method of measuring clearances is with a dial indicator. This is also a quick method that does not require any disassembly.

Web Deflection

Crankshaft web deflection, determined by web gauge or inside micrometre is the most important indication of the foundation, grout, frame soft foot or issues with alignment of main bearings. If a situation occurs where a bearing become more worn than the adjacent one, the effect will be shown as a change in pattern of deflections. When the crank is turned from BDC to TDC the weight of the running gear causes the crank web and the crank pin to bend in such a manner that the distances between the webs decreases and continue to decrease until the bearing is no longer in contact with the journal.

The deflection when the crankshaft is approaching TDC will then go from its normal positive reading to zero and then to negative reading at which point the assembly is supporting the weight without the assistance from the lowered main bearing. Thus any change from

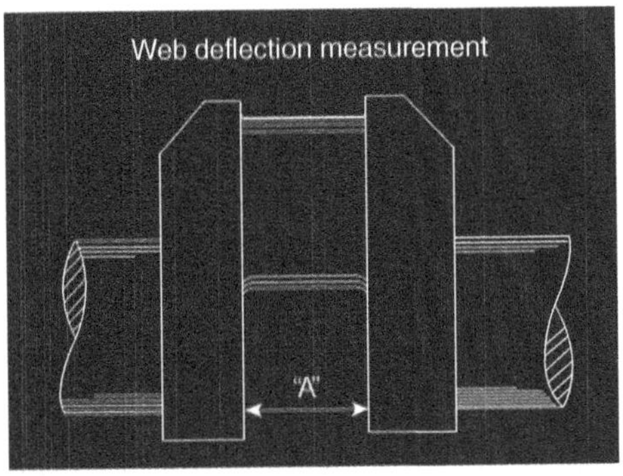

Figure 14.12 Web deflection measurement.

natural deflection can be related to main bearing misalignment and is proportional to the difference in height of the bearings. Misalignment of bearings takes place for the following reasons.

- Damage or wipe-out of the main bearing.
- Loose compressor foundation bolt leading to vibration.
- Crack in the bearing saddle.
- Loose main bearing bolt leading to damage of main bearing.

Foundations under reciprocating equipment may deteriorate over time. When this happens, and the foundation subsides, the web deflections in the machinery crankshafts can go beyond OEM specifications. Excessive web deflections can cause crankshaft fatigue failures. A common approach to restoring the crankshaft web deflections caused by foundation problem is a very costly affair if re-grout of the machine is to be done. Not only is this type of foundation work costly, it will normally require significant downtime and therefore lost production.

Piston Rod Run-out

Piston rod run-out is a measure of how well the cylinder is lined up with the frame and, also, how well the internals of the compressor (piston, rod, crosshead) are lined up with each other and the frame and cylinder combination. This runout is checked and adjusted with all parts "cold"; the ambient temperature should be in the 60–68 °F range. Piston rod run-out must be rechecked on site any time the piston or rod is serviced.

Horizontal Run-Out

Horizontal rod run-out readings can be used as a direct indication of horizontal alignment from the crosshead through the distance pieces to the cylinder. Horizontal rod run-out is measured by placing a dial indicator pickup on the side of the rod. Horizontal rod run-out should be the same whether the unit is cold or hot. Excessive horizontal rod run-out is corrected by realignment of components involved. This may include cylinders, heads, distance pieces, crossheads, crosshead guides, rod, and piston.

Vertical Run-Out

This reading generally indicates how well the compressor piston and rod are in line with the cylinder and frame. Normal expected cold vertical rod run-out is not an indication of misalignment. For perfect alignment, it is the result of the difference between the normal cold running clearance of the piston to bore and the crosshead-to-crosshead guide, which in large cylinders causes the piston center line to lay below the crosshead centreline.

Vertical rod run-out is measured by placing a dial indicator on the top or bottom of the rod. As the rod is moved through the entire stroke length, it will "run-out" by an amount equal to the ratio of the stroke-to-rod length times the Δ drop, i.e., half the difference between the running clearances.

What is an acceptable vertical rod run-out figure? API-618 2nd Edition requires 0.00015 in/in of stroke at normal operating temperatures. Although this figure is deemed by most manufacturers as being too stringent, especially where large cylinders are involved. It is felt by some that 0.001–0.002 in/in of stoke may be perfectly acceptable for normal operating vertical run-out, especially where large cylinders are involved. This amount of run-out creates just enough movement in the packing rings to keep them from sticking in cups, especially in dirty gas applications.

Water Jackets

After years of operation of the compressor, water jackets become dirty and fouled, particularly if jacket water quality is not good and the jacket cooling effect may have deteriorated considerably. The inside of the jacket walls may be covered with sand, lime, or magnesium, which interferes with normal heat transfer and cooling. The solution to this problem is to clean not only the cylinder jackets but also the water cooling passages in the front and rear heads. Chemical cleaning is recommended at every fourth or fifth overhaul for normal water conditions.

Machine Monitoring

Essentially, the objectives of any reciprocating compressor monitoring system are threefold:

- Operating condition monitoring,
- Early fault detection and diagnosis, and
- Efficiency monitoring and improvement.

Operating Condition Monitoring

Continuous operational condition monitoring implies that machine condition data are collected and evaluated in real time. Of special interest here are data that could rapidly change and are tell-tale signs of unexpected degradation of the condition of the machine. Vibration amplitude excursions indicate rapid changes and will give an early warning of disaster. A timely shutdown will minimize possible consequential damage and the resulting extended downtime. The important parameters that are monitored are compression ratio, cylinder loading, rod load, rod reversal, peak pressure, and discharge temperature.

Compression Ratios

This is the ratio of absolute discharge pressure to absolute suction pressure (Pd/Ps). Analysis of compression ratios helps in that, it should not cause rod loads to exceed manufacturer limits.

Improper Cylinder Loading

Cylinder loading is an area of compressor analysis that sometimes has been left to flow charts and curves using ideal and theoretical assumptions. There are programs on the market today that create operating curves based upon actual operating conditions that can save hundreds of thousands of dollars for corporations; lost dollars of which the corporations are not even aware. The measurement of volumetric efficiencies and the determination of true rod loads can be greatly enhanced to save machines, production and money.

Rod Load

The maximum rod load reached in compression and tension during the stroke should not exceed equipment manufacturer (EM) limits, and should go from compression to tension for a short period of time. Rod reversal allows lubrication to both sides of the cross head pin. Rod load can be based upon internal cylinder or line pressures and should consider effects of inertia. It is important to identify the method used by the EM to set the limits and make any comparisons on same basis.

Rod Load Reversal

There are two types of reversals. The first, piston or crank angle reversal is the physical reversing in direction of the piston, which happens at both TDC and BDC positions. The second type of reversal is pressure reversal. Pressure reversal occurs as the internal cylinder pressure goes from more pressure on the head-end side to more pressure on the crank-end side of the piston. Without pressure reversal, lubrication of both sides of the crosshead pin may not take place. This lubrication is necessary and lack of lubrication will cause early failure.

Valve failures can also produce a non-reversal situation if the compressor is already running close to a non-reversal condition. It is necessary for the rod load to go from compression to tension for a short period of time.

Peak pressure

Seen in the compressor can be useful in detecting fluid build up in the compressor or anything else that is reducing the top dead center (TDC) clearance and causing higher compression. It can also be used to detect flow restrictions.

Discharge Temperature Delta

This reading is the actual discharge temperature minus the theoretical discharge temperature. High DTD generally indicates re-circulation of gas caused by leaking valves or rings. A general rule of thumb is that the DTD should range from 10 to 40 °F higher than theoretical. This value should be trended. With ring leakage, DTD can go to 100 °F or more. Valve leakage results, typically, in DTD of 40–50 °F.

Condensed Liquids

Entrapped in the system similar results are obtained when condensation of liquids occurs in the system. These liquids are formed when pressure differential and temperatures are just right to cause the gas to condense. This typically occurs on the suction side due to lower temperatures. If condensation is heavy enough, vibration will become audible as fluid is slammed through the valves by the piston. If allowed to increase, the results can be almost as bad as detonation within the cylinder. If fluids in the cylinder are from a cooling system leak that has reached large enough proportions, the trace may be similar.

Early Fault Recognition and Diagnosis

The early recognition of faults is an important factor when it comes to detecting defects and avoiding consequential damage. Where a fault is signaled at an early level, the machine in question may possibly continue in operation for a limited period of time. However, the development of such a situation must be closely observed, and a shutdown for maintenance should be scheduled to intervene as required.

Faults arising or produced can be diagnosed by referring to machine data collected for monitoring and early-detection purposes. Various analysis are carried out in the process, localizing the fault in terms of place and operational effect, so that an estimate can be arrived at regarding necessary replacement parts and anticipated period of the outage. Trend analysis is particularly important in this context. Trends are calculated, for example, by hourly averaging of recorded machine data, and provide a means of tracing the (long-term) development of machine state.The important parameters that are monitored for fault detection and diagnosis are described below.

Temperature

Machine temperatures are a valuable indication of machine condition and are a primary tool for reciprocating compressor condition monitoring. The primary temperature measurements include cylinder discharge temperature, valve temperature, packing temperature, crosshead pin/big end bearing temperature, and main bearing temperature. Cylinder discharge temperature is one of the protection parameters recommended by API 618 (1995) since leaks in rings and valves result in recompression of gas that will raise the discharge temperature. Packing case temperature or packing leak off temperature can give an indication of packing leakage, while main bearing temperature measurement has been proven effective in preventing major damage due to main bearing failures.

Suction and discharge cylinder valves are subject to ring breakage or malfunction due to dirt or solids accumulation. The resulting backflow of gas causes an increase in the temperature of the valve, valve covers and cylinder body. Temperature rise can also be caused by prolonged unloaded operation of the cylinders. It is usually recommended that this be limited to a maximum of a few minutes.

Measuring the gas temperature in the valve pocket is the simplest and most cost-effective method of valve condition monitoring. The gas temperature in the valve chamber can be measured with a temperature probe, observing the upflow of the suction valve and the downflow of the discharge valve. If there is an obvious increase in temperature at one valve, one can assume that there is damage. This method has the advantage of being inexpensive and simple to install in virtually all types of valves.

Rod Drop Monitoring

Both the rider bands and piston rings contact the cylinder liner. However, the rider bands are designed to carry the full weight of the piston, while the piston rings are intended only to provide sealing between the piston and bore. The piston rings and their mounting grooves are purposely engineered with clearance to allow the rings to "float" without carrying any of the vertical load of the piston's weight. The rider bands are intentionally designed to seat fully within their mounting grooves so they can support the piston's weight.

The piston rod assembly drops during operation due to normal rider band wear. In good operating conditions, the rate of wear can be very low, especially in lubricated cylinder applications. However, adverse conditions such as improper cylinder lubrication, component misalignment, or abrasive particles or moisture in the process gas can increase the wear rate. These may sometimes sudden reduction in rider band life.

In a reciprocating compressor proximity probes are typically located under the piston rods, and are used to measure the rod position, which can be converted to rider band wear (Figure 14.13). Trending this displacement measurement allows one to get an indication of potential rider band wear over time.

Rod Run-Out

Whereas Rod Drop is a measurement of rod position, Rod Run Out is a measurement of the rod's actual dynamic motion as it travels back and forth on its stroke. The main purpose of a rod position monitor is to constantly measure both the vertical and horizontal position of the piston rod during operation and to monitor rod flexing in both axes.

The amount of Rod Run Out is highly dependent on the cylinder alignment with the Crosshead. Due to inherent looseness in the Crosshead and thermal growth of the machine, higher readings of Rod Run Out are allowed in the vertical direction. In the event of a malfunction such as a cracked piston rod attachment, a broken crosshead shoe, or even a liquid carryover to a cylinder, the operating rod run out will increase significantly. The horizontal direction allowances are much less and high readings are attributed to misalignment.

Also knowing the crosshead clearance and the maximum movement expected at the piston, you can find the maximum displacement expected at the pressure-packing case proximity probe and can create alarm set points accordingly. The probe position changes continuously as the piston travels through its stroke. Eddy Probes can make this measurement while the machine is operating. This provides a highly accurate measurement of the actual dynamic motion of the rod under full load conditions.

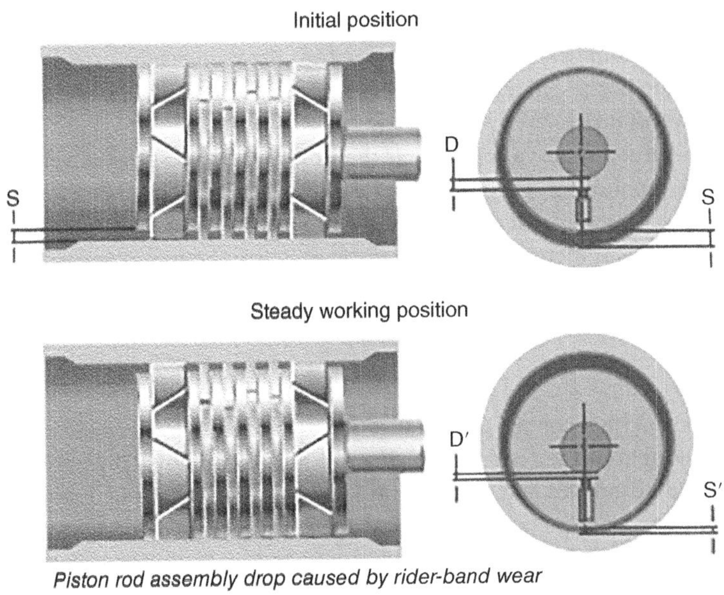

Figure 14.13 Rod drop measurement.

Vibration

Traditional means of vibration monitoring are employed to gauge the overall mechanical condition of reciprocating compressors and to effect precautionary shutdowns so as to avoid consequential damage. To that end, vibration sensors or accelerometers are mounted in the crank or crosshead slide areas

Frame Vibration

Crankcase vibration has been used as a basic protection parameter for decades. Most compressor OEMs specify a crankcase velocity alarm/shutdown level and the most common configuration is to mount the transducers on each end of the crankcase about halfway up from the baseplate in line with a main bearing and shall capture the transmitted vibration from the main bearings.

These velocity sensors can be installed on the frame to measure overall machine vibration. These measurements can be useful for evaluating the forces caused by piston rod loading (gas and inertial forces) and moment forces imparted to the main bearings by the crankshaft because of its asymmetric design. The magnitude for a frame velocity signal is usually low (less than 0.3 in/s). However, at low frequencies, even small amplitudes of measured velocity correspond to large amounts of displacement.

Crosshead Vibration

The other primary vibration measurement that has proven to be effective is to measure acceleration on the crosshead or distance piece of each cylinder. Malfunctions such as liquid carryover, loose piston nuts, loose crosshead attachment valve problems, clearance problems, and many others can be identified with this measurement. Of all the vibration measurements that could be made, this is probably the most effective vibration protection measurement available. If a machine is undergoing catastrophic distress, it will typically be picked up on the crosshead accelerometers.

Cylinder Vibration

This vibration measurement deals with valve operation and movement, acoustic effects and the analysis of various leak scenarios. Acceleration sensors can be installed on the cylinder to detect impulse events associated with the compressor cylinder valves. It is especially important to ensure that recommended sensor mounting practices are used for accelerometers.

p–V Diagram Analysis

Pressure velocity (PV) analysis is a technique that has proven to be very effective in assessing the condition of reciprocating machine. Dynamic pressure transducers are used to measure the pressure inside the cylinder over the course of the stroke. This allows the analyst to evaluate the condition of the rings, valves, and packing, while at the same time calculating the dynamic rod load, which is the source of the forces and moments. This requires that a pressure transducer be installed in the cylinder, either on a temporary basis using a valved port or for an online measurement using a permanent transducer installation. *p–V* diagram analysis serves to analyze the condition of the packing. Increase in leakage cause discrete changes in the indicated pressures and the instantaneous timing of the attendant pressure excursions. Continuous online monitoring of reciprocating compressors through dynamic

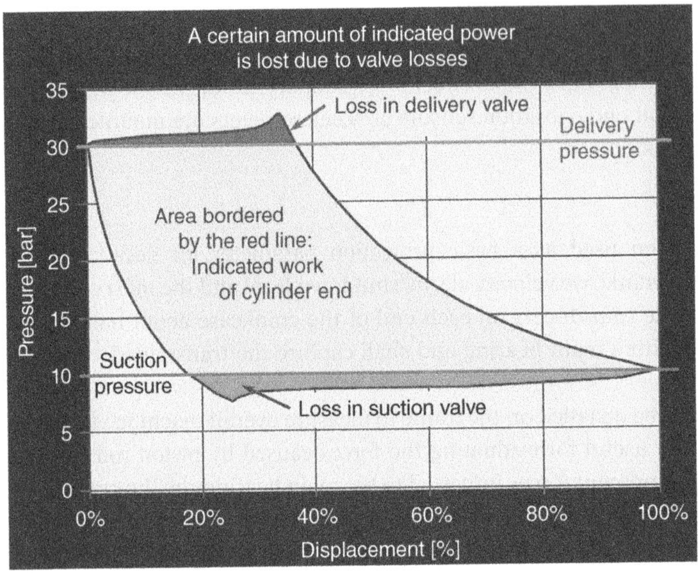

Figure 14.14 *P–V* diagram of a reciprocating compressor.

analysis allows malfunction diagnosis, improving the time and cost to repair and provides continuous protection from catastrophic events such as rod or cylinder rupture. The Pressure × Volume ($P \times V$) diagram is the best way to represent the energy cycle in a reciprocating machine.

Efficiency Monitoring

Efficiency is another indicator of the condition of a machine. The primary purpose of efficiency monitoring in reciprocating compressors is to record changes in process or machine parameters that influence energy transfer to the compressed gas. The efficiency of a reciprocating compressor may be determined, for example, by finding how indicated compression power relates to the drive power required – for example, the power consumption of an electric motor. How much gas is compressed and how much power is required for that compression are both derived directly from the P–V diagram. Hence, understanding the *P–V* diagram is critically important and forms the foundation for a deeper understanding for how a reciprocating compressor works.

The P–V diagram is a plot of the pressure of the gas versus the volume of the gas trapped in the compression chamber. In Figure 14.15, P_S, suction pressure, represents the pressure of the gas at the inlet to the compressor cylinder. P_D, discharge pressure, represents the pressure of the gas at the outlet from the compressor cylinder. V_1 represents the maximum volume of gas trapped in the compression chamber and V_3 the minimum. The difference between V_1 and V_3 is known as piston displacement, or how much volume is displaced in one stroke length of the piston:

This completes the basic reciprocating compression process. One cycle around the P–V diagram represents one revolution of the crankshaft and two stroke lengths of the piston,

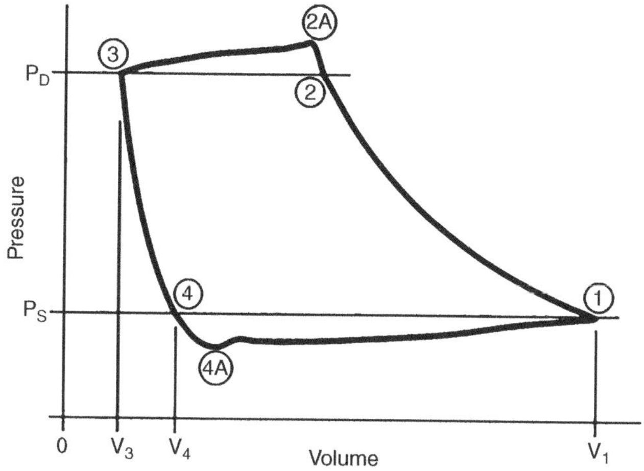

Figure 14.15 *P–V* diagram.

one from Positions 1–3 and another from Positions 3–1. Four events make up the PV diagram – compression, discharge, expansion and suction.

Capacity: The volume of gas compressed by this P–V diagram is the difference in volume between Positions 1 and 4:

$$V_1 - V_4 = \text{Capacity}$$

This volume is influenced by the compression ratio (R_C) and the magnitude of V_3. Compression ratio is:

$$R_C = \frac{P_D}{P_S}$$

PD and PS are with absolute pressures.

As compression ratio increases the capacity decreases as less gas is drawn into the compression chamber and compressed. As V_3 increases there is more gas that must expand thus reducing capacity. V_3 is referred to as the fixed clearance volume.

Volumetric efficiency: Volumetric efficiency (*VE*) is the percentage of stroke that fills with gas at suction pressure and suction temperature. In equation form

$$VE = \frac{V_1 - V_4}{V_1 - V_3} \times 100\%$$

VE represents the capacity

- *VE* is NOT suction valve open time. The suction valves do not have to be open for the full *VE*.
- A higher number for *VE* does not mean it is "better" as might be the case for energy efficiency. *VE* simply represents capacity.
- The influence of *VE* on energy efficiency is through the relationship of *VE* to average piston velocity (average velocity of gas through valves).

Power: The power required to drive a reciprocating compressor can be divided into three pieces, adiabatic, valve loss and friction. The power required to compress a volume of gas is represented by the area enclosed by the *P–V* diagram, or:

$$\text{Work} = \int P\,dV$$

Inefficiency in the *P–V* diagram is the pressure drop incurred in moving the gas from the inlet flange of the cylinder into the compression chamber and in moving the gas from the compression chamber to the outlet flange. Overcoming this pressure drop requires energy.

Compression Efficiency

Compression efficiency is defined as:

$$\text{EFF} = \frac{AP}{BP} \times 100\%$$

Where
EFF = Compression efficiency
AP = Adiabatic power
BP = Brake power

Figure 14.16 is a plot of compression efficiency versus compression ratio for a given compressor cylinder when compressing two different gases, hydrogen and nitrogen:

Note how efficiency increases as compression ratio increases. The efficiency curve will have this shape for any reciprocating compressor cylinder. Also note how much higher the efficiency is for hydrogen (with a very low mole weight of two) as compared to natural gas (with a medium mole weight of 18). Everything else being equal, compressing hydrogen will have one ninth (11, 89% less) the valve loss power, hence much better efficiency.

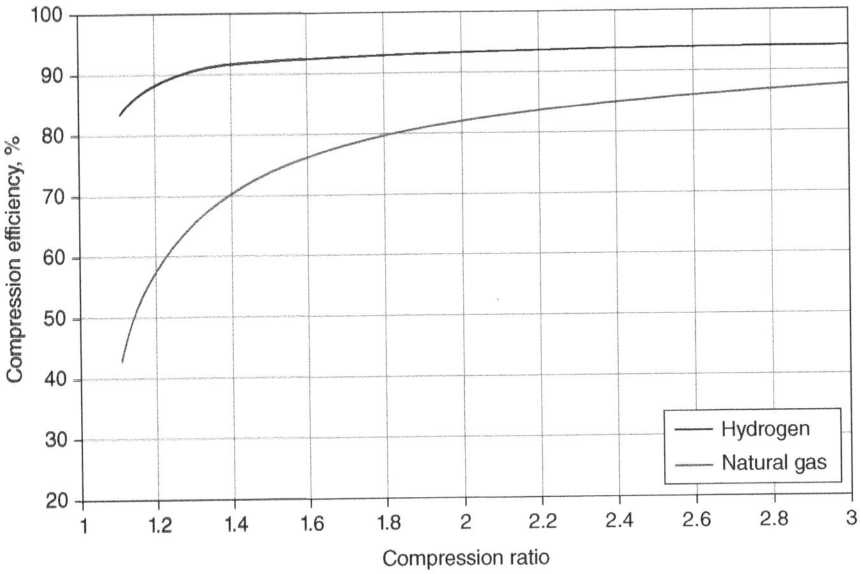

Figure 14.16 Plot of compression efficiency versus compression ratio.

Varying Conditions

Figure 14.17a shows how the P–V diagram changes when the discharge pressure increases with the suction pressure staying constant. The volumetric efficiency (capacity) decreases and the power (area enclosed by the P–V diagram) increases. The P–V diagram with the higher compression ratio will have a better efficiency.

Figure 14.17b shows the discharge pressure decreasing. Here the volumetric efficiency increases and the power decreases. The diagram with the lower discharge pressure, therefore lower compression ratio, will have a lower efficiency.

Figure 14.17c shows the suction pressure increasing. The volumetric efficiency increases, but it is not at all clear how the power changes just by looking at the P–V diagram as it gets shorter but also wider.

The diagram with the higher compression ratio will have a higher efficiency. In fact, as the pressures increase or decrease the power changes.

Figure 14.18 shows that as compression ratio decreases (suction pressure increases with the discharge pressure constant) the power will increase to a point but then decrease. The P–V diagram changes in height and width at different rates as the pressures change leading to the curve.

While power changes following a curve capacity always increases with decreasing compression ratio (Figure 14.19):

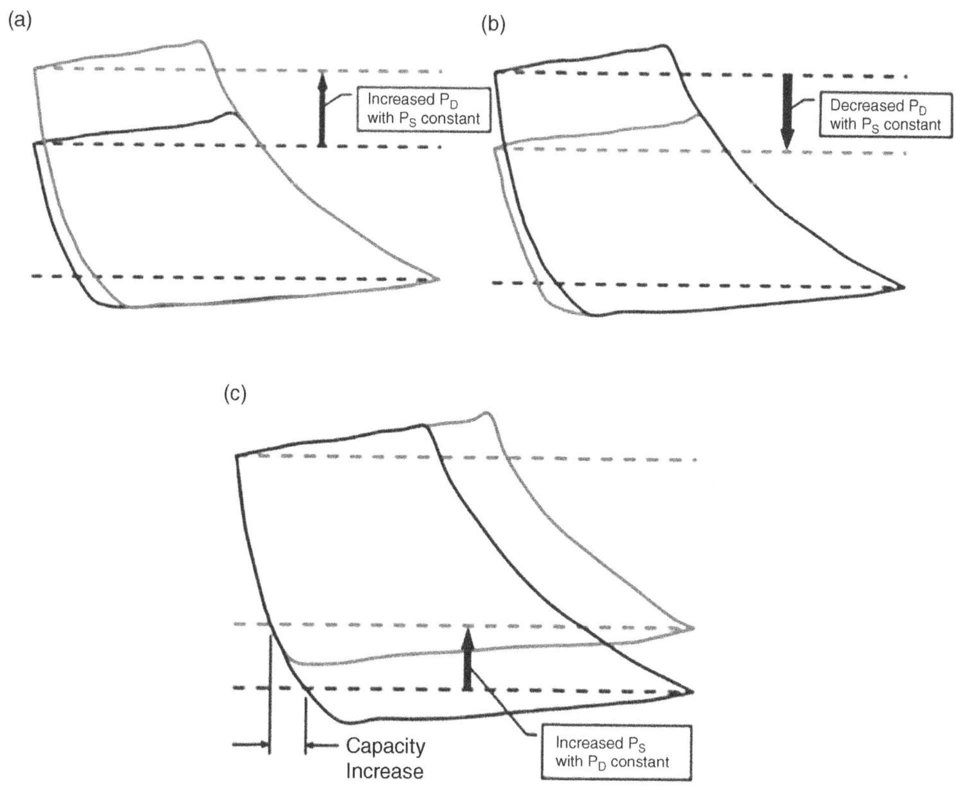

Figure 14.17 (a)–(c) are *P–V* diagrams showing the changes that result from increased discharge pressure, decreased discharge pressure and increased suction pressure.

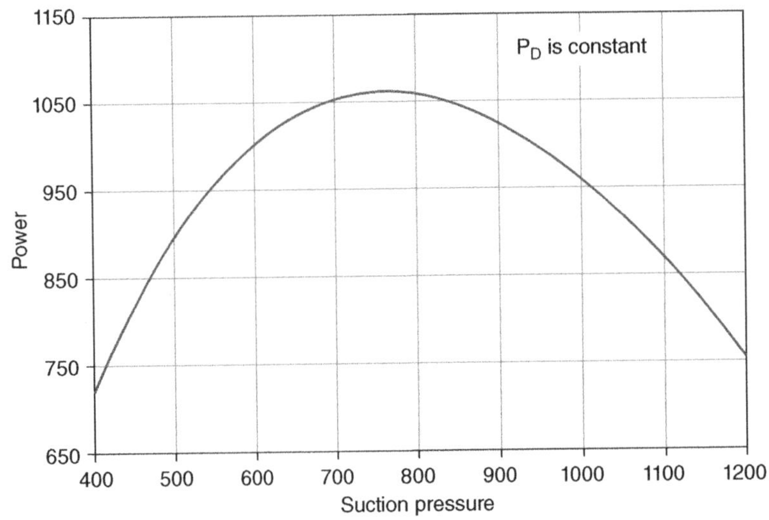

Figure 14.18 Plot of power versus suction pressure with constant discharge pressure.

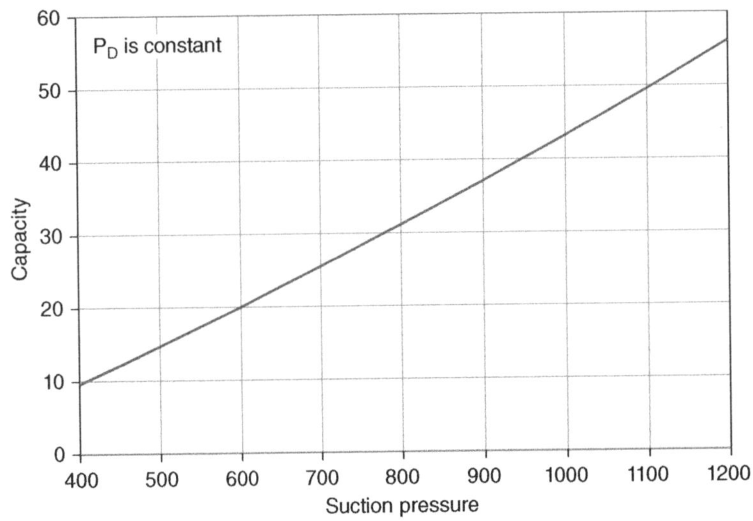

Figure 14.19 Plot of capacity versus suction pressure with constant discharge pressure.

Conclusion

Reciprocating compressors are used in many applications because they can compress air or gas in a wide range of pressure ratios and flowrates. They can offer excellent operational flexibilities for different services. However, beside many advantages, there are some disadvantages and areas of concern. For example, pulsation is a major challenge for nearly any reciprocating compressor. It can impose many issues, including severe vibration, piping failures (due to fatigue), reduction in compressor efficiency, and errors in flow measurements. Also, the cylinder valves and unloaders can be sources of operational problems and reliability issues. Through condition monitoring, you get an early indication of developing problems in reciprocating compressors just like in other critical mechanical assets. This can help you plan outages for repairs and hold production stoppages to a minimum.

15

Lubrication Related Failure in Machinery

Introduction

According to industry experts, as few as 10% of bearings reach their L10 life before failing. So what happened to the other 90%? As many as 60–80% of all bearing failures (catastrophic, functional, and premature) are lubrication-related, whether it's poor lubricant selection, poor application, lubricant contamination or lubricant degradation. Then you can start to understand the tremendous cost saving opportunities of eliminating not only catastrophic and functional failures which impact production, but to extend the life of many oil- and grease-wetted components through precision lubrication.

To understand why so many failures are in fact related to lubrication, or more accurately, poor lubrication, one must look at typical root causes of failure. Water is one of the main culprits that shorten bearing life or contaminate oil. Also, thermal cycling caused by temperature changes tends to draw airborne contaminants into the bearing housing. Damaged or worn seals allow product to enter the housing and create severe problems.

Using the wrong oil or additives, foaming, particles, moisture and under or over lubrication contribute to operational failures. Improper oil levels and incorrect oil addition along with sudden volume loss can lead to operational failures as well. Also, filtration systems that are both on and offline are essential, as is proper equipment for draining oil. These failure modes arise largely from improper maintenance and operations, meaning that a strong knowledge of how machines work and how to apply lube is the best defense against them. Lubrication excellence programs are essential to preventing lubrication failures and extending the life of machines. Many of these failures can be a product of a lack of knowledge, best practice and maintenance, all of which can be prevented with a strong program.

Taking proactive measures to ensure that the lube is performing its best can go a long way in extending the lifespan of a machine. In fact, a proper failure mode prevention program can make a machine last three to eight times longer. For this reason, it is important to take note of all of the failure modes of lubricant and lubrication programs.

Root Cause Failure Analysis: A Guide to Improve Plant Reliability, First Edition. Trinath Sahoo.

Lubrication Related Failure in Sump and Circulating System of Turbomachinery

Water Contamination

Water in oil will dramatically decrease the effectiveness of the lubricant and therefore reduce service life. Typical recommendations, depending on the source, are to maintain water at 100 or 200 parts per million (ppm) or less in the lubricating oil. This can be a challenge in some applications but should still be a target.

In circulating oil applications, dewatering can be done by centrifuging so that the water content of the oil can be maintained at a suitably low level. One of the major routes for water to gain access to the bearing housing is intake of moist air into the housing when the pump is shut down and cools. This moist air cools as the pump cools, potentially releasing condensation into the bearing housing.

Temperature also plays a major role in how much water an oil can hold. Damaging levels of water, or free water, begin to occur in some mineral oils at between 400 and 500 ppm at 60 °C (140 °F). Free water may form at 200 ppm at 51.7 °C (125 °F) in the same oil. Therefore, it is important to know the saturation point of an oil at a given temperature to begin to determine a valuable set point for effective lubrication maintenance. By the time water becomes visible, damage is already occurring to both the oil and the surfaces of the equipment and components. Figure 15.1 shows moisture in oil and its effect.

Particle Contamination

Particle contamination can occur as a result of intake from the surroundings, improper cleaning of the bearing housing during maintenance cycles, and corrosion products from the high water content in the oil.

Even though ball bearings are made of hardened steel, they are surprisingly susceptible to failure due to contamination of particles. Particulate contaminants, whether those particles are soft or hard, large or small, will cause some damage to bearings. The reason is the very high contact pressures that are present in the small contact zones between the rolling

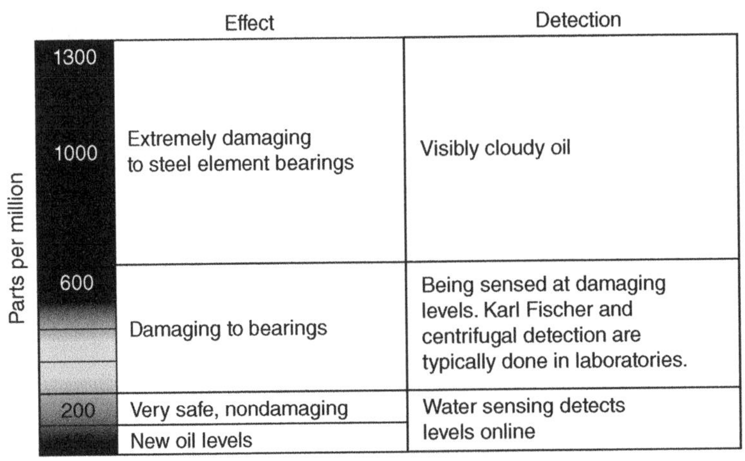

Figure 15.1 Moisture in oil its detection and its effect.

elements and the raceway. Pressures of around 200 000 lb/in^2 between the ball and the race in an angular contact ball bearing are not uncommon. Over-rolling contaminant particles with such high contact pressures result in dents in the raceway.

The presence of contamination shortens bearing service life in two ways. Every time a rolling element passes over a dent, contact pressure increases at the edge of the dent. Higher stresses result in shorter fatigue life. The second mechanism is wear. Although the balls do roll in a ball bearing, because of the curvature of the balls and races, some sliding occurs as well. The sliding portions of the contact, when contamination particles are present, can result in wear of the surfaces.

Also thermal cycling caused by temperature changes tends to draw airborne contaminants into the bearing housing. Damaged or worn seals allow moisture-laden contaminated air to enter the housing and create severe problems. Thermal cycling occur when bearing housings "breathe" as a result of rising temperatures which causes gas volume expansion and dropping temperatures at night cause gas volume contraction.

Another source of contamination is airflow across pumps generated by fans, blowers which is sufficient to create a pressure differential between the bearing housing and the surrounding atmosphere. This will increase the intake rate into the bearing housing as much as 10 times, leading to higher levels of contamination. Steam quenching and high-pressure washing can significantly increase the rate of intake of moisture into bearing housings. Care should be taken to avoid direct spray or steam around vents, oilers, and seals. Many vents, oilers, and seals are designed to prevent this intake – but they do need to be specified.

Lower particle counts significantly extend the life expectancy of equipment. For example, by reducing contamination levels from ISO 21/18 to ISO 14/11, life of a 50 gal/min pump could be extended by a factor of 7.

Heat

Elevated operating temperatures are a major contributor to oil oxidation. Combined with air, particulate, and water contamination, the chain reaction of oil oxidation begins. Additives are affected first, followed by the base stock, which leads to machine and component surface wear and fatigue. For every 18 °F (7.8 °C) oil operating temperature, the oxidation rate doubles. Oil operating at 75 °C (167 °F) will last 100 times longer than oil operating at 130 °C (266 °F). (Figure 15.2 shows oil life vs type.)

Air Entrainment

Air entrainment is the primary source of oxygen in the oxidation failure of oil. New oil can contain as much as 10% air at atmospheric pressure. Splash-lubricated gearboxes, bearing housings using flinger rings or slingers, and compressors are all aeration-prone applications. Excessive aeration has a negative effect on acid number, oil color, film strength, and viscosity. In addition, air entrainment can lead to accelerated surface corrosion, higher operating temperatures, and oil varnishing.

Oil Degradation

The primary causes of oil degradation are high heat, air entrainment, and mixing incompatible fluids. Increased viscosity (thickening) is one of the results of this degradation. This usually

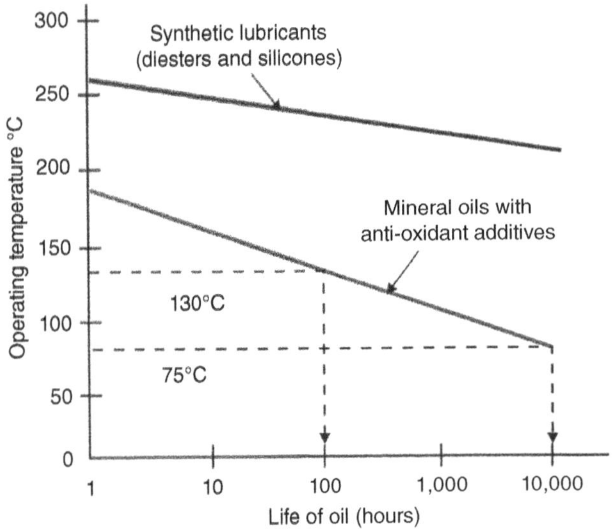

Figure 15.2 Oil life varies with base oil type and temperature.

happens overtime and varies by the combination of these elements. The viscosity of the oil increases where this fluid film acts as a solid and allows the ball to roll without metal-to-metal contact. When the viscosity is "wrong," the load-carrying ability of the lubricant is negatively affected. Additionally, if the oil degrades to a point where it is too thick to penetrate between these surfaces, the oil supply may not be adequate to prevent sacrificial contact.

Lubrication Related Failure Specific to Reciprocating Compressor

Though lubrication of crankcases and bearings require splash lubrication, the lubrication system for compressor cylinders and packing require relatively small amounts of oil at higher pressures. For cylinder lubrication, a force-feed lubrication system that uses a positive-displacement pump that must be capable of accurately delivering, monitoring, and protecting the oil flow to each of the required lubrication points is needed. If the system fails or does not work properly, the compressor units will be seriously damaged in a short period of time during operation. These are once-through systems, and the volume of oil delivered at each point needs to be just enough for proper lubrication. Therefore, the rate of lubrication at each point is critical, and over lubrication must be avoided. Excessive oil volumes can cause fouling of valves, gumming of the packing, and accumulation in the downstream piping system.

Factors Affecting Cylinder Lubrication

In compressor cylinders, operating temperature has an important role because of its effect on oil viscosity and oil oxidation, as well as on the formation of deposits. At higher temperatures, oil viscosity is reduced, and hence higher-viscosity oils are required to maintain adequate lubrication films.

Oil on discharge valves and valve chambers is heated when it comes in contact with hot metal surfaces and is continually swept by the heated gas after the compression cycle. This is a severe oxidizing condition, and the compressor oils oxidize to an extent that depends on the conditions to which they are exposed. The oxidation products formed are deposited mainly on the discharge valves, which are the hottest parts. The deposits on discharge valves may get struck in the valve seating area, leading to leakage of hot, high-pressure gas back into the cylinders. This high-temperature gas heats intake gas on the suction stroke, leading to high discharge temperatures.

The oil fed to compressor cylinders is subjected to oxidizing conditions as a result of high temperatures encountered when the oil leaves through the discharge valves, where temperatures are highest. Hence, to minimize the formation of deposits on the valves, oil fed to the cylinders should be kept to a minimum. It also helps to reduce excessive oil carry over to down stream equipment. The Compressed Air Institute suggests the use of 1 pint of oil for each 6 million square feet of area swept by the piston in air compressors. These are general recommended starting points and may need to be adjusted (up or down) based on gas and operating conditions in the compressor.

Moisture is an important factor because condensation occurs in the cylinders during idle periods when cylinders cool below the dew point of the air. The water thus formed tends to displace the oil films and come in contact with the metal surfaces, leading to rusting. The rust, when stuck between the surfaces of piston and cylinder, will result in excessive wear. Also, rust tends to promote oil oxidation and contribute to the formation of deposits. If this potential exists, the oils should be selected with good rust-inhibiting qualities and fortified with effective additives that will adhere to metal surfaces.

Rate of Oil Feed

The amount of oil fed to the compressor cylinders should be sufficient to provide lubrication and effectively seal the piston against leakage. Oil feeds above this amount are wasteful, cause oxidation, and tend to increase oil carryover to distribution lines.

Excessive Oil Feed. All oil fed to the cylinders is subjected to oxidizing conditions. Under prolonged heating, even the best-quality compressor oils will oxidize to some extent. Therefore, feeding more oil than is actually needed results in increasing the amount of oxidation products formed. Because the highest temperatures are encountered on discharge valves and in discharge passages and most of the oil fed to the cylinders eventually leaves through the discharge valves, it is here that deposits tend to accumulate. To prevent or minimize trouble from deposits, an oil especially suitable for compressor service that permits using very low rates of oil feed should be used. Feed rates for compressor cylinder lubrication are typically shown in drops per minute.

Cylinder Oil Feed. In the lubrication of double-acting compressor cylinders, one of the most important factors is the rate of oil feed. The likelihood of over lubrication is greater than that of supplying too little oil. Many problems associated with compressor operation can be overcome by preventing excessive lubrication. Proper control of the supply of oil to the cylinders is the most effective means of preventing the formation of objectionable deposits around valve ports, in ring grooves, and on cooler surfaces.

Most people believe that just maintaining or adding lubricant to rotating equipment provides effective lubrication. There are many different methods to achieve the goals of your lubrication management program. Points below discuss how to design a lubrication program to improve overall plant reliability.

Crankcase Oil Oxidation. In a reciprocating compressor as oil is broken up into fine spray or mist by splash, a large surface of oil is exposed to the oxidizing influence of warm air. Hence oxidation takes place at a rate that is a function of the operating temperature and the ability of the oil to resist chemical change. Oil oxidation is accompanied by a gradual increase in viscosity and, eventually, by the deposition of insoluble products in the form of gum or sludge. These deposits may accumulate in oil passages and restrict the flow of oil to bearings.

Lubrication Program Management

The goal of every lubrication program should be to ensure that all equipment receives and maintains the proper levels of lubrication such that no equipment ever fails from inadequate or improper lubrication. For this to happen, we must follow the six R's of lubrication – right lubricant, right condition, right location, right amount, right frequency, and right procedures – for each piece of equipment.

Right Type

Two major factors in the selection of an oil-based lubricant are the correct viscosity and additives in the formulation. To get the right type of lubricant, you refer to the OEM manual and contact the OEM if you have any questions. With old equipment, the manual may be outdated, and better lubricants may be available. When in doubt, use your lubricant supplier along with the OEM manual. Most manuals are written for ideal conditions, but these guidelines do not address the real environment in which the equipment operates. To determine which type of lubricant is best for an application, one must understand the current situation. Look at application/environmental factors such as speed, temperature, load, vibration, moisture, and dust. Consider that

- Temperature determines lubricant base oil type.
- Speed determines the required viscosity (at operating temperature).
- Load, vibration, and moisture determine the additive package.

As temperatures or speeds increase, the viscosity of a lubricant will decrease. *Viscosity*, the measure of a fluid's resistance to flow, is essential to protect equipment. You need a formulation that will provide a sufficient film layer of lubrication to reduce friction.

When selecting the appropriate industrial lubricant, keep in mind the application environment. For example, if you expect that a bearing will be subjected to significant amounts of water contact, then choose a lubricant with low water wash out and high corrosion resistance properties. In situations where the bearings are operated at low speeds and extreme pressure, you can increase equipment reliability by choosing a higher base oil viscosity lubricant with high load capabilities. Selecting the right lubricant in each scenario can have considerable payoff.

Choosing the right lubricant is even more essential if you are operating in an aggressive environment, where there is typically high moisture and corrosion and extreme temperatures. It is even more critical to equipment life to select a lubricant that will maintain a sufficient film of lubrication to reduce friction, resist load and wear, and prevent corrosion.

Right Quality

Once the right type of lubricant has been selected, it is important to select a high-quality lubricant. Quality is both the ability of the lubricant to meet OEM specifications, based on performance on ASTM tests, and the cleanliness of the fluid in which it is delivered. You can have the highest-quality lubricant, but if it is not handled properly during delivery or storage, it will not perform the way you expect it to.

Right Amount

Too much lubricant in a system can be as destructive as too little, which is also a major failure mode. It is important to understand the various parameters surrounding the operation of any given bearing to properly select relubrication intervals. Over greasing will lead to increased operating temperatures, resulting in energy losses and eventual bearing failure. Similarly, using too little grease will not allow the grease to properly carry the load applied to it, which will also result in bearing failure. In splash-lubricated systems, enough oil needs to be splashed up for cooling and lubrication. Too high an oil level will cause churning, which overheats the oil; too low a level will not provide proper oil cooling and lubrication for bearings and gear teeth. Spur helical, bevel, and spiral bevel gears are lubricated with the gears dipping into the oil at twice the tooth depth. The OEM will provide information on the correct oil level.

Right Place

Once we have selected the right type of lubricant and the correct quantity to add, we need to apply it at the proper location. Adding the wrong oil to a lubrication point is not uncommon. This situation will usually go undetected until a problem occurs. With an oil analysis program, early-stage detection is more likely, thus helping to avoid possible equipment damage. All lubrication points should be properly labeled as to the lubricant to be added. Lubricant manufacturers provide lube tags for proper identification of the proper lubricant to be used at the lube point.

It is a good practice to use separate containers for different lubricant types. Mixing lubricants with different additive packages is not recommended. Normally, each lubricant supplier color-codes its tags by lubricant types.

Right Time

Once we have established our program with the right type, quality, amount, and place, we need to establish proper lubrication intervals. This timeframe is known as the *regreasing frequency*. This calculation is more complex. The machine's operating conditions must be collected along with some additional bearing information to derive regreasing intervals.

Temperature is known to affect both grease and oil, so naturally one of the first correcting factors that must be collected is operating temperature. The hotter the environment, the more frequently the grease must be replenished. The ambient contamination (how dusty the environment is) and moisture (how humid the environment is) work in a similar manner. The dirtier and wetter the environment, the more often the bearing must be regreased.

Right Lubrication Procedures

Once the correct lubrication type, quality, amount, and place and time have been determined, procedures should be put in place to maintain a lubrication program. This will ensure that the proper lubrication procedures for each piece of equipment throughout the plant will be followed by all maintenance personnel. Putting together a lubrication plan should become part of the maintenance standard operating procedures.

These factors should include

Reviewing storage and handling conditions

- Maintaining records of the correct lubricant type for each application.
- Determining the proper amount of lubrication per day and frequency of relubrication.
- Tracking mean time between failures (MTBR)/mean time between maintenance.

By maintaining careful records, you can see patterns over time in performance and adjust accordingly.

Important Points to Be Considered for Developing an Effective Lubrication Program

Lubricant Selection

When choosing a lubricant for a particular piece of equipment, the equipment manufacturer's operation and maintenance manual should be consulted. The operation and maintenance manual will usually outline the required characteristics of the appropriate lubricants as well as a recommended schedule for replacement or filtering. If the maintenance manual is not available or is vague in its recommendations, lubricant manufacturers and distributors are other sources. Several factors need to be considered in the selection of the proper lubricant. The most important things you need to know are

1) Operating speed
2) Load (type and magnitude)
3) Bearing type (if applicable)
4) Temperature
5) Method of lubrication
6) Operating environment

All the pertinent information on the equipment and unusual operating conditions should be provided to the lubricant manufacturer or distributor so that a lubricant with the proper characteristics can be chosen. Whenever possible, lubricants should be purchased that can be used in several applications. By limiting the number of lubricants onsite, the chance of mixing different lubricants or using the wrong lubricant is minimized.

Storage and Handling of Lubricants

The important issues that affect lubricants as a result of improper storage and handling are as follows:

1) Pollutants such as dust and water may enter the lubricant if adequate precautions are not taken. These pollutants then get into the lubricated equipment.

2) If the lubricants are not labeled properly, lubricant identification become difficult. As a result, an unsuitable lubricant may be added to a machine, causing damage to the equipment.
3) Mixing some lubricants together may cause equipment breakage.
4) Damaged packaging can result in leaks that can affect the environment.

Lubricant Degradation During Storage

Lubricants can degrade in storage mainly for the following reasons:

1) Contamination, most frequently dirt and water.
2) Exposure to excessively high temperatures.
3) Long-term storage.

Entrained water promotes base oil degradation and additive depletion. Free water provides a place for microbial contamination that is corrosive and harmful to lubricant properties.

Emulsified water has a tendency to also impair air-release properties of oil. When air fails to detrain (release air to the headspace), a common consequence is oil oxidation. Solvents, fuels, and other incompatible lubricants are harmful contaminants for lubricating oil.

Proper Storage Environment

Whether your lubricants are stored indoors or outside, environmental conditions can reduce their shelf life. Fluctuating temperatures may cause a reaction known as *thermal siphoning*, in which air moves in and out of the container's headspace and the atmosphere. Moisture and airborne particles travel with the air, resulting in contamination and degradation of the lubricant. The fluids are also vulnerable to contamination by dust and dirt. Extreme hot or cold temperature can lead to chemical degradation.

Accurate Labelling

Are the lubricants clearly labeled and easily distinguishable? Lubricants are formulated with many performance additives and base stocks to match the lubrication requirements of the equipment in which they are used. Some lubricants are incompatible with each other because of differences in additive chemistry that lead to undesirable chemical reactions. If these oils are mixed, insoluble material may form and then deposit onto sensitive machine surfaces. This may lead to mechanical problems and shorter equipment life. Also, if incorrect lubricant are used, it may not develop an adequate film thickness and may cause equipment damage and health and safety issues. Proper labeling helps ensure that the appropriate lubricant is always used in the right component.

Oil Sampling

One of the most important milestones of an oil analysis program is sampling of the oil. The way a sample is collected, the frequency, the accessories used, and the procedures followed all dictate how informative the oil samples will be and, subsequently, dictates how beneficial the results will be. This is why it is of great importance to take the sample correctly. If the oil sample is not representative of the oil in the machine, then the results will not accurately reflect the condition of that component. The result will be an incorrect diagnosis based on incorrect analysis because of a poorly taken sample.

Strategic Sampling Considerations
- Sampling location
- Sampling hardware
- Sample bottles
- Sample procedure

Data Density and Data Disturbance

The oil sample should represent the body of oil about which data are required in order to increase the effectiveness of oil and machine decisions. This will happen by maximizing data density and minimizing data disturbance. Taking a sample that maximizes data density and minimizes data disturbance is of vital importance when selecting a sampling location.

The first goal is to maximize data density. The sample must be taken in such a way as to ensure that there is as much information per milliliter of oil as possible. This information relates to such criteria as cleanliness and dryness of the oil, depletion of additives, and the presence of wear particles generated by the machine. However, maximizing data density depends on the nature of the data you desire. For example, if you want to assess the effectiveness of a system's filter, you must collect a representative sample before and after filtration. The difference between the two samples is reflected in the differential particle count across the filter. Depending on the results, a decision to retain or change the filtering practices can be made.

The second goal is to minimize data disturbance. The sample should be extracted in such a way that the concentration of information is uniform, consistent, and representative. It is important to make sure that the sample does not become contaminated during the sampling process. This can distort and disturb the data, making it difficult to distinguish what was originally in the oil from what came into the oil during the sampling process. Failing to sample from a running machine, where the oil is not hot or well mixed, is a common source of data disturbance. Ideally, the machine should be operating at normal load and speed in its typical environment when a sample is taken; otherwise, particles and moisture can settle before the sample is taken, causing data disturbance.

Strategic Sampling Considerations

The following questions should be answered when designing a sampling program:

- Where is the best location to draw an oil sample to ensure that the correct information is collected?
- What are the best tools for drawing a sample from a specific location?
- Who will be responsible for pulling the sample, and how consistent will the sample be each time it is drawn from the specific location?
- What are the procedures followed to collect the sample?
- How clean is the sampling hardware?

Locations for Sampling

Sampling Port Location Troubleshooting problems in oil analysis is greatly assisted by the installation of several sampling ports in various locations to isolate individual components. Using multiple sample ports provides an analytical edge for both discovering potential component failures and analyzing the root causes. Sample ports are classified into two

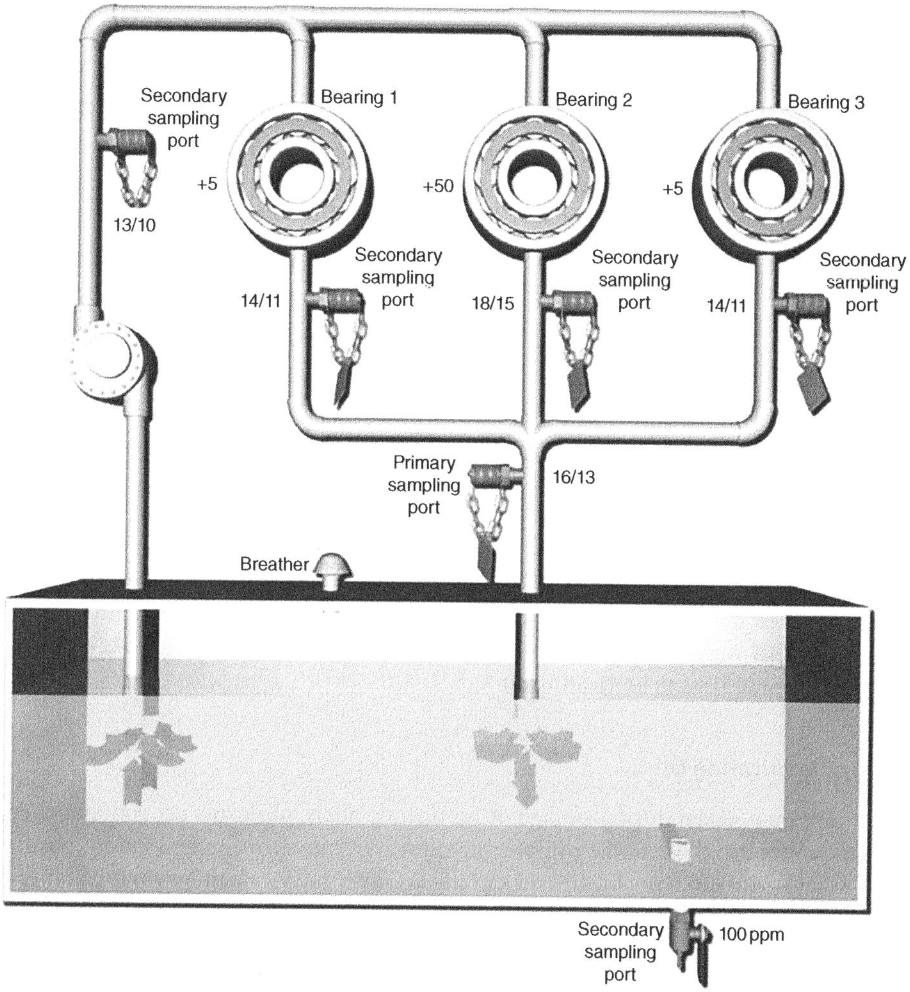

Figure 15.3 Sampling port locations on a lube oil system that feeds three sets of bearings.

categories, primary and secondary. Figure 15.3 shows sampling port locations on a lube oil system that feeds three sets of bearings.

Primary Sampling Ports The primary sampling port is the location where routine oil samples are taken. The oil from this sample location is usually used for monitoring oil contamination, wear debris, and the chemical and physical properties of the oil. Primary sampling locations vary from system to system but are typically placed on a single return line prior to entering the sump or reservoir.

Secondary Sampling Ports Secondary sampling ports can be placed anywhere on the system to isolate upstream components. This is where contamination and wear debris contributed by individual components will be found.

Consider a lube oil pump that feeds three sets of bearings (Figure 15.3). The return for the three bearings combines into a single return line before entering the sump. The primary sample port is on the single return line after all three bearing lines join and before the oil enters the sump. The secondary sampling locations are immediately downstream of the pump (upstream of all three bearings) as well as downstream of all three bearings.

The sample drawn from the primary sample port informs us that something has changed in the system, but it does not show what has changed. This is why secondary sample ports are used. They pinpoint the precise location of the problem. Secondary sample ports also can be used to monitor the general performance of filters. The primary port will show what is going into the filter, whereas the secondary ports show what is coming out. This procedure enables a filter change based on condition, long before the differentiated pressure indicator shows that a filter is in bypass.

Contamination Control

The easiest way to control contamination is to avoid using practices that have a risk of exposing the lubricant to a contaminant(s). Contamination of newly commissioned storage and handling equipment includes preservatives, paint, moisture, rust particles, and fabrication debris such as dust and dirt. Before storage and handling, the equipment should be thoroughly cleaned, made scale-free, and treated internally with a protective coating that is compatible with the lubricants used. The equipment should be carefully dried and cleaned before it is charged with a lubricant to be stored. Lubricants in storage are most prone to become contaminated with water from headspace condensation. Dissolved, emulsified, and free waters pose potential risks to lubricating oil.

Handling of Lubricating Oil

Although all care is taken during storage of lubricants, often lubricants get contaminated between storage and end use in the equipment. Many root causes of contamination causing machine shutdown are mainly due to a lack of attention in the distribution of a lubricant.

Even though the right lubricants are used consistently, many cross-contaminations occur because the same dispensing equipment is used for multiple lubricants. Ideally, dispensing equipment (such as pumps, transport containers, funnels, grease guns, filter carts, and transfer carts) should be labeled using the same system applied to the lubricants, matching each piece of equipment to the lubricant with which it is to be used. The same labeling system should be applied to machine lubrication points as well. Plastic jugs are preferred. Metal jugs can rust. Galvanized steel jugs can result in the transfer of zinc into the oil, and zinc can be poisonous with respect to some machine parts (e.g., turbine plain bearings). It is preferable to use jugs with a lid to prevent pollution by dust between storage and use in the equipment. When the jugs are not being used, cover them with a rag to prevent pollution by dust (Figure 15.4). Use a different grease gun for each type of grease. The nozzles of the grease guns must be wiped off after use. Grease nipples on the machine must be wiped off before packing them with grease.

Prefiltering of Lubricants

Whenever the concentration of solid contaminants in the lubricant exceeds the QC target limits for allowable contaminants, the lubricant should be filtered before use. Filtering lubricants prior to use can be done easily and relatively inexpensively. Filter elements should have high dirt-holding characteristics, low back pressure limits. Figure 15.5 shows the lubricant pre filtration pump.

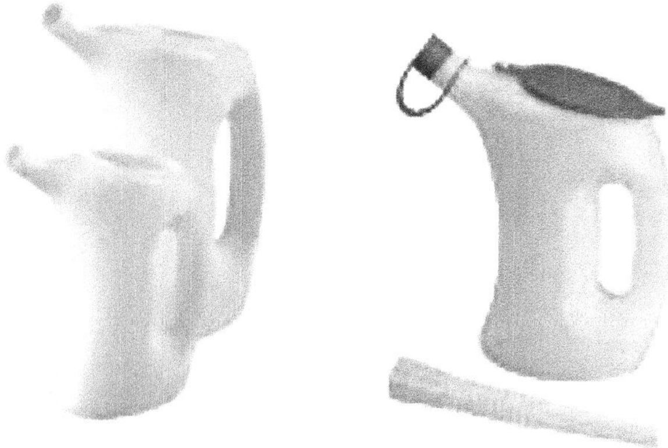

Figure 15.4 Dispensing equipments for lubricants.

Use of Sealable, Cleanable Top-Up Containers

After the effort of testing and pre-filtering the products to ensure that they are capable of performing in the desired manner, the end user certainly wants to preserve the improved quality state of the lubricant while it is being transported around the plant in the possession of the lubrication technician operator. This can be challenging. Any surface that is wetted with oil becomes a magnet for atmospheric contaminants. Funnels, new oil cans, dipsticks, transfer container, and so on all eventually become grossly contaminated if they are not isolated from the atmosphere following use. Even the top up containers if remain open may attract moisture and dirt. Figure 15.6 shows sealable and cleanable oil-handling containers which prevent dirt to enter the oil.

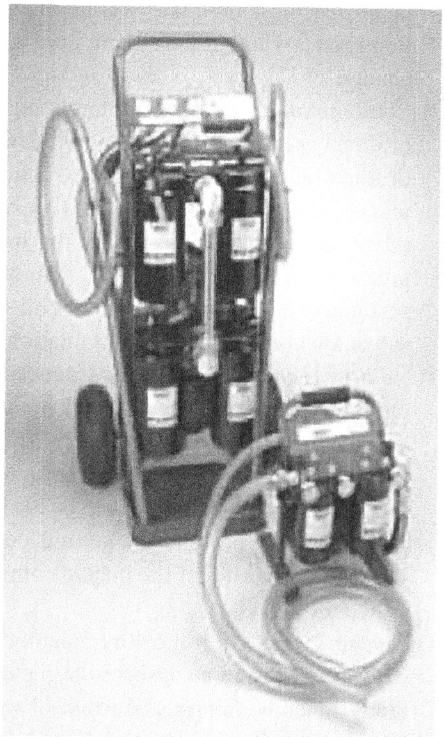

Figure 15.5 Lubricant pre filtration pump.

Understanding Oil Analysis: How It Can Improve Reliability

Oil analysis is performed to determine both fluid and machine integrity by measuring the elements found in the four main categories: physical fluid properties, fluid contaminants, additive metals, and wear debris. This process involves analyzing a lubricant's chemical and physical properties in order to detect lubricant or equipment issues before major problems develop. Hence, oil analysis serves as a valuable preventative maintenance tool, allowing an accurate scientific look at the lubricant's service condition as well as the condition of the operating equipment.

Figure 15.6 Sealable and cleanable oil-handling containers with colour coding.

The four primary focus areas of oil analysis tests reveal information about lubricant condition, contamination, additive elements, and wear debris

1) *Lubricant condition*. The condition of the lubricant is monitored with tests that quantify the physical properties of the oil to ensure that it is serviceable. Metals and debris associated with component wear are measured to monitor equipment health. Lastly, some tests target specific contaminants commonly found in oils. Assessment of the lubricant condition reveals whether the system fluid is healthy and fit for further service or is ready for a change.
2) *Contaminants*. Increased contaminants from the surrounding environment in the form of dirt, water, and process contamination are the leading cause of premature machine degradation and failure. Increased contamination alerts you to take action to save the oil and avoid unnecessary machine wear. Similarly, the presence of grease contaminating an oil system may be indicated by the increased presence of aluminum or barium if the grease contains metallic soaps. Although contamination is commonly associated with substances entering a component's oil system from an outside source, wear metals themselves are also a form of contamination.
3) *Additive elements*. These are substances that have been added to the oil to impart particular characteristics needed for specific applications. A measure of additive elements can provide the necessary information to determine whether the oil is still chemically able to perform its job. For example, the additive calcium (Ca) is used as a detergent; if oil analysis shows calcium levels to be too low, it would be correct to surmise that the oil will not be able to prevent sludge and deposits. The additive metals category focuses on chemical properties of the oil and compares them with baseline metals previously established for the oil.
4) *Machine wear*. An unhealthy machine generates wear particles at an exponential rate. The detection and analysis of these particles assist in making critical maintenance decisions. Machine failure as a result of worn-out components can be avoided. Remember, healthy and clean oil leads to the minimization of machine wear. Combinations of wear

debris can identify components within the machine that are wearing. For a proper analysis of wear debris/metals, the component materials incorporated into the particular system must be known. Results from wear-debris testing can indicate whether components in the system are operating in a normal state, are nearing failure, or have already failed. This is accomplished not only by assessing the type of debris/metals present but also by looking at their relative concentration in the sample, which can be indicative of specific component failures.

It is imperative to select the proper blend of tests to monitor the machine's lubricant condition, wear debris, and contaminants in order to meet the goals of successful oil analysis.

Trending

Trending or trend analysis is the most accurate way to assess the condition of an oil or machine and involves a data set of three or more variables. Variables represent the instances the oil sample was taken, and in order to establish a trend, oil would have to be sampled at least three times. Although manufacturers publish general guidelines for wear rates and contamination limits, the guidelines can be ambiguous or left to interpretation. Trending provides the most accurate method for determining normal values for interpreting oil analysis results and making recommendations.

Trending allows analyzers to look for abrupt changes over time. At least three samples must be on record for a trend to be determined. In the case where there is an inadequate amount of data, wear rates provide a guideline. For machines with a trend report, significant changes within the report will aid in determining potential problem areas.

Conclusion

A comprehensive oil analysis provides reliability professionals with information on a past or imminent failure. By taking advantage of oil analysis, owners can get a clear picture of the equipment's operating state, whether components are wearing at an acceptable rate or are showing signs of accelerated deterioration. By using this information, appropriate measures can be taken to repair troublesome components before an unexpected and costly failure occurs.

16

Steam Traps Failure

The job of the steam trap is to remove condensate, air and carbon dioxide from a piping system as quickly as they accumulate with minimum steam loss. Once the steam has released its valuable latent heat and condensed, the hot condensate must be removed immediately to prevent water hammer. Any air in the steam system consumes part of the volume that steam would otherwise occupy. The temperature of the air/steam mixture falls below that of pure steam. Air is an insulator and clings to pipe and equipment surfaces, resulting in slow and uneven heat transfer. When carbon dioxide is present, the steam pushes it to the walls of the heat transfer surface, where it reduces heat transfer. Carbon dioxide also dissolves in the condensate to form carbonic acid, which may corrode piping and equipment.

This chapter reviews the basic types of steam traps,their operating principles, advantages and limitations, installation requirements and offers advice on possible failure causes that may be encountered in steam trap operation.

Over the years, three basic trap types have evolved and have been classified according to their mode of operation. Certain types of traps may combine two working principles in their operation. Within the scope of this book, however, the predominant condensate discharge principle shall designate the trap type. The three types of traps normally used are:

- **Thermodynamic traps**- Traps that are actuated by the principles of thermodynamics and fluid dynamics.
- **Mechanical traps**- Traps that are actuated by a float, responding to changes in condensate level.
- **Thermostatic traps**- Traps that are actuated by temperature sensitive devices, responding to changes in condensate temperature.

Thermodynamic Trap

In a thermodynamic trap (Figure 16.1), cool condensate at process pressure enters the trap body, raises a disc, and flows over the seat and out the discharge. Hotter condensate (near the steam temperature) flashes as it flows under the disc. This high-velocity flash steam creates a low-pressure area on the bottom of the disc. The pressure of the flash steam above the disc combined with the lower pressure below it force the disc to seat.

The disc remains closed until the flash steam above it condenses and its pressure drops to the inlet pressure. When the downward force on the disc is reduced, the disc again lifts up and discharges fresh condensate, which produces flash steam again, and the cycle repeats.

Root Cause Failure Analysis: A Guide to Improve Plant Reliability, First Edition. Trinath Sahoo.

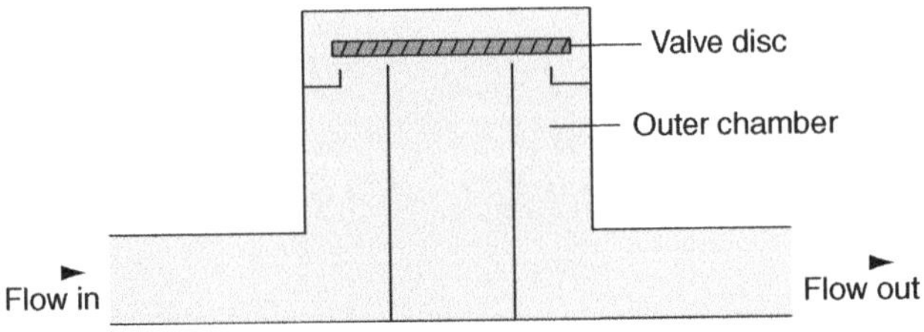

Figure 16.1 Thermodynamic trap.

These traps operate intermittently. Thermodynamic traps find application in steam main lines, tracers, constant-pressure applications, high-pressure lines and superheated lines.

Advantages of this type of trap include: it is small and lightweight; it does not freeze; it can resist water hammer and vibration; it is corrosion-resistant since most parts are made from stainless steel; it is unaffected by pressure and temperature fluctuations; a non-return valve (NRV) is not required because the disc prevents return flow; and it can operate in superheated steam. Its major drawbacks are that it cannot work with low inlet pressure and high discharge pressure, and its operation is noisy.

Float Trap

The simplest type of mechanical steam trap is a float (or float-and-thermostatic) trap (Figure 16.2), which operates on the principle of the difference in the densities of steam and condensate. A float trap consists of a chamber, a hollow spherical float and a valve. Condensate enters the trap's chamber and raises the float, which is directly attached to the discharge valve such that the higher the condensate level, the more the valve opens. As the valve opens, the condensate flows out of the chamber. When there is no condensate flow or when steam enters the trap, the float drops and closes the valve. The flow of condensate from this type of trap is continuous. If air enters with condensate, it will reduce the efficiency of the steam trap.

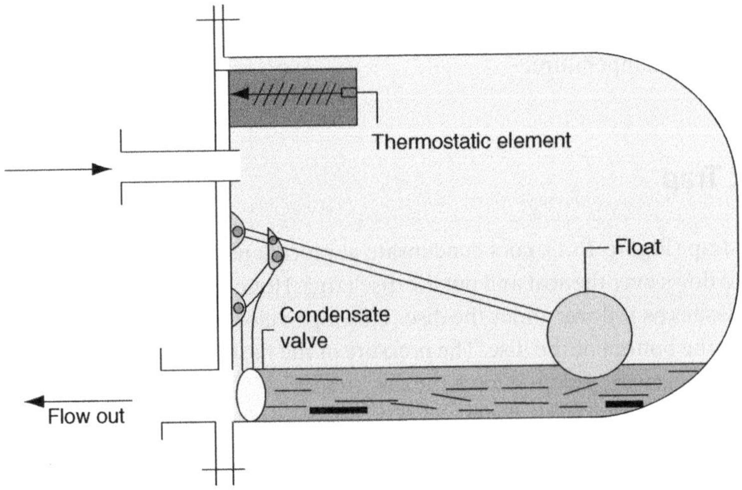

Figure 16.2 Float steam trap.

Because the condensate discharge valve is always submerged in condensate, it cannot discharge noncondensible gas. For this reason, the condensate collection chamber is fitted with a thermostatically controlled valve in the vapor space above condensate level. When steam flow is turned on, the condensate pushes the air toward the trap and out through open air vent valve. After a sufficient amount of air has collected, the low temperature of the air causes the thermostatic valve to open and allow the air to leave. Float traps discharge condensate at steam temperature, and are unaffected by sudden or wide load or pressure changes. They can be used where large condensate loads demand immediate removal and where steam locking occurs. Typical applications include heat exchangers with high and variable condensate loads, reboiler jacket pans, and batch operations that require frequent startup of an air-filled system.

Float traps are available with a wide range of orifice designs and sizes to handle various inlet pressures, and they can handle rapid pressure changes. They have a high capacity for handling non condensibles. The float trap's limitations include: It is susceptible to damage by water hammer. It cannot be used in superheated steam. The float can be damaged by freezing. Water expands as it get hotter, and the water's lower density limits the float trap's maximum operating pressure.

Float traps are vulnerable to dirt. The discharge valve is located near the bottom of the trap, where large dirt particles may accumulate. Also, float traps can operate steadily in a partially open position, which allows debris to become lodged between the plug and seat of the discharge valve. The protection against dirt is to buy a trap that has an integral strainer, or to install a strainer ahead of the trap.

Inverted Bucket Trap

The inverted bucket trap contains an inverted cylinder that is open at the bottom (resembling an upside-down bucket) and is connected directly to the condensate discharge valve mechanism (Figure 16.3). The bucket has a small hole on the top through which air escapes. As condensate enters the trap, it accumulates inside the body and the bucket. The water level

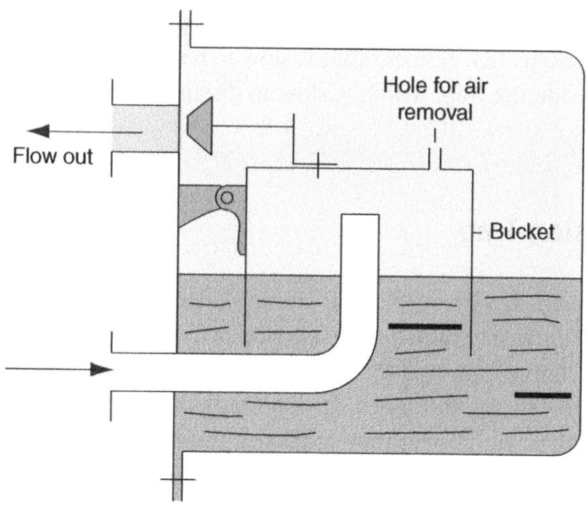

Figure 16.3 Inverted bucket trap.

climbs faster outside the bucket than inside, which gives the bucket buoyancy and causes it to rise, closing the valve.

Condensate will gradually fill up the bucket until its buoyancy is not sufficient to hold it up and it sinks. This pulls the valve off its seat and allows the water to flow out. As soon as enough water has left the bucket, its buoyancy is restored, and it again rises and closes the valve. These traps discharge condensate intermittently at steam temperature. Typical applications are high-pressure steam mains, shipboard systems, and continuous processes with small quantities of non-condensable. Its main advantages are that it can withstand water hammer and it can operate at very high pressures.

However, the small discharge hole in the top of the bucket discharges air slowly. The water seal, which must always be maintained to ensure proper operation, may be lost if there is an abrupt drop in steam pressure or in the presence of superheated steam. The inverted bucket trap cannot discharge condensate continuously as rapidly as it forms; it will always lose some steam. If the bucket vent gets blocked, the trap locks shut.

Inverted bucket traps are resistant to dirt becausethe discharge valve is located at the top of the trap, away from dirt that settles in the bottom of the trap body. The discharge valve opens abruptly and fully, so dirt carried in the condensate does not become lodged in the valve seat.

Inverted bucket traps are most likely to fail in the open position. They fail in the open position if they run dry, because then the bucket cannot float. (This problem is most likely to occur if the steam is superheated.) They may also fail from misalignment of the internal mechanical linkage. When they fail in the open position, the size of the discharge orifice is the only factor that limits steam loss.

Thermostatic Traps

Thermostatic traps respond to changes in temperature and therefore discriminate very well between steam and cooler non condensable gases. They can rapidly purge air from a system, especially on a cold start-up, and can be installed in various positions. Most frequently, actuation is by means of a bimetallic element or a bellows like capsule filled with a vaporizing liquid. Thermostatic traps respond slowly to changing conditions even though the cause is usually misunderstood. It is not the heat sensitive element that is slow to respond. Rather it is the heat energy in the condensate inside the trap, which is slow to dissipate, that causes the time delay.

Thermostatic Metallic-Expansion Trap

A thermostatic metallic-expansion trap utilizes the sensible heat of the condensate. Generally, the body of the trap is made of steel and the inside tube is made of copper. Condensate enters the tube, flows past the valve, and is discharged through the outlet port (Figure 16.4).

When steam enters, the copper tube heats up and its expansion shuts the valve. As the steam condenses and the temperature drops, the copper tube contracts, the valve opens, and the condensate is removed.

The metallic-expansion trap can be adjusted to open at a predetermined temperature within the operating range. Its main applications are steam tracing lines and condensate lines.

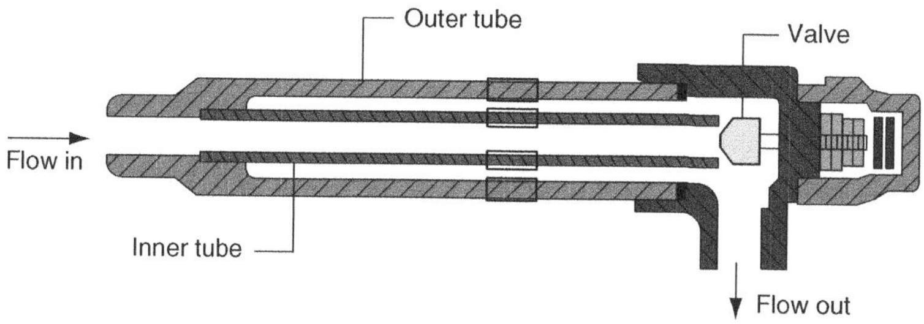

Figure 16.4 Metallic-expansion trap.

This type of trap allows the discharge of non-condensable at start-up. It is not affected by superheated steam, water hammer or vibration, and it can resist freezing.

However, its elements are subject to corrosion, it is not suitable where condensate must be discharged as soon as it forms, and it works best if the pressure, and thus the temperature, of the steam are fairly constant and predictable.

Thermostatic traps are prone to leakage because they open and close relatively slowly. This provides time for erosion by high velocity steam and water while the sealing surfaces are barely separated. The partially open discharge valve may trap dirt, preventing tight closure. The piping layout of thermostatic traps may be critical. The trap does not open until the condensate cools somewhat. If the condensate does not cool quickly enough, condensate may back up into the steam equipment or steam line. Therefore, install thermostatic traps so that they are surrounded by air that is substantially cooler than steam temperature. With low pressure steam, the trap may have to be installed at a certain distance from the equipment or steam lines. Do not insulate the trap or the pipe that leads to it.

Balanced-Pressure Thermostatic Trap

In this design (Figure 16.5), the thermostatic element consists of one or more bellows filled with a liquid that boils at a temperature lower than water. When steam enters the trap chamber, the liquid inside the bellows is heated, then the element expands and closes the valve.

The bellows is subjected to the system pressure and temperature, so the boiling point of the internal material increases as the surrounding steam pressure increases. Thus, the trap automatically compensates for changes in system pressure.

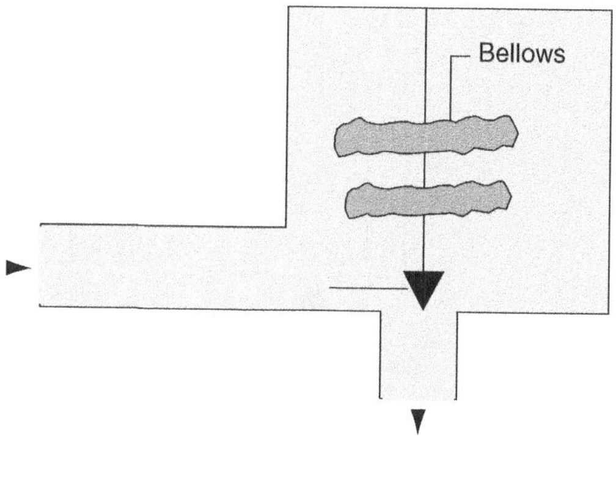

Figure 16.5 Balanced-pressure thermostatic trap.

These traps are used in plants where a batch process requires rapid discharge of non condensibles at start-up and in steam tracing lines, and as the air vent valve for float traps. They should not be used for superheated steam or where water hammer is a possibility. When superheated steam enters the trap body, the liquid inside the bellows boils vigorously and expands greatly, resulting in a high internal pressure that the thin and fragile bellows elements cannot withstand.

The balanced-pressure trap is small and light weight, and provides maximum discharge of non condensibles at start-up. It is self-adjusting to all pressures within its operating range.

However, it can be damaged by water hammer and superheated steam, and freezing may destroy the delicate bellows. It is not suitable for use where condensate must be discharged as soon as it forms.

Bimetallic Trap

The bimetallic trap (Figure 16.6) consists of a strip of two dissimilar metals with different coefficients of thermal expansion joined together. When cold, the bimetallic strip is flat, allowing the valve to open. When steam enters the chamber, the strip is heated and takes aoval shape, causing the valve to close.

On sensing the temperature of the steam/condensate, owing to dissimilar growth of bimetal, the above strips tend to warp and effectively push the disc down against the seat to stop the flow of fluid through the discharge port.

Similar to liquid expansion trap it is possible in a bimetallic trap, to set the discharge temperature of the condensate by adjusting the position of the lock nut at the top of the deck of the bimetallic strips.

It is used in constant-pressure steam main lines and tracing lines.

This trap is small and lightweight, and provides maximum discharge of noncondensibles at start-up. It can withstand freezing, water hammer, high pressure, and superheated steam.

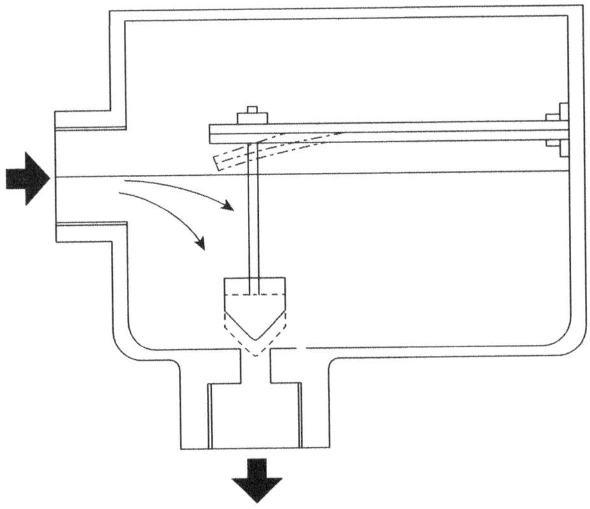

Figure 16.6 Bimetallic steam trap.

It discharges condensate well below steam temperature to reduce flash steam. However, it responds slowly to load and pressure changes, and cannot discharge condensate at steam temperature.

Selection Criteria

From the standpoint of efficiency, ability to block steam flow is the main consideration in selecting traps. In addition, you need to consider other characteristics, as described below.

Capacity range. Float, bucket, and thermostatic traps block steam efficiently from zero condensate flow up to their maximum rated capacity. Disc traps adapt to different drainage rates, but they are limited to small capacities because of the way they operate.

System pressure. Inverted bucket traps are available for any pressure. Float traps are limited in pressure by the possibility of crushing the float. Bellows type or encapsulated thermostatic traps are limited in pressure by the possibility of crushing the thermostatic element. Orifice and disc traps are limited in pressure by the erosion that occurs when high-pressure steam passes through narrow passages. Some disc traps require a minimum pressure drop between the steam side and the condensate side to operate properly, typically 10 PSI or more. In addition, disc traps are vulnerable to back pressure because proper operation requires steam to be ableto exit from the trap at high velocity.

Venting a cold system at start-up. Thermostatic traps provide rapid venting of cold systems. Thermostatic elements are included in F&T traps for cold system venting.

Inverted bucket traps do not vent air rapidly because the smallness of the vent hole in the bucket limits the flow of gases through the trap. To compensate for this, a thermostatic elements can be fitted to the bucket that increases the size of the vent hole when the bucket is cold. This added complication reduces the reliability of the trap, all other things being equal.

Disc traps vent a cold system very slowly because the trap is closed by air in the same way as by livesteam. Orifice traps are poor for venting a cold system because of the typically small size of the orifice. You can vent a cold system by using separate airvents, which gives you greater latitude in selecting trap types.

Venting a warmed-up system. F&T traps, inverted bucket traps, and orifice traps all do a good job of venting non-condensible gases from a system that is operating at normal temperature. If a thermostatic trap is kept flooded by condensate, air never has a chance to reach the thermostatic element, so the trap cannot vent.

Vulnerability to freezing. Inverted bucket traps and float traps remain partially filled with water when they are idle, which invites freezing damage. Of the two, float traps are more vulnerable because they contain a float ball and a thermostatic element (in F&T traps) that are easily crushed by ice expansion. Bimetallic thermostatic traps, disc traps, and orifice traps are less likely to be harmed by freezing. Installation practice is a major factor in avoiding freeze damage, both for traps and for other steam equipment. The basic principle is to completely drain the portions of the system that may be exposed to freezing temperatures. Competent steam fitters have a variety of techniques for accomplishing this. Major trap manufacturers publish guidance in avoiding freezing in steam systems.

Operation with Superheated Steam. Superheated steam can be a problem for several trap types when they are used in drip legs. If superheated steam reaches an inverted bucket trap, the steam will rush through and keep the trap dry, causing the trap to remain in the open position. Superheated steam may cause a disc trap to chatter and pass steam. Superheated steam may burst the expanding element of a bellows trap. Superheated steam may increase the loss through orifice traps. In principle, you could use any of these types with superheated steam, provided that you install them so that liquid condensate always forms a head of the trap.

Vulnerability to water hammer. A slug of condensate propelled by high steam pressure has enough energy to crush trap components, especially thermostatic elements and float balls. Float trapsand bellows-type thermostatic traps are sensitive to water hammer.

Service Life: A long lasting steam trap is obviously desirable. Regardless of design, however, every steam trap's life is shortened as steam pressures and temperatures increase. Hot condensate is a particularly difficult liquid to handle. It can be both corrosive and erosive and, under certain conditions, it can cause cavitation. It can destroy valves and seating faces in a matter of days in extreme circumstances. The frequency with which a trap must open and shut also obviously influences its life span. Opposing these destructive forces are the designer's skills, hardened materials and the level of special concern a plant has for selecting and sizing traps suitable for the conditions they will experience.

Ease of Checking: Considering the difficulties inherent in detecting whether a steam trap is performing properly, it is surprising more consideration isn't given to this criterion. A trap that has a crisp open-close cycle can quickly be judged to be healthy with a simple listening device. Traps that modulate or have a slower and softer open close cycle often leave the steam trap checker uncertain about their condition. Is it capable of shutting off tightly. As a group, traps that normally provide modulating control such as the float and thermostatic or the bimetallic thermostatic-are more difficult to check than other trap types.

Sensitivity to Backpressure: Traps discharging in to closed condensate return systems will experience varying amounts of back pressure, depending on the return system's design and the number and condition of other traps discharging into it. Most bimetallic traps will discharge condensate at progressively cooler temperatures as they experience increasing back pressures. Thermodynamic traps tend to decline in efficiency as back pressures exceed 50% of the inlet pressure.

Responsiveness to Changing Loads: Not all trap types accommodate themselves quickly to the changing condensate loads typical of process applications. Mechanical and thermodynamic traps are very responsive, but thermostatic traps must first cool slightly before they can openwider to pass a greater amount of condensate.An adequate "cooling leg" is required in front of thermostatic traps to assure good system efficiency.

Resistance to Shock, Vibration & Water Hammer: Despite the system designer's best efforts, all steam systems tend to experience some level of vibration, shock, or water hammer. Start-ups, pressure changes or changing loads, are the periods that generally are the hardest on equipment such as steam traps. Not all traps however, are equally vulnerable to damage from these causes. Thermodynamic traps and bimetallic thermostatic traps are generally very rugged. The bellows in thermostatic traps and the closed float in some mechanical traps are fragile and damage-prone. These damage producing conditions are most frequently seen in process applications.

Discharge Mode (cyclic or modulating and continuous): Does a trap that has adistinct open and shut cycle provide inherent advantages or disadvantages relative to one that has a continuous and modulating discharge? The cycling thermodynamic disc and bucket traps are easier to check for proper performance, and perhaps better at passing dirt particles. On the other hand the continuous draining float trap is especially responsive to rapidly changing condensate loads and it does not contribute to pressure surges in the return system. The judgment as to which is the superior trap depends on the relative value that the plant operator assigns to these various characteristics.

Ease of Maintenance: All steam traps fail in time. It is a matter of management philosophy whether they are repaired or scrapped and replaced. While steam traps are available in both repairable models and throw-away versions, the longer economic view favors repair ability. This preference has encouraged several manufacturers to design traps that significantly simplify the maintenance task.

Common Problems of Steam Traps

Steam leakage. The valve seat of a steam trap is subjected to erosion and corrosion. When the seat is damaged, the valve will not seat completely and the trap may leak live steam.

When the steam trap is oversized, it may waste significant quantities of steam. Even a bimetallic trap that is calibrated to be fully open with minimum sub-cooling may blow steam if its load is reduced. A properly working thermodynamic trap may fail to close if the condensate pressure is too low.

Improper sizing. A trap that is undersized will cause condensate to interfere with the heat transfer efficiency because the condensate forms a thin film on the heat-transfer surfaces. Traps are typically specified several times larger than required using a safety factor to calculate the trap capacity. A trap having too much excess capacity wastes money, acts sluggishly and generates high back-pressure that may significantly reduce the life of the trap.

Dirt. Steam condensate often contains particles of scale and corrosion products that can erode trap valves. If the particles are large enough, they can plug the discharge valve or jam it open. To prevent this, a strainer must be installed upstream of each trap. The strainers must be cleaned when the system is first started up and when any steam piping is replaced.

Noise. With the exception of thermodynamic traps, most traps tend to operate relatively quietly. In some cases, traps may produce a slight audible sound as condensate flashes into steam downstream of the trap valve. Noise in a steam system is usually caused by lifting condensate up vertical return lines, water hammer, or a failed trap that leaks live steam to the condensate returnline.

Air binding. When the trap is connected to the plant by a long length of small diameter horizontal pipe, condensate holds up in the steam space and cannot flow to the trap. To prevent air binding, the piping to the traps should be of a larger diameter pipe and shorter length, which allows a higher flowrate. Another method to prevent air binding is to put a vent valve at a high point in the system.

Steam locking. When the trap is connected to the plant by a long length of small-diameter horizontal pipe, steam may prevent the condensate from reaching the trap. Condensate cannot get to the trap unless it can displace the steam. To prevent steam locking, it is necessary to put the trap as close as possible to the equipment or the line to be drained. If the trap is just below the equipment or line, a balance pipe may be installed between the two to serve as a vent and prevent steam locking. Traps can also be fitted with a steam-lock release valve.

Water hammer. Condensate lying in the bottom of a steam line can cause water hammer. Steam traveling at a very high velocity produces waves as it passes over this condensate. If condensate quality increases, the high-velocity steam pushes the condensate and creates a dangerous slug when it changes direction. This is called water hammer. When the high-velocity condensate comes to a halt, the kinetic energy is converted to pressure energy, and this sudden pressure increase destroys the trap mechanism of float traps and balanced pressure thermostatic traps. To prevent water hammer, robust traps like thermodynamic or inverted bucket traps may be used, or the pipe may be re-oriented.

Freezing. If the steam system is shut down with a sufficient quantity of condensate remaining in the trap, and the ambient temperature falls below 0 °C, then freezing inside the trap will occur. Float traps and balanced-pressure thermostatic traps are destroyed by freezing. If freezing is likely, thermodynamic or bimetallic traps, which are not affected by freezing, should be used. Opening the drainvalves after shutdown is another option.

Loss of prime. This problem relates to inverted bucket traps. The inverted bucket trap starts functioning only when there is some quantity of water inside the trap. If there is a sudden pressure drop or superheated steam enters the trap, the prime is lost and the trap won't operate. To prevent this, a check valve may be put in the trap inlet line.

Backpressure. Small levels of back pressure typical of a properly designed condensate return system are not generally a problem. It is the elevated levels of back pressure found in the inadequate return system that creates drainage problems.

Corrosion. All steam systems and their associated components suffer from the effects of corrosion. Corrosion attacks valve components, fittings and steam traps. Over time all these items succumb. The primary defense is a carefully monitored and maintained boiler feed water treatment system that controls the gases (oxygen and carbon dioxide) which promote corrosion. Carbon dioxide by itself is not corrosive, but it can combine with free hydrogen to form carbonic acid which is corrosive. A principal reason stainless steel is used extensively in steam traps is to resist the effect of corrosion and prolong the life of the trap.

Corrosion can be best controlled by proper boiler water treatment but any piping arrangements that interfere with good drainage increases the potential for corrosion problems.

Sharing of Traps

Do not use a single trap to serve more than one item of equipment. Condensate drainage is sensitive to discharge pressure, so if two units discharge to a common line, the discharge from the higher pressure unit can block the flow of condensate to the other unit. This problem is especially severe with modulating equipment. Sharing traps makes it difficult to diagnose problems in traps or the equipment they serve.

Installation

Proper operation of a steam trap depends on the way it is installed. As we have seen, installation practice is important for freeze protection, and it may be a factor in providing adequate surge capacity. Piping the trap to maintain a water column ahead of the trap may be necessary for satisfactory operation and longevity of thermostatic and disc traps.

Strainers

As we have seen, dirt is a potential problem with steam traps. It can keep the valve from closing completely (all types, except orifice traps) and it can clog the trap passages (especially orifice and disc traps). The general solution to these problems is to install a strainer ahead of the trap. Figure 11 shows a strainer that is properly installed. The strainer itself is an item that requires periodic inspection and cleaning. Unless the steam system has a particular dirt problem, inspection is needed only at long intervals. This leads people to forget about them. Make sure that your maintenance schedule includes such multi year inspections. To simplify the installation, you can get traps with integral strainers in most types and sizes. However, these do not eliminate the need to clean out the strainers periodically.

Maintenance of Steam Traps

Trap Checking Techniques

Checking a steam trap is the process of observing its performance and comparing it with the performance characteristics that one has learned are typical for a healthy trap of the same type. If the performances are similar, the trap may be judged to be OK. If there are differences, it can be concluded that either the trap is faulty or the system in which the trap is installed has a problem.

There are three basic techniques used in observing the performance characteristics of a steam trap. They are:

- **Sight**-visually observe the discharge pattern;
- **Sound**-listen to the functioning of the valve mechanism and the flow of fluid through the seat;
- **Temperature**-determine the trap's temperature.

Each of the checking techniques has its limitations and seldom can a conclusive opinion be reached on the basis of a single type of observation. Experienced steam trap checkers invariably try to use all three techniques. Some will use more expensive checking equipment than others in this checking process, but the quality of their results does not seem to vary much.

Sight

Watching the discharge pattern of a steam trap is probably the most reliable method of determining whether it is working properly. Unfortunately, many traps discharge into closed condensate return systems without test "T"s and the discharge pattern is not visible. Under these conditions trap performance appraisal is limited to sound and temperature monitoring. An experienced eye begins to distinguish the lazieraction and white appearance of flash steam from the more transparent jet-like discharge of live steam that can be seen right at the trap's outlet. The best clues corne with watching a steam trap that has a normal crisp off/on cycle, such as a disc or bucket trap. If the discharge vapor has any velocity during the closed period of the trap's cycle, it can be reasonably assumed that it is leaking steam and should be replaced. While most thermostatic traps can have a cyclic mode discharge, it tends to be slower and less definite than either the disc or bucket types. Thermostatic traps, especially bimetallic traps, can also have a continuous discharge pattern. Float traps that are designed for continuous modulating drainage are harder to diagnose, especially in the early stages of failure.

Sound

Listening to the sound of a steam trap functioning with the bit of a long handled screw driver (held against the trap and its handle pressed against an ear) can provide important information as to whether it is doing its job properly. Because of the high background noise that often surrounds a steam trap, the screw driver or industrial stethoscope is giving way to the ultrasonic listening device with ear phones. These help to screen out the normal ambient background noise and permit more precise identification of a noise pattern in an individual trap. Listening to a trap which is designed to have an open and closed cycle action can quickly reveal if the trap is operating in this manner. It can also reveal if there is any significant leakage past the valve when it is in its closed position. But the value of listening to float traps or thermostatic traps operating in a modulating mode in an effort to determine if they are leaking steam is questionable.

Temperature

Every individual who checks steam trap performance employs some sort of temperature sensing means. A wet finger (obviously dangerous), a squirt bottle, a surface-contacting pyrometer, or an infrared sensor all have the same initial objective: Is this trap hot? If not, it can be deduced that little or nothing is flowing through it and the trap is not performing one of its major functions-draining condensate. This does not necessarily mean that the trap has failed. It first must be determined if some obstruction-either up or downstream of the trap-has blocked flow in the pipe. Once that issue is resolved, it is logical to proceed with steps investigating the trap for blockage by pipeline dirt or failure of its internal mechanism.

When it has been determined that a steam trap is hot, the next question is whether it is as hot as it should be. Steam traps designed to discharge condensate every close to steam temperature such as disc, bucket or float traps will have a surface temperature about 5–10% below the temperature of steam in the system. Unfortunately, it does not follow that if temperatures are observed in this range that the trap is healthy. It only means it is doing half its job properly, i.e., draining hot condensate. It is difficult to determine by temperature alone if a trap is leaking steam when it is designed to operate at close to steam temperatures. If measured temperatures are much below the expected range, the trap can be suspected of backingup more condensate than normal. In general, there will be a significant difference in temperatures between that observed upstream of a trap and that observed downstream. These temperatures will be directly related to the pressure in the system at the point of measurement. If the temperatures measured up and downstream of a trap are the same, it can only be deduced that the pressure on either side of the trap is the same. As this is an abnormal condition, it can be concluded that a system problem exists that may or may not be caused by a faulty trap discharging into a common return. More analysis is necessary.

Diagnosis

A combination of temperature measurement (to determine drainage failures) and ultrasonic technologies (to determine leakage failures) has proven to be very effective to diagnose trap failure. High-quality temperature measurement at the trap inlet will reveal whether the trap is backing up condensate – which usually indicates some form of drainage failure. Identifying leakage failures is more difficult. The movement of fluid in a closed system develops ultrasonic noise. An ultrasonic testing instrument recognizes the noise created by flowing condensate or steam, or both, and can be used to determine a trap's operating condition based on the difference between the ultrasonic profiles of leaking steam and discharging condensate.

Some instruments recognize the difference in amplitude between the ultrasonic levels generated by steam and condensate, providing information about the trap's condition but not necessarily the source of the leak. Some instruments provide an audible output that requires the tester to make a subjective judgment, the accuracy of which depends on the experience of the tester. Any measurement should be compared to a reference standard to identify the source of the ultrasonic noise and determine the condition of the trap. At least one company makes an ultrasonic testing instrument that indicates trap condition by comparing the actual ultra-sonic readings to empirical data that reflect known values.

Knowing what failures exist and what actions to take to correct them requires an accurate diagnosis of every steam trap's actual condition. Otherwise, cost-recovery opportunities to correct leakage failures might be missed, or maintenance funds might be wasted replacing properly functioning traps. Consider implementing routine testing using diagnostic instrumentation that has been independently validated by a recognized authority in order to accurately assess the condition of your plant's steam traps.

While there are many steam trap users who routinely provide "preventive maintenance" by replacing trap elements annually, this not often practical. In fact it can prove costly and ineffective since traps can fail or leak in between these routines and many traps will work for years before the elements need to be replaced.

Steam Trap Audit

It is often more cost effective to establish a routine steam trap audit. As part of any predictive maintenance routine, knowledge of the system is critical. For this reason, before inspection begins, a map or some diagram of the location of all the steam traps and valves in a facility should be available. All traps should be tagged and coded and referenced on the diagram. In addition, the trap inventory should include the trap type, size, manufacturer, and application.

To improve on inspection routines, it is recommended that some form of record keeping/data collection be employed to provide information about the steam system overtime. This is useful in spotting potential areas of recurrent problems, possible clues about misuse of traps, data about costs and savings incurred. There is commercially available steam management software available which can be quite useful in maintaining accurate trap records.

As mentioned above, every trap should have a tag with a corresponding identification code. During the inspection procedure, trap condition should be noted. All poorly operating traps should be documented in a non-compliance report and a follow up procedure should be planned. Include a digital photograph of the trap in your report. The follow up procedure should include such items as trap number, condition and date of repair. As part of a quality assurance procedure, all repaired traps should be scheduled for re-test. A comprehensive report is recommended to describe the results of a steam trap survey. The report should include items such as the number of traps tested, the number found in good condition and the number of faulty traps. A cost analysis should be included as well. The cost analysis should indicate the gross amount of savings, the repair costs and the net savings for the survey.

Once the record keeping has been put in order, various methods of inspection should be considered. The most common are visual inspection, acoustic stethoscopes, temperature and ultrasonic testers. Ultrasonic testers translate the high frequency emissions of a trap down into the audible range where they are heard through head phones and seen as intensity increments on a meter. Some units have frequency tuning to filter out additional signals and to tune into the sounds of steam and condensate while others have on-board recording and data logging so that users can record the sounds of steam traps and data log important test information.

Any steam system, no matter how diligent the operation, can leak; any trap can potentially waste steam. If performed properly, a routine, planned program of steam trap inspection and repair can continually pay for itself and contribute to a company's bottom line in terms of productivity, quality and energy savings. A rule of thumb states that if there has been no steam trap survey or maintenance program, upwards of 50% of a system's traps can be blowing live steam. If a survey is performed annually, this figure drops to about 25%. A bi-annual survey will reduce this even further to less than 12%.

Conclusion

Steam Traps are an important element of any steam system. Faulty steam traps not only waste energy, they can contribute to pipe erosion due to poor water quality and Contaminants allowed to pass downstream. Faulty steam traps can negatively affect product quality in various processes such as paper, food or chemicals, and even add to environmental pollution. Hence proper steam trap maintenance is essential for a steam system. Steam Traps should be inspected routinely. The frequency of inspection is often determined by application. As an example, steam systems used just for facility heating are routinely inspected annually while systems that utilize steam as part of a manufacturing process might be inspected biannually or annually depending on the impact steam has on the process.

17

Proactive Measures to Avoid Failure

Critical equipment can fail at the time of peak production requirement which may impact company bottom line, may sometime rack up downtime and drain a maintenance budget. Developing a proactive maintenance program help organizations identify potential faults before they disrupt service, as a result unplanned outages and associated costs are eliminated. Proactive maintenance is a preventive maintenance strategy that works to correct the root causes of failure and avoid breakdowns caused by underlying equipment condition. It helps to see machine failures as something that can be anticipated and eliminated before they develop. Although proactive maintenance requires in-depth planning, top level buy-in and investment, it ultimately helps organizations avoid over or under maintaining assets and offers a balance of costs, reliability, up time and asset health.

The purpose of the proactive maintenance tasks in the failure management strategy is to

1) Prevent failures before they occur
2) Detect the onset of failures in sufficient time so that the failure can be mitigated before it occurs.

What Are Proactive Maintenance Tasks

Proactive maintenance tasks are divided into following categories.

Planned-Maintenance Tasks

A planned-maintenance task is performed at a specified interval, regardless of the equipment's condition. The purpose of this type of task is to prevent functional failure before it occurs. Many times this type of task is applied when no condition-monitoring task is identified or justified, and the failure mode is characterized with a wear-out pattern. Planned maintenance can be divided into the following two subcategories:

1) **Restoration Task**. A scheduled task that restores the capability of an item at or before a specified interval (age limit) to a level that provides a tolerable probability of survival to the end of another specified interval. For the case of scheduled restoration of a diesel engine, rebuilding the fuel injectors would be an example.
2) **Discard Task**. A scheduled task involving discarding an item at or before a specified age limit regardless of its condition at the time. Note that the terms "restoration" and "discard" can be applied to the same task. For example, if the piston ring and rider ring of a

Root Cause Failure Analysis: A Guide to Improve Plant Reliability, First Edition. Trinath Sahoo.

compressor are replaced with new ones at fixed intervals, the replacement task could be described as scheduled discard of the cylinder rings or scheduled restoration of the compressor.

Condition-Monitoring Tasks

A condition-monitoring task is a scheduled task used to detect the potential onset of a failure so that action can be taken to prevent the functional failure. A potential failure is an identifiable condition that indicates that a functional failure is either about to occur or is in the process of occurring. Condition-monitoring tasks should only be chosen when a detectable potential failure condition will exist before failure.

Combination of Tasks

Where the selection of either condition-monitoring or planned-maintenance tasks on their own do not seem capable of reducing the risks of the functional failure of the equipment, it may be necessary to select a combination of both maintenance tasks.

Failure-Finding Tasks

A failure-finding task is a scheduled task used to detect hidden failures when no condition-monitoring or planned-maintenance task is applicable. It is a scheduled function check to determine whether an item will perform its required function if called upon. Most of these items are standby or protective equipment. An example would be testing the safety valve on a boiler.

Risk based Inspection – Risk based inspection is the process of developing a scheme of inspection of pressure vessels, piping, structures and machinery based on knowledge of the risk of failure. The essential process is a risk analysis. This is the combination of an assessment of the likelihood (probability) of failure due to flaws damage, deterioration or degradation with an assessment of the consequences of such failure. The information gained from this process is used to identify (i) the type of damage that may potentially be present, (ii) where such damage could occur, (iii) the rate at which such damage might evolve, and (iv) where failure would give rise to danger. A suitable inspection scheme will be developed at a frequency that provides adequate confidence about the condition, taking account of the damage mechanisms and the reliability of the inspection techniques used.

Servicing and Routine Inspection

These are simple tasks intended to: (i) maintain the failure rate and failure pattern as predicted by performing routine servicing (i.e., lubrication, filter changing, other replenishment tasks); (ii) spot accidental damage; (iii) spot problems resulting from ignorance or negligence or (iv) respond to unsatisfactory condition monitoring techniques results. They provide the opportunity to maintain the general standards of maintenance at a satisfactory level. These tasks are not based on any explicit potential failure condition. Servicing and routine inspection may also be applied to items that have relatively insignificant failure consequences, yet should not be ignored (minor leaks, drips, etc.).

One-Time Changes

One-time changes are used to reduce the failure rate or manage failures in which appropriate maintenance tasks are not identified or cannot effectively and efficiently manage the risk. The basic purpose of a one-time change is to alter the failure rate or failure pattern through:

- Equipment redesigns or modification, and
- Operational improvements, or both.

One-time changes most effectively address equipment failure modes that result from the following failure mechanisms:

1) Faulty design, materials, or both
2) Improper fabrication, construction, or both
3) Mis-operation
4) Maintenance errors

These failure mechanisms often result in a wear-in failure characteristic, and thus, require a one-time change. When no maintenance strategy can be found that is both applicable and effective in detecting or preventing failure, a one-time change should be considered. The following briefly describes each type of one-time change:

Equipment Redesign or Modifications- Redesign or modifications entail physical changes to the equipment or system. An example would be mitigating main engine bearing damage caused by no or low lube oil flow by adding a low pressure sensor in the lubricating oil pump outlet that would start a standby pump to maintain lube oil flow to the main engine bearings.

Operational Improvements. Operational improvements are: changes in the way equipment is operated, modifications in the way maintenance is performed on the equipment, or both. Operational improvements usually entail changing the operating context, changing operating procedures, providing additional training to the operator or maintainer, or any combination thereof. For example, in the case of a centrifugal pump running near to it's best efficiency point. If part load running is envisaged then a bypass line from Discharge to suction to be provided.

A proactive maintenance plan can help make the above programs operate more efficiently and limit some of the major expenses that can emerge from putting a plan into action. Predictive and preventive maintenance approaches work together to influence proactive maintenance, as it is more of an approach-based strategy rather than an individualized program for asset maintenance.

Evolution of Different Type of Maintenance

Since the Industrial Revolution breakdown maintenance has been prevalent for upkeep of machinery. At about 1940, the concept of inspecting and lubricating machinery became fairly common. It was not until about 1950, however, that the concept of Preventive Maintenance came along. PM was a dramatic improvement over Breakdown Maintenance, and it remains a key ingredient to reliability systems today. Systematic planning and scheduling came about in the early 1960s as the early computers allowed automation of these processes. Today, CMMS systems allow very effective planning and scheduling of resources.

During the early 1970s, the concept of Predictive Maintenance became publicized. Using actual trend data to determine when machinery may have a problem had some distinct advantages over the time-based approach of PM. As the years have passed since 1970, the focus has been on better and better diagnostic evaluation tools to determine exactly what machine faults may exist at the component level. The role of the analysts in the reliability Team is defining exactly what is wrong with a machine so that spares can be ordered, outages may be planned, and the proper tools and skilled workers can be assembled for the repair activity. In the late 1980s, the Reliability Engineering programs became available. Reliability engineers were sought after by enterprises looking to increase production and minimize costs. The factors that impact reliability are; design imperfections, material deficiencies, operating abuse, and improper application. The first step in implementing reliability engineering is formation of a Reliability Team consisting of representatives from production, maintenance, and engineering personnel to address the following issues:

- Root cause failure analysis,
- Equipment brand consolidation,
- Equipment upgrade and redesign,
- Lubricant consolidation,
- Materials upgrade and standardization,
- Spare parts consolidation and inventory reduction,
- Preventative maintenance modification, improvements, and elimination,
- Maintenance and repair work methods improvements
- The skills needed by maintenance personnel

At around 1990 Proactive maintenance become popular. Proactive maintenance is a preventive maintenance strategy that works to correct the root causes of failure and avoid breakdowns caused by underlying equipment conditions. The purpose of proactive maintenance is to see machine failures as something that can be anticipated and eliminated before they develop. Creating a proactive maintenance program helps organizations find hidden inefficiencies. It largely depend on condition monitoring program efficacy and it must be aggressively pursued to allow more focused maintenance effort and the elimination of unwarranted emergency repairs and costly downtime. Effective machinery condition information is required to efficiently plan maintenance during scheduled downtime. Maintenance downtime should be scheduled to assure the equipment performs as expected when needed instead of becoming the next "crisis of the day". Predictive and preventive maintenance approaches work together to influence proactive maintenance, as it is more of an approach-based strategy rather than an individualized program for asset maintenance. Proactive maintenance depends on rigorous machine inspection and condition monitoring. In mechanical machinery it seeks to detect and eradicate failure root causes. Condition monitoring is the part of preventive and predictive maintenance relating to the detection of failures.

Condition Monitoring Technologies

Condition Monitoring systems allow management of individual assets based on their actual condition. There is a wide array of available measurement technologies from portable systems to continuous on-line systems with permanently mounted transducers. Through effective monitoring, the technologies allow fault identification, diagnosis, and sometimes prognosis. Predictive technologies available are vibration, oil analysis, infrared thermography, motor current signature analysis (MCSA), ultrasonic detection and many others.

Vibration

Vibration analysis has been proven to be the most successful predictive tool when used on rotating equipment, both in increasing equipment availability and reliability. Vibration analysis detects repetitive motion of a surface on rotating or oscillating machines. The repetitive motion may be caused by unbalance, misalignment, resonance, electrical effects, rolling element bearing faults, or many other problems. The various vibration frequencies in a rotating machine are directly related to the geometry and the operating speed of the machine. By knowing the relationship between the frequencies and the types of defects, vibration analysts can determine the cause and severity of faults or problem conditions. The history of the machine and the previous degradation pattern is important in determining the current and future operating condition of the machine. The presence of trained vibration specialists with equipment to conduct analysis will form the basis of a strong vibration program.

Vibration analysis monitors the following conditions on rotating machines:

- Cracks, pits, and roughness in rolling element bearing components.
- Unbalance of rotating machine parts.
- Shaft misalignment.
- Coupling problems.
- Bends, bows, and cracks in shafts.
- Excess sleeve bearing wear.
- Loose parts.
- Misaligned or damaged gear teeth.
- Deterioration caused by broken or missing parts.
- Deterioration caused by erosion and corrosion.
- Resonance of components.

Integrating oil and vibration analysis can yield early detection and trending of numerous equipment problems. Detecting the faults is the first step in the diagnostic process. Thermography can also be used to verify vibration analysis on rotating equipment. Thermography can detect any overheating of bearings due to insufficient lubrication, misalignment, and other causes.

Oil Analysis

One of the basic technologies of condition based maintenance is lubricating oil analysis. The reason for this is that lube oil analysis is a very effective tool for providing early warning of potential equipment problems. One of the leading causes of premature bearing failure is improper lubrication practices. These include, use of improper lubricant, excessive amounts of lubricants or not enough lubricant and contaminants in oil. The goals of oil monitoring and analysis are to ensure that the bearings are being properly lubricated. This occurs by monitoring the condition of both the lubricant and the internal surfaces that come in contact with the lubricant.

Oil condition monitoring requires an examination of the physical, chemical and additive properties to maintain oil stability as needed by the operating equipment. Routine oil monitoring can provide information about additive breakdown, viscosity changes, water content, coolant leakage and wear particle contamination. Equipment condition monitoring with oil analysis include: fluid contaminant analysis and control, fluid property analysis, and fluid wear debris analysis. These fluid parameters are typically monitored using on-site screening and laboratory analysis. Typical tests performed are:

- Particle count
- Viscosity
- Water
- Elemental spectroscopy
- FTIR spectroscopy
- Total acid number
- Wear density (by exception from particle count)
- Analytical ferrography (by exception from wear density).

Viscosity is the single most critical physical property of oil. An increase in viscosity is more tolerable than a decrease in viscosity. Viscosity limits are usually enveloped with an upper and lower limit. The specification sheets from lubricant vendors along with the baseline laboratory results of new oil will provide the viscosity limits.

The Neutralization Number or Total Acid Number (TAN) corresponds to the acidity of the oil. This acid can be produced from oil degradation. Acid numbers vary depending upon the additives in a lubricant.

Another key elements in oil analysis is the International Standards Organization Code (ISO 4406). This solid contaminant code is a standard for expressing the cleanliness levels of liquids and is internationally recognized. The code is based upon the number of particles per unit of volume greater than 5 and 15 μm in size. These two sizes are significant because the concentration at the lower range gives an accurate assessment of the "silting" condition of fluid and the 15 μm range reflects the "wear" type particle concentration. The ISO code is represented by the 5 μm range number, over (slash), a 15 μm range number. This provides the user with valuable information about the hard particle contamination in a fluid. We can now tell if new oil is more contaminated than the recommended cleanliness level and if so we can filter it down to the necessary cleanliness code. Through trending analysis, this cleanliness level will provide information on in-service lubricants as they degrade or ingress contaminants.

Water contamination is another critical component of oil analysis. Water contaminationis expressed in PPM (parts per million) of water content. One percent (1%) of water equals 10 000 PPM. The lower the percentage (or PPM) of water the less contamination is present and better protection of the asset will result.

The tribology program for used and in-service oil will focus on controlling both the root causes of machine and lubricant degradation (proactive maintenance) as well as detect and describe abnormal lubricant and machine condition (predictive maintenance). These can be satisfied through on-site screening of all used and in-service oils with interacting laboratory test for bench marking and exception analysis. Most experts in tribology recommend that oil analysis be performed on all new oils. This testing should include the following:

- On-site particle count using pore blockage (10 μm screen)
- On-site viscosity
- On-site water screen.

The laboratory tests listed below should be performed on the "critical" fluids. These are the lubricants that are used in the equipment that is tested on a more routine basis than 60 or 90 days. The following tests will provide baseline information used for trending and analysis:

- Off-site Karl Fischer
- Off-site elemental spectroscopy
- Off-site FTIR spectroscopy
- Off-site total acid number (TAN).

The new oil ISO code and water count should be at the World Class level or better. If the new oil does not meet the necessary criteria, then it will need to be filtered down to that cleanliness level or the contract with the vendor will need to be reviewed.

Thermography

Thermal measurement technology measures absolute or relative temperatures of key equipment parts or areas being monitored. Abnormal temperatures indicate developing problems. Temperature and thermal behaviour of plant components are the most critical factors in the maintenance of plant equipment. For this reason, temperature is frequently considered the key to successful plant maintenance and is the most measured quantity. There are two types of equipment used in this technology, contact and non-contact. Contact methods of temperature measurement using thermometers and thermocouples are still commonly used for many applications. However, non-contact measurement using infrared sensors has become an increasingly desirable alternative over conventional methods.

Infrared thermography is the technique of producing an image of invisible infrared light emitted by objects due to their thermal condition. Thermography works by revealing the thermal patterns present with all equipment and processes. The invisible thermal or infrared energy radiates from hot and cold areas. Thermal imaging systems "see" the radiation by focusing the infrared radiation on a detector or detector element. This radiation is then converted into electrical energy, amplified and processed into a visible image and then presented on a view finder or monitor as a thermogram.

More sophisticated cameras can actually measure the apparent temperatures of any object or surface in the image. The cameras can also produce colour images that make interpretation of thermal patterns easier.

Some applications of thermography are introduced below.

Quickly measure and compare height signatures of each piece of equipment on the inspection route without disrupting operations.

- Monitor and measure bearing temperatures in large motors or other rotating equipment.
- Identify hot spots on electronic equipment.
- Identify leaks in sealed vessels.
- Find faulty insulation in process pipes or other insulated processes.
- Find faulty terminations in high power electrical circuits.
- Locate overloaded circuit breakers in a power panel.
- Identify fuses at or near their current rated capacity.
- Identify problems in electrical switchgear.
- Capture process temperature readings.
- Identify steam trap leaks.
- Temperature of brick furnaces.

Motor Current Signature Analysis (MCSA)

MCSA helps in quantifying the severity of electrical faults, static eccentricity, and dynamic eccentricity in the motor under operating circumstances. These data are acquired using a clamp-on current probe, special filters, a digital signal analyzer, and (normally) an expert system to evaluate the results. Data are acquired with the unit under normal operating conditions (at least 70% load). MCSA capable of identifying: phase unbalance bad rotor bars faulty collector rings dynamic eccentricity static eccentricity. The rotor bar/end ring faults are easily quantified through this technique, whereas vibration analysis is a more qualitative assessment of these types of problems. When clarification is required, MCSA is an excellent means of gaining an understanding of the severity of the electrical fault. Identifying these problems in large motors (above 150 Hp) is especially critical, since eccentricity causes electromagnetic unbalance and associated alternating stresses on the stator elements. These alternating stresses degrade the stator over time and can result in loose iron, insulation breakdown, and shorts.

Ultrasonic Emissions Evaluation

Ultrasound occurs at frequencies far above the audible range of humans. While most humans hear in the range from 20 Hz to 20 kHz, the ultrasonic range is typically accepted to be from 20 to 100 kHz. Much like vibration, low frequency sound energy travels long distances while high frequency ultrasound energy dissipates very quickly. Unlike vibration which can be highly directional in nature, ultrasound propagates radially in all directions. Due to the quick dissipation of high frequency energy, the ultrasonic intensity will be greatest at the source. This characteristic makes ultrasonic emissions evaluation ideal for locating the precise source of problems. Fault conditions associated with each application generate dynamic pressure in a particular frequency range. Using ultrasonic instruments, the maintenance personnel may identify the presence of a problem in the equipment and then pinpoint its source. Most equipment involves the use of head-phones and filters which make the ultrasonic frequency range audible to the human ear.

Applications for airborne ultrasonic include:

- Pressurized gas, air leaks
- Vacuum leaks
- Boiler tube, Heat Exchanger leaks
- Steam traps
- Valve seat leaks
- Bearing faults
- Compressor valve leakage.

Proactive Inspection Program for Static Equipments

This step contains two major points:

- Development of inspection programmes addressing the types of damage that inspection engineer should detect, and the appropriate inspection techniques to detect the damage.
- Reducing risk through inspection, discusses the application of RBI tools to reduce risk and optimise inspection programmes.

Inspection influences risk, primarily by reducing the probability of failure. Many conditions (design errors, fabrication flaws, malfunction of control devices) can lead to equipment failure but in service inspection is primarily concerned with the detection of progressive damage.

The purpose of an inspection programme is to define and perform those activities necessary to detect in-service deterioration of equipment before failures occur. An inspection programme is developed by systematically identifying:

- What type of damage to look for,
- Where to look for,
- How to look for the damage (inspection techniques),
- When or how often to look for.

Equipment Modification Factor

The equipment modification factor identifies the specific conditions that can have a major influence on the failure frequency of the equipment items. These conditions are categorised in to following items.

1) The technical module that examines the material of construction, the process environment and the inspection programme.
2) Mechanical consideration that vary from item to item.
3) Universal condition that affect all equipment items in the facility.
4) Process influences that can affect equipment integrity.

Effectiveness of the facility's inspection program is to identify and monitor the operative damage mechanisms prior to failure. The RBI approach considers by definition that items are monitored and if an inspection points out a critical state for an item, preventive measures are systematically taken (repaired, changed). So we don't consider only the failure likelihood but the likelihood that an undetected damage state becomes critical.

Risk Based Inspection

Risk based inspection (RBI) is recognized an effective method to reduce risk and raise the level of safety in many industries. Whilst in traditional deterministic approach the inspection frequencies and modes are fixed, in RBI for each item of equipment, the inspection interval depends on the risk level. RBI is a method for using risk as a basis for prioritising and managing the effort of an inspection program to rationally allocate inspection resources. The term of "inspection" is understood as a systematic procedure used to assess equipment technical conditions. It is usually performed on a fixed periodical basis. In an operating plant or installation, a relatively large percentage of risk is associated with a small percentage of the equipment items. Typically, about 80% of risk of equipment's failure is associated with only 20% of equipment (Figure 17.1). RBI allows shifting inspection and maintenance resources to provide a higher level of coverage on the high-risk items and an appropriate effort on lower risk equipment. RBI ensures a higher safety level without increasing the costs.

The RBI method defines the risk of operating equipment as the combination of two separate terms: the likelihood of an undetected failure and the consequence of such a failure. It is worth to notice, that such approach of term "risk" is also used in safety standards as well as in other disciplines.

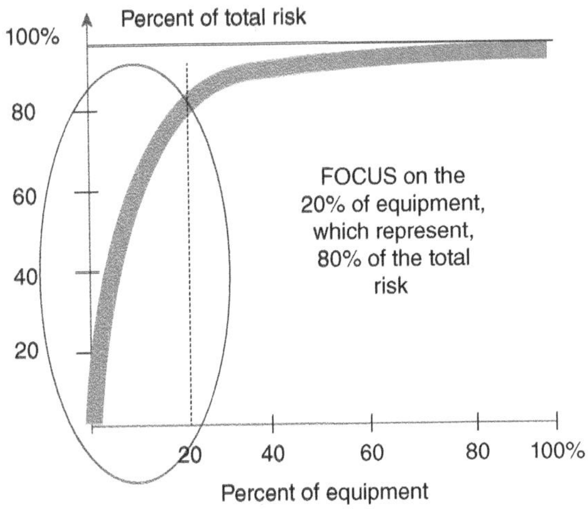

Figure 17.1 Typical risk plot "total risk vs. quantity of equipment."

The likelihood of failure assessment takes into consideration such criteria as:

- The damage mechanisms applicable to the item analysed.
- The inspection history of the item.
- The effectiveness of the previous inspection.

The assessment of failure consequences follows these steps:

- Scenarios definition in which failure (i.e.: leak) progress into undesirable events.
- Estimation of the physical effect of each scenario.
- Adverse effect on people, equipment, environment, productivity as a result of the outcome.

The detailed method to assess consequence and likelihood depends directly upon the facility type. Furthermore, the level of detail of the method is fitted to the future use of the result, the available data for the analysis, the need of accuracy of the result: a range of probability/consequences or a formal probability/consequences. In general, main steps of RBI study are as follow:

- Preliminary analysis,
- Failure probability assessment,
- Consequence evaluation,
- Risk ranking, and
- Development of Inspection program.

Condition Monitoring and The Internet of Things

In the future, everything will be connected to everything else. The Internet of Things is simply a grand name for devices being able to connect and communicate, independently of user interaction. This allows processes such as condition monitoring to run in the background while machines are operating, communicating between two or more machines, even between machines in different locations, in different countries.

The overwhelming benefit of this facility of communication is the ability for constant monitoring, with machines able to both provide and receive data, and running speeds, vibrations and temperatures to be constantly compared and evaluated. The more data that is available, the further into the future analysts and engineers are able to see. The future is looking very bright for condition monitoring with internet of things. With each new tool in the condition monitoring analyst or engineer's toolbox, there is actually less investigative work to carry out, as more and more information is on hand to assist in detecting and correcting changes in the running of machines. With IoT-based predictive maintenance, the continuous flow of data to the cloud gives engineers greater accuracy to identify operating conditions and errors.

Equipment Condition Assessment and Decision Making

Condition monitoring data should be trended and analyzed. Trained analysts should use analysis software where appropriate. There should be a process in place to generate and distribute equipment condition reports including a technology-integrated status of anomalies for equipment owners to review. It should be clear who is making decisions to act, based on indications of equipment condition. Such decisions should be made in a timely manner among personnel taking and analyzing data, owners of equipment, and personnel responsible for assigning, planning, and performing maintenance tasks.

Also, best practice plants create an equipment condition report with all condition indicators for each important piece of equipment. The report clearly identifies degraded conditions or adverse trends. The report is distributed to all work groups. The plant corrective action programme is used to document deficiencies discovered during condition monitoring activities. All sources of condition data are utilized, to take corrective action based on condition data.

Operator-Based Maintenance

Even though offline condition monitoring programme is in place, internet of things in predictive technology is gaining ground in industry, still LOOK, LISTEN and FEEL by operator plays an effective condition monitoring tool which identifies just in time the key indicators to machine failures and maintenance requirements. The concept here is to look, listen and feel the machines to know what service work need to be done on the machine and when. This does not mean to wait until it breaks and fix it. What it does mean is to collect data while the equipment is running, establish threshold limits either from OEM information, vendors, or best practices, and schedule the maintenance activities accordingly. Using equipment operators to perform routine maintenance activities help reduce downtime as envisaged by Total productivity Maintenance concept.

Conclusion

The proactive maintenance approach makes it possible to perform maintenance only when necessary, based on gathered condition data. The transition to proactive maintenance can be challenging, with issues of culture change, budget, lack of training resources, and full-time employees to learn and perform analysis. There are a few ways to help organizations get started,

1) Focusing on failure modes: Prioritization of plant systems and failure modes with impact on capacity and availability.
2) Receiving buy-in from leadership: Business decision of where to invest maintenance resources and technologies.

3) Technologies selection: Infusion of available maintenance technologies with wireless capabilities.
4) Enhancing employee skill set: Increase in the core competency of the maintenance team.

By implementing proactive measures organizations become empowered to Identify the root causes of machine failure, resolve problems before they become failures and extend the life of machinery and reduce downtime. Equipment for which proactive maintenance is considered to be ineffective or too costly. Redesigns, modifications, and operational improvements are considered.

Index

Printed and bound by CPI Group (UK) Ltd, Croydon, CR0 4YY
17/09/2026
14964322-0005